STUDIES IN IMPERIALISM

general editor John M. MacKenzie

Established in the belief that imperialism as a cultural
phenomenon had as significant an effect on the dominant
as on the subordinate societies, Studies in Imperialism
seeks to develop the new socio-cultural approach which
has emerged through cross-disciplinary work on popular
culture, media studies, art history, the study of education
and religion, sports history, and children's literature. The
cultural emphasis embraces studies of migration and race,
while the older political, and constitutional, economic
and military concerns will never be far away. It will
incorporate comparative work on European and
American empire-building, with the chronological focus
primarily, though not exclusively, on the nineteenth and
twentieth centuries, when these cultural exchanges were
most powerfully at work.

Geography and imperialism, 1820–1940

This book examines the ways in which European
imperialism was facilitated and challenged by
geographical theory and practice in the period 1820–1940.
It adds to current multi-disciplinary debates on the
complex cultural, ideological and intellectual bases of
European imperial conquests and colonisations by
reference to geographical science.

These authors examine maps and surveys, exploration and
travel, the activities and debates of metropolitan and
provincial geographical societies, and a range of written
and visual representations. The use of geographical
knowledge as a tool of imperial propaganda is evaluated,
together with its contribution to imperial debates on race,
environmental perception and management.

The book explores imperial photography, the cartography
of decolonisation and environmental management and
conservation in the British empire. The work of societies
such as the Royal Geographical Society, the Royal Dutch
Geographical Society and the Société Géographie de Paris
is evaluated.

Morag Bell is Reader in Geography, Robin Butlin
is Professor of Geography and Michael Heffernan
is Senior Lecturer in Geography, at Loughborough
University of Technology.

STUDIES IN IMPERIALISM

Geography and imperialism
1820–1940

edited by Morag Bell,
Robin Butlin and Michael Heffernan

MANCHESTER UNIVERSITY PRESS
Manchester and New York

Distributed exclusively in the USA and Canada by
ST. MARTIN'S PRESS

Published by Manchester University Press
Oxford Road, Manchester M13 9NR, UK
and Room 400, 175 Fifth Avenue,
New York, NY 10010, USA

Distributed exclusively in the USA and Canada
by St. Martin's Press, Inc.,
175 Fifth Avenue, New York, NY 10010, USA

British Library Cataloguing-in-Publication Data
A catalogue record for this book is available from the British Library

Library of Congress Cataloging-in-Publication Data
Geography and imperialism, 1820–1940 / edited by Morag Bell, Robin
    Butlin, and Michael Heffernan.
        p.   cm. — (Studies in imperialism)
    ISBN 0–7190–3934–7
    1. Imperialism—History.   2. Colonies—History.   3. Geopolitics.
    I. Bell, Morag.   II. Butlin, R. A. (Robin Alan), 1938–   .
    III. Heffernan, Michael J., 1959–   .   IV. Series: Studies in
    imperialism (Manchester, England)
    JC359.G46   1995
    910—dc20   94–41730

ISBN 0 7190 3934 7 *hardback*

First published in 1995
99 98 97 96 95   10 9 8 7 6 5 4 3 2 1

Photoset in Trump Medieval by
Northern Phototypesetting Co. Ltd., Bolton
Printed in Great Britain by
Biddles Ltd, Guildford and King's Lynn

CONTENTS

[v]

CONTENTS

FIGURES

[vii]

TABLES

GENERAL INTRODUCTION

In the nineteenth century, many Europeans argued that geographical knowledge was the key to imperial power. The emergent discipline should therefore be a central tool of the expansion of commercial and political influence. Although the basic proposition was seldom contested, the nature and uses of such knowledge were often subjected to controversy, and its wider dissemination frequently met with apathy, particularly in Britain. By the end of the century, after several decades of geographical enthusiasm on the continent of Europe, many British geographers were beginning to castigate their fellow citizens for an inadequate response to the varied and practical forms of geographical information by then available, for official and private ignorance, a lack of societies and pressure groups, and above all a laggard approach to geographical education at all levels.

There are a number of reasons for the recognition accorded to geography as both practical discipline and popular study. It was seen as an accessible and bridging science, capable of offering insights into other more technical branches of scientific knowledge. Related as it was to travel, exploration, photography, and the adventure tradition of popular culture, it was considered to be capable of substantial cultural penetration, both feeding off and promoting the imperial chauvinism of the age. It was thought that this also made it an ideal school discipline, with attractive texts, and one, moreover, capable of stimulating the interest of both males and females. Despite the lengthy controversy surrounding the entry of women to the London Royal Geographical Society, by the last decades of the nineteenth century, women travelled, published, were highly active in geographical societies, and became professional geographers. It was also seen as a key source of official knowledge, feeding geopolitical ambition in Europe as well as elsewhere, and ultimately it was adopted by the fascist states of Europe as a discipline with intense propaganda value.

Many of these issues are touched upon in this book. The first thrust of imperial geography, exploration, is placed within its cultural context, an activity highly modified by both the stated objectives and the self-seeking publicity surrounding such expeditions. The discipline and its institutional forms are also seen as a clearing house for the dissemination of environmental knowledge, for example in relation to desiccation theory and forestry, as a prompter of the remarkable reconstruction of the world through the European lens of the photographer in the second half of the nineteenth century, as a source of provincial pressure groups, major publications projects, and imperial cartography. There are contributions on the role of women in the evaluation of landscape and as intermediaries between metropolis and imperial periphery, while several chapters examine the significance of geographical information to the Dutch, the French and the Italians at different periods in their jockeying for European and global power. This is a volume which reflects the growing potential of studying the inter-disciplinary and multi-faceted character of geographical activity in the nineteenth and twentieth centuries and which opens up fresh empirical and interpretative landscapes.

J. M. M.

LIST OF CONTRIBUTORS

David Atkinson is Lecturer in Geography at St David's University College, Lampeter. He is currently completing a doctoral thesis on 'Geography and geopolitics in fascist Italy' and was awarded an Italian government scholarship for this work in 1992–93. His research interests include the history of geopolitics and the cultural and historical geography of post-Risorgimento Italy.

Morag Bell is Reader in Geography at Loughborough University, and author of *Contemporary Africa* (1986). Her main research interests are in the geography of cultural and environmental relations in Britain and Southern Africa since the late nineteenth century and the contemporary history of western development thought.

Robin Butlin is Professor of Geography at Loughborough University, and author of *Historical Geography: Through the Gates of Space and Time* (1993). His principal field of research is historical geography, with reference to the historical geographies of empire, of the Holy Land, and of rural England, Ireland and Western Europe in the early modern and modern periods.

James Cameron is Dean of Education at the Northern Territory University in Australia. He combines an interest in teacher education with the history of British settlement of Australia, and has published extensively in both fields. His publications include *Ambition's Fire: The Colonization of Pre-Convict Western Australia* (1981), and the *Atlas of Northern Australia* (1982). He is currently working on a major study of Sir John Barrow.

Richard Grove is a Fellow of Churchill College, Cambridge, co-editor of *Conservation in Africa* (1987), and the author of many articles on the history of conservation. His current research interests include the history of anti-desiccation forestry projects since the seventeenth century and the history of environmental thought.

Michael Heffernan is Senior Lecturer in Geography at Loughborough University. His research interests focus on the cultural and political geography of France and the French overseas empire during the nineteenth and twentieth centuries and on the history of European geographical and environmental thought since the Enlightenment.

Cheryl McEwan is Post-Doctoral Research Tutor in the Department of Geography, University College of Swansea. Her research interests include the historical and cultural geography of the British empire, women and histories of geographical thought, and nineteenth-century travel literature.

LIST OF CONTRIBUTORS

John M. MacKenzie is Professor of Imperial History at Lancaster University. His principal research interest is the culture of British imperialism. He is the author of *Propaganda and Empire, The Empire of Nature,* and *The Orientalism Debate, History, Theory and the Arts* (in press); and editor of *Imperialism and Popular Culture, Imperialism and the Natural World* and *Popular Imperialism and the Military.*

James Ryan is a Departmental Lecturer at the School of Geography, University of Oxford. His doctoral research is on *Photography, Geography and British Imperialism, 1840–1914.* His research interests lie in the cultural dimensions of British imperialism and the cultural history of geographical knowledge.

Jeffrey Stone began his career as an Administrative Officer in the Provincial Administration, Northern Rhodesia. He returned to Britain in 1964, to the Geography Department of the University of Aberdeen, where he is now Senior Lecturer. His interests include the history of cartography, with particular reference to Scotland and Africa.

Paul van der Velde is a member of the International Institute for Asian Studies at Leiden in The Netherlands, and from 1988 to 1993 worked at the Institute for the History of European Expansion at Leiden University. His research interests are in the history of the Royal Dutch Geographical Society, the travel writings of Jacob Haafner (1754–1809), and the life and work of P. J. Vath (1815–95).

INTRODUCTION

Geography and imperialism, 1820–1940

Morag Bell, Robin A. Butlin
and Michael Heffernan

The important thing [about European imperialism] was to dignify simple conquest with an idea, to turn the appetite for more geographical space into a theory about the special relationship between geography on the one hand and civilised or uncivilised peoples on the other (Edward W. Said, 1978).

The attempt to build an empire or to develop its resources without geography is like building a house without first considering the climate or the locality, or seeing the character of the soil on which you must plant its foundations (Douglas W. Freshfield, President of the Royal Geographical Society, 1917).

In recent years, new perspectives have been added to our understanding of the European imperial impulse.[1] The familiar interpretation of imperialism as an economic process of capitalist expansion into the non-capitalist world, an assessment associated with early twentieth-century commentators such as Lenin and Hobson,[2] has been modified by recent nuanced accounts of the complex cultural, ideological and intellectual bases of European conquest and colonisation. The deep-seated, structural economic forces which underpinned European imperial expansion cannot be ignored but they need no longer be interpreted as anonymous, monolithic and inevitable. European imperialism was sustained – perhaps even defined – by the beliefs and convictions of a wide range of people including those in executive power and others, both men and women, beyond the realms of formal decision-making who sought to influence the political process through their writings and other activities. The imperial visions and ambitions of these people were diverse; they frequently clashed with their equivalents in neighbouring countries and with opposing schools of imperialist thought in their own political arena. Imperialism was, in short, a complex and contested discourse involving different individuals and interest groups

[1]

whose moral, ideological and even psychological values can be pro-
fitably investigated to extend our understanding of the imperial
process.[3]

Concern with the discursive nature of the imperial mind and with the
moral and intellectual bases of overseas expansion has been stimulated
in part by a growing realisation that 'colonial' attitudes and assumptions, particularly those concerning race and religion, have persisted
into the 'post-colonial' era and represent one of the most serious
obstacles to global economic development. Despite the dismantling of
the great colonial empires, the relationship between former colonisers
and colonised is still frequently characterised by barely concealed mis-
trust, suspicion and mutual incomprehension. In times of crisis – the
latest international recession and the Gulf War are both illustrative –
these latent, unexamined hostilities re-surface with undiminished
intensity.[4] Understanding the tenacity of these belief-systems within
the west requires careful excavation of their historical roots and
detailed, contextual examination of the intellectual climates in which
these attitudes developed.

An important and complex question arises; in what ways was
European imperialism facilitated, sustained or perhaps challenged by
European science and scholarship during the nineteenth and twentieth
centuries? The age of the great colonial empires which began in the
eighteenth century was an era of dramatic scientific and technological
development in Europe. In consequence, some historians have argued
that innovations in science, medicine and technology determined not
only the enormous growth of European commercial and economic
power during the Industrial Revolution but also the pace and direction
of European imperial expansion in the same period.[5] Without the
Gatling gun and quinine, the conquest and occupation of the tropical
world by small numbers of European soldiers and settlers would have
been impossible. Other writers have moved beyond the 'hardware' of
imperial control to consider the more diffuse ways in which particular
facets of European science and scholarship were influenced by, and in
turn, influenced, the process of imperial expansion. The history of
entire disciplines – including anthropology, archaeology, astronomy,
history, philology, medicine, the natural sciences and physics – have
now been critically re-assessed in the light of the imperial assumptions
and values underlying their origins and development.[6] It has even been
argued that the entire structure of European knowledge, the very epistemological foundations on which the modern academy has been
erected, is largely the outcome of unexamined, universalist (and hence
imperialist) preconceptions about the world and its inhabitants.
Through their claims to universal validity, professional practioners

within the various branches of European science saw themselves as contributing towards a higher imperial goal – to understand the world and its contents.[7] In practice, they conspired in the imposition of its methods and philosophies in regions where knowledge and understanding previously took different forms.[8]

Modern European science and European imperialism thus marched arm-in-arm: both were supremely ambitious, universalising projects concerned to know all, to understand all and, by implication, to control all. In his path-breaking survey of British and French attitudes to the Middle East since the Enlightenment, Edward Said dates the emergence of this totalising scientific-imperialist impulse to the turn of the eighteenth century. By this point, Said argues, the relationship between European scientific advance and the process of European imperial expansion had become so intimate and mutually dependent that the two seemed virtually synonymous, at least in the minds of some imperialists. The exploration of the Pacific by Captain James Cook and his French rival, Louis Antoine de Bougainville, can be seen in these terms – as major episodes in European imperial expansion couched principally in the language of science.[9] For Said's purpose, however, the great Napoleonic survey of Egypt carried out during the three-year French occupation of the Nile basin represents a perfect illustration of the fusion of science and imperialism. The French republican forces which set sail from Toulon for Alexandria in 1798 were accompanied, at Napoleon's insistence, by a small 'army' of leading scientists and scholars especially recruited to undertake a vast survey of the region. Although their work was rudely cut short by the British blockade which ended the French occupation in 1801, enough had been completed on the archaeology, architecture, ethnography, topography, history and natural history of the region to fill twenty-three huge volumes of the *Description d'Égypte*, published by successive French governments between 1809 and 1828.

The most important characteristic of the Egyptian survey was its exhaustive, encyclopaedic thoroughness. It was explicitly designed to catalogue everything that was known, or could be known, about the region. For Said, this survey marked a critical turning point in the European representation and intellectual appropriation of the Orient: by conquering Egypt militarily through force of arms and intellectually through the power of European scholarship, the Napoleonic state was seeking, in a supremely ambitious imperial way, to possess not only the strategic advantages, land, resources and people of late eighteenth-century Egypt but also the region's history and civilisation. This was much more than mere territorial conquest; by uncovering and *re*-presenting Egypt, modern scientifically sophisticated France claimed authority over Egypt's soul and identity.[10] The Napoleonic survey of Egypt was thus an early example of science as a co-operative regional synthesis. Each member of the expedition contributed his specialist expertise to construct a complete portrait of Egypt, past, present and future. This notion of regional synthesis, virtually always incorporating a strong historical perspective, was to become the defining characteristic of European geography as it developed during the nineteenth and early twentieth centuries.

Geography illustrates better than any other 'imperial science' the soaring proprietorial ambition of the European imperial mind. Unlike other disciplines, particularly in the pure sciences, geography's philosophical and epistemological status and subject matter have always been pragmatically defined. While other sciences have modestly claimed expert authority over this or that body of knowledge or material, geography has frequently – and rather immodestly – set itself apart from this specialising process and championed instead the cause of a general science of the earth and its inhabitants. Notwithstanding the discipline's occasional energetic striving for recognition as a physical science, geographers saw themselves, particularly in the last century, as part of a 'pan-discipline' whose task it was to unite, partly for educational purposes, the otherwise disparate branches of the sciences and humanities. Geography's function was to incorporate into its own research data, material from other disciplines and to re-present it in neat, regionally specific digests.[11] The principal geographic 'tool' was the map. By representing the huge complexity of a particular physical and human landscape cartographically in a single image, geographers provided the European imperial project with arguably its most potent device. The European exploration, mapping and topographic surveying of Africa, Asia and Latin America during the eighteenth and nineteenth centuries – generally with state support – was self-evidently an exercise in imperial authority. To map hitherto 'unknown' regions (unknown, that is, to the European) using modern techniques in triangulation and geodesy was both a scientific activity dependent on trained personnel and state-of-the-art equipment and also a political act of appropriation which had obvious strategic utility to occupying military forces.[12] The British Survey of India (1818–82), staffed by skilled surveyors, expanded rapidly in the nineteenth century due to its military and political significance.[13]

Throughout the nineteenth century, exploration and map compilation in far-flung regions remained a source of enormous public interest. To many, the geographer was the ideal macho hero (the notion of a female geographer was almost a contradiction in terms) pitting himself against the elements and hostile indigenous peoples in remote and threatening environments.[14] However, as geography developed into a professional discipline with a prominent position in schools and universities, it began to expand beyond its original focus on exploration and topographical map-making to assume intellectual authority over a wide range of regionally specific environmental, economic, social, political and cultural evidence. Maps remained critically important, of course, but they became increasingly sophisticated and capable of representing not only the material reality of the physical and human landscape but also the 'hidden' features of a locality's economic, social or political characteristics through the skilful deployment of cartographic symbols.[15] By the end of the nineteenth century, at the high point of European imperialism, geography had become, in the words of Thomas Richards, 'unquestionably the queen of all imperial sciences ... inseparable from the domain of official and unofficial state knowledge'.[16] For Sir Harry H. Johnston, explorer, African colonial administrator and prominent Fellow of the Royal Geographical Society (RGS), geography had pride of place in the British educational system as 'a science worthy of the greatest deference from its sister sciences, as it is really the Eldest Sister of the Band'. It should also be a compulsory subject for all aspiring politicians, diplomats and civil servants.[17] For Johnston, the detailed geographical description of a region, complete with authoritative and regularly updated topographical and thematic maps, allowed that region to be known, understood and therefore possessed in the most complete way. By dividing the world into regions and ordering the burgeoning factual information about the world into regional segments, geography offered one solution to the yearned-for objective of classifying and understanding the human and environmental characteristics of the entire globe. It thus perfectly reflected the extraordinary optimism of the imperial age: the world could, at last, be visualised and conceptualised as a whole.

Even politically radical geographers on the left often shared the imperialist and universalist aspirations which lay at the heart of modern geographical inquiry. The prolific French anarchist geographer Elisée Reclus saw French imperialism in North Africa in a positive light and his extraordinary idea for an enormous terrestrial globe to occupy pride of place at the Paris Exposition Universelle in 1900 represents a perfect illustration of geography as the all-encompassing science of global representation.[18] Patrick Geddes, the Scottish Edwardian polymath, demonstrated an even more expansive enthusiasm for geographical education as a pathway not only to global understanding but also to a fuller comprehension of the entire cosmos. In 1902, Geddes proposed using the Reclusian terrestrial globe alongside another, equally vast, orb depicting the celestial view from earth as the centrepieces of his ambitious Institute of Geography. Geddes's Institute, the plan of which is shown on the front cover of this volume, was to include a lofty Tower of Regional Survey – a modification of his famous Outlook Tower – from which visitors could perceive the surrounding local landscape. There would also be a huge lecture theatre and congress hall, massive panoramas of typical scenery from around the world and dozens of exhibition rooms each devoted to the physical and human characteristics of different countries and regions. Visitors to the Institute – sadly never built – would emerge with memorable visual impressions of every spatial scale from the local through the regional and the global to the cosmic.[19]

The contemporaneous and symbiotic relationship between geography and imperialism at the end of the nineteenth century has been widely acknowledged. Several writers have argued that geographers – both the amateur explorers and the professional university-based scholars – were the essential midwives of European imperialism. They provided both the practical information necessary for overseas conquest and colonisation and the intellectual justification for expansion through their increasingly-elaborate 'theoretical' writings on geopolitics and the impact of climatic and environmental factors on the evolution of different races.[20] It has also been suggested that, for much of the late nineteenth and early twentieth centuries, the teaching of geography in schools was little more than a thinly disguised form of racial and imperial propaganda.[21] Recent work on the historiography of geography has extended and modified this assessment.[22] As geography developed into a large and heterogenous academic pursuit so its relationship within the politics of imperialism became more elaborate. By the early decades of the twentieth century, it was difficult to see the geographical community, even within a single country, as a cohesive group. Various, often mutually exclusive, forms of imperial engagement were advocated by different schools of geographical thought. A few politically radical geographers, such as Prince Kropotkin, began to develop a serious critique of European imperialism which challenged the values and assumptions of their fellow geographers.[23] By the inter-war years, geography and imperialism had become such broad and contested intellectual and ideological arenas that the discipline could no longer be defined unproblematically as a science of empire. Insofar as individual geographers remained committed to European empires, many eschewed aggressive, exploitative interventions overseas in favour of a liberal, benign and developmental colonial ethos which emphasised the sacrifices and responsibilities of western scientists and colonial administrators.

With the onset of decolonisation following World War Two, some geographers actively supported the anti-imperial cause on a variety of moral, political and environmental grounds. A noteworthy example is the great French desert geomorphologist and North Africanist, Jean Dresch, who worked hard to convince his reluctant comrades in the French Communist Party to support the Algerian cause during the war of independence.[24] By the 1960s and 1970s, in a conscious attempt to rid the discipline of its unsavoury imperialist connotations and resulting lack of academic rigor and respectability, geography entered an abstract world of mathematical modelling and quantitative spatial science. In the aftermath of widespread social protest across the western world around 1968, many of its practitioners also recaptured the western fascination with a recently defined 'Third World' through a more politically engaged orthodoxy in which the geographer's science was employed to champion the rights of the world's poor and dispossessed.[25] In line with this more recent tradition and a broader 'post-colonial' cultural critique within the social sciences, western geographers continue to engage in critical debate over the persistent eurocentrism underpinning intellectual thought and political practice. Within the context of a New World Order they explore, once again, the ethics of research in, and the politics of engagement with, the ex-colonial periphery.[26]

European imperialism and European geography have therefore developed together, but in complex and contested ways. There is considerable scope for examining in greater detail the precise mechanisms involved in this relationship beyond those suggested by a broadly coincidental chronological development. The essays in this volume examine different facets of this relationship from the late-eighteenth to the mid-twentieth century, from the origins of the modern colonial empires through to decolonisation. The book is not intended to trace geography's evolution within the educational system. Neither is it a general history of geographical thought and practice over this period. Rather, the essays consider those discourses of geographical inquiry both within and beyond the academy which were implicated in the operation of European cultural and political power in the expanding colonial world. Each chapter in the book can be read as a separate essay but they are arranged in broad chronological order to convey some sense of the changing nature of the geo-imperialist discourse.

The essays explore a number of separate themes within this unfolding discourse. The first theme addresses the technical aspects of geographical inquiry and the complex role this technology played in the construction and dismantling of overseas empires. Chapters by James Ryan on imperial photography, by David Atkinson on the cartographic representations of Africa produced by Italian geographical theorists, particularly during the Fascist era, and by Jeffrey Stone on the cartography of decolonisation in Swaziland, deal specifically with the techniques underlying the crucial visual and cartographic aspects of geographical representation within the imperial context. Ryan deals particularly with the photographic archive of the Royal Geographical Society (RGS), founded in 1830 and the third most senior of the geographical societies that sprang up around the world during the 1800s (after Paris and Berlin, established in 1821 and 1828 respectively). The RGS was easily the largest and wealthiest of these societies and its prominent role in Ryan's account demonstrates implicitly its critical part in promoting the cause of empire. The impact of the geographical societies as pressure groups advocating various kinds of imperial engagement in different periods and national contexts forms another of the book's major themes. Five chapters consider the different ways in which institutionalised geography became woven into the fabric of state imperial power. James Cameron considers the role of prominent figures within the RGS in political debates about exploration to Australia in the early nineteenth century. The contribution of the RGS to issues of environmental management and conservation in the expanding British empire is outlined by Richard Grove with reference to the early and middle years of the nineteenth century. John Mackenzie discusses the British provincial geographical societies in Edinburgh, Manchester and other port cities as active agents in British 'municipal imperialism'. The role of the Dutch Geographical Society as an imperial pressure group in The Netherlands is reviewed by Paul van der Velde while Michael Heffernan details the campaigns of the Société de Géographie de Paris (SGP) to extend the French empire during and after the Great War. Most of these chapters touch upon the book's third major theme, the contribution of geographical knowledge to popular visions and interpretations of empire within Europe. While the geographical societies as educational agencies, encouraged a particular version of geography within a desired imperial curriculum, this promotional and educational role was also carried out beyond the confines of geographical institutions by powerful and dedicated individuals. Robin Butlin illustrates this with reference to the historical geographies of the British empire produced in the form of educational textbooks and informational compendia for civil servants and administrators by university-based academics such as H. B. George and Sir Charles Lucas. This is also the subject of chapters by Morag Bell and Cheryl McEwan dealing with the impact in Britain of influential individuals – Violet Markham and Mary Slessor – whose travels and popular geographical writings on south and west Africa respectively, contributed new perspectives to Britain's colonial relations and to the gendered nature of imperial power.

INTRODUCTION

Notes

1 Imperialism as a term only acquired its modern geopolitical meaning late in the nineteenth century. 'Europe' and 'European' refer here both to the imperial states of Europe itself and to those settler European colonies which themselves became imperial or 'sub-imperial' powers during the nineteenth and twentieth centuries, particularly the United States, South Africa and Australia. The imperialism of other, non-European modern powers, most notably Japan, is not considered.

2 V. I. Lenin, *Imperialism: the highest stage of capitalism*, Moscow, 1966 (originally published 1917); J. A. Hobson, *Imperialism: a study*, London, 1902. For more recent discussions in this tradition, see A. Brewer, *Marxist theories of imperialism: a critical survey*, London, 1980; B. J. Cohen, *The question of imperialism: the political economy of dominance and dependence*, London, 1973; N. Etherington, *Theories of imperialism: war conquest and capital*, London, 1984; B. Warren, *Imperialism: pioneer of capitalism*, London, 1980.

3 See, as early examples of studies on the cultural dimensions of the European imperial impulse, Henri Baudet, *Paradise on earth: some thoughts on European images of the non-European world*, New Haven, 1965; Philip D. Curtin, *The image of Africa: British ideas and action, 1780–1850*, London, 1965; Victor G. Kiernan, *The lords of humankind: European attitudes towards the outside world in the imperial age*, London, 1969; Bernard Smith, *European vision and the South Pacific, 1768–1850*, Oxford, 1960. The work of Edward W. Said, especially his *Orientalism*, London, 1978 and, more recently, the much-criticised *Culture and imperialism*, London, 1993 have been extremely influential. See also Rana Kabbani, *Europe's myths of Orient: devise and rule*, London, 1986; Timothy Mitchell, *Colonising Egypt*, Cambridge, 1988; Paul Carter, *The road to Botany Bay: an essay in spatial history*, London, 1987; Ronald Hyam, *Empire and sexuality: the British experience*, Manchester, 1990.

4 See, on the Gulf War, Philip M. Taylor, *War and the media: propaganda and persuasion in the Gulf War*, Manchester, 1992. For an intriguing analysis of the Gulf War as spectacle, see Jean Baudrillard, *La guerre du Golfe n'a pas eu lieu*, Paris, 1991 and, for an angry response, Christopher Norris, *Uncritical theory: postmodernism, intellectuals and the Gulf War*, London, 1992

5 Daniel R. Headrick, *The tools of empire: technology and European imperialism in the nineteenth century*, Oxford, 1981; and, Daniel R. Headrick, *The tentacles of progress: technology transfer in the age of imperialism*, Oxford, 1988; Michael Adas, *Machines as the measure of men: science, technology and ideologies of Western dominance*, Ithaca, 1989.

6 D. Arnold, *Imperial medicine and indigenous societies*, Manchester 1988; T. Asad (ed.), *Anthropology and the colonial encounter*, London, 1973; Peter Bowler, *The invention of progress. Victorians and the past*, Oxford, 1989; L. H. Brockway, *Science and colonial expansion: the role of the British botanic gardens*, New York, 1979; G. Leclerc, *Anthropologie et colonialisme: essai sur l'histoire de l'africanisme*, Paris, 1972; Roy MacLeod and M. Lewis (eds.), *Disease, medicine and empire: perspectives on Western medicine and the experience of European expansion*, London, 1988; Daniel Nordman and Jean Pierre Raison (eds.), *Sciences de l'homme et conquête coloniale: constitution et usages des humanités en Afrique (XIXe.–XXe. siècles)*, Paris, 1980; Lewis Pyenson, *Civilizing mission: exact sciences and French overseas expansion, 1830–1940*, Baltimore, 1993; Edward W. Said, 'Representing the colonized: anthropology's interlocutors', *Critical Inquiry*, 15 (1989), 205–25; George W. Stocking, *Victorian anthropology*, New York, 1987.

7 The debate about the role of science in the European imperial project remains intense. See, for example, Lewis Pyenson, 'Why science may serve political ends: cultural imperialism and the mission to civilize', *Berichte zur Wissenschaftsgeschichte*, 13 (1990), 69–81; P. Palladino and M. Worboys, 'Science and imperialism', *Isis*, 84 (1993), 91–102.

8 See Samir Amin, *Eurocentrism*, London, 1988.

9 Smith, *European vision*; D. R. Stoddart, *On geography and its history*, Oxford, 1986, esp. pp. 28–40; Paul Carter, *The road to Botany Bay*.

10 Said, *Orientalism*, pp. 73–92 and Martin Bernal, *Black Athena: the Afroasiatic roots of classical civilization*, 2 vols, London, 1987–92. See also R. Anderson and I. Fawsey (eds), *Egypt in 1800: scenes from Napoléon's Description de l'Égypte*, London, 1988; Charles Coulson Gillispie and M. Dewachter (eds), *Monuments of Egypt: the Napoleonic edition*, Princeton, 1987. The maps accompanying the text of the survey have been analysed by Anne Godlewska, 'The Napoleonic survey of Egypt: a masterpiece of cartographic compilation and early nineteenth-century fieldwork', *Cartographica*, 25, 1–2 (1988), monograph 38–9. The 'success' of the Egyptian survey spawned subsequent French scientific-imperialist expeditions to Morea (southern Greece), Algeria and Mexico: see Numa Broc. 'Les grandes mission scientifiques françaises au XIXe. siècle (Morée, Algérie, Mexique) et leurs travaux géographiques', *Revue d'Histoire des Sciences*, 34, 1 (1981), 319–58; Anne Godlewska, 'Traditions, crisis, and new paradigms in the rise of the modern French discipline of geography 1760–1850', *Annals, Association of American Geographers*, 79, 2 (1989), 192–213; Michael J. Heffernan, 'An imperial utopia: French surveys of North Africa in the early colonial period', in Jeffrey Stone (ed.), *The mapping of Africa*, Aberdeen, 1993; and Gary S. Dunbar, ' "The compass follows the flag": the French scientific mission to Mexico, 1864–1867', *Annals, Association of American Geography*, 78, 2 (1988), 229–40.

11 The protean and often contested nature of geographical thought and practice is admirably surveyed in David N. Livingstone, *The geographical tradition: episodes in the history of a contested enterprise*, Oxford, 1992.

12 On the relationship between cartographic knowledge and power, including imperial power, see J. B. Harley, 'Maps, knowledge and power', in Denis Cosgrove and Stephen Daniels (eds), *The iconography of landscape: essays in the symbolic representation, design and use of past environments*, Cambridge, 1988, pp. 277–312; Denis Wood, *The power of maps*, London, 1993.

13 See, for example, D. Kumar, 'Problems in science administration: a study of the scientific surveys in British India 1757–1900', in P. Petitjean, C. Jami and A. M. Moulin (eds), *Science and empires*, Dordrecht 1992, pp. 269–80.

14 On public perceptions of the explorer, see Felix Driver, 'Henry Morton Stanley and his critics: geography, exploration and empire', *Past and Present*, 133 (1991), 134–66, and Beau Riffenburg, *The myth of the explorer: the press, sensationalism and geographical discovery*, London, 1993. On hunting as a masculine imperial sport, see J. M. MacKenzie, *The empire of nature: hunting, conservation and British imperialism*, Manchester, 1988; and W. Beinart, 'Empire, hunting and ecological change in southern and central Africa, *Past and Present*, 128 (1990), 162–86. There were, however, a number of women travellers and explorers, see Dea Birkett, *Spinsters abroad: Victorian lady explorers*, London, 1989; and Sara Mills, *Discourses of difference: an analysis of women's travel writing and colonialism*, London, 1991. See also the chapters in this volume by Bell and McEwan.

15 Arthur H. Robinson, *Early thematic mapping in the history of cartography*, Chicago, 1982.

16 Thomas Richards, 'Archive and utopia', *Representations*, 37 (1992), 104–35, quotation on p. 106; and also Thomas Richards, *The imperial archive: knowledge and the fantasy of empire*, London, 1992, p. 13.

17 Quoted in memorandum from Johnston to Arthur R. Hinks, dated 11 November 1916, reporting his speech to the Conjoint Board of Scientific Societies on behalf of the Royal Geographical Society, Royal Geographical Society Archives, Hinks correspondence (arranged by year). On Johnston, see R. Oliver, *Sir Harry Johnston and the scramble for Africa*, London, 1957.

18 On Reclus and the great globe, see Gary S. Dunbar, *Élisée Reclus: historian of nature*, Hamden, Connecticut, esp. pp. 105–9. See also M. Fleming, *The anarchist way to socialism: Élisée Reclus and nineteenth-century European anarchism*, London, 1975; H. Sarrazin, *Élisée Reclus ou la passion du monde*, Paris, 1985; and Beatrice

Giblin, 'Élisée Reclus (1830–1905)', in T. W. Freeman (ed.), *Geographers: bio-bibliographical studies*, 3 (1979), 125–32.

19 Patrick Geddes, 'Note on draft plan for Institute of Geography', *Scottish Geographical Magazine*, 18, 3 (1902) 142–4. See also J. G. Bartholomew, 'A plea for a National Institute of Geography', *Scottish Geographical Magazine*, 18, 3 (1902), 144–8. On Geddes, see H. Meller, *Patrick Geddes: social evolutionist and city planner*, London, 1990 and Ian Stevenson, 'Patrick Geddes 1854–1932', in T. W. Freeman (ed.), *Geographers: bio-bibliographical studies*, 2 (1978), 53–66.

20 See, for example, Yves Lacoste, *La géographie, ça sert, d'abord, à faire la guerre*, Paris, 1981; Brian Hudson, 'The new geography and the new imperialism', *Antipode*, 9 (1977), 12–19; and Richard Peet, 'The social origins of environmental determinism', *Annals, Association of American Geographers*, 75 (1985), 309–33.

21 See, for example, J. A. Mangan, *'Benefits bestowed'?: education and British imperialism*, Manchester, 1988; J. A. Mangan (ed.), *The imperial curriculum: racial images and education in the British colonial experiences*, London, 1993; Manuela Senudei, 'De l'empire à la décolonisation à travers les manuels scolaires', *Revue Française de Science Politique*, 16, 1 (1991), 56–86; and John M. MacKenzie, *Propaganda and empire: the manipulation of British public opinion, 1880–1960*, Manchester, 1984. The last named correctly notes that the same could be said of history teaching. On this see F. J. Glendenning, 'Attitudes to colonialism and race in British and French history schoolbooks', *History of Education*, 3, 2 (1974), 57–72 and F. J. Glendenning, 'School history textbooks and racial attitudes 1804–1911', *Journal of Educational Administration and History*, 5 (1973), 33–4.

22 See Anne Godlewska and Neil Smith (eds), *Geography and empire: critical studies in the history of geography*, Oxford, 1993; Felix Driver, 'Geography's empire: histories of geographical knowledge', *Environment and Planning D: Society and Space*, 10 (1992), 23–40; and David N. Livingstone, *The geographical tradition*.

23 On Kropotkin, see Myrna Margulies Breitbart, 'Peter Kropotkin, the anarchist geographer', in D. R. Stoddart (ed.), *Geography, ideology and social concern*, Oxford, 1981, pp. 134–53; and Olga Alexandrovskaya, 'Pyotr Alexeivich Kropotkin 1842–1921', in T. W. Freeman (ed.), *Geographers; bio-bibliographical studies*, 7 (1983), 57–62.

24 On Dresch's prominent opposition to the war, in which he joined forces with Jean-Paul Sartre, Aimé Césaire, Charles-André Julien and others, see Danièle Joly, *The French Communist Party and the Algerian war*, London, 1991, esp. pp. 123–9.

25 The most important text heralding this shift was David Harvey, *Social justice and the city*, London 1973, though this was concerned primarily with the contradictions and inequalities within the urban environments of the 'developed' world. See also Richard Peet (ed.), *Radical geography: alternative views on contemporary social issues*, London, 1978.

26 See, for example, Stuart Corbridge, 'Third World Development', *Progress in Human Geography*, 16 (1992), 584–95; James Sidaway, 'In other words: on the politics of research by "First World" geographers in the "Third World" ', *Area*, 24 (1992), 403–8; David Slater, 'On the borders of social theory: learning from other regions', *Environment and Planning D: Society and Space*, 10 (1992), 307–27; Michael J. Watts, 'Development I: power, knowledge, discursive practice', *Progress in Human Geography*, 17 (1993), 257–72.

Acknowledgements

The ideas for this collection originated in a conference on Geography and Empire, hosted by the Royal Geographical Society in May 1990. We are grateful to the society for its generous hospitality on that occasion and for financial support provided by the Nuffield Foundation. The ideas enshrined in several chapters were initially presented at this conference and benefited from the discussions which followed.

CHAPTER ONE

Agents and agencies in geography and empire: the case of George Grey

James M. R. Cameron

Early in the afternoon of Monday 14 November 1836, George Grey and his friend Franklin Lushington met Sir John Barrow, President of the Royal Geographical Society (RGS), to offer to explore any part of Australia that Barrow or his society nominated.[1] Grey and Lushington, both twenty-four years old and still lieutenants, had recently graduated from the senior officers' course at Sandhurst and were looking for fresh challenges. Neither knew much about exploration. They knew even less about Australia, although Grey was well aware, from his brief flirtation with the systematic colonisers, of the glory that had attached to Charles Sturt for his epic voyage along the Murray River, the mouth of which was now being considered as the site for the capital of the proposed Wakefieldian colony of South Australia. They did know from their discussions with Captain John Washington, RN, the secretary of the RGS, that its council was to meet later that afternoon and that the whole society was to meet that evening for the first time since June. They also knew that the society was active in the promotion of exploration and amenable to the approach of young men prepared to offer their services.

Barrow, always impressed when approached by enthusiastic and adventurous young naval and army officers who were prepared to risk their lives in the pursuit of geographical knowledge, recommended that the council accept Grey's and Lushington's offer. The council agreed. So, too, did the general membership before it settled down to hear highlights from Captain Robert FitzRoy's recent survey of south Atlantic and Pacific waters in the *Beagle*.[2] Thus was set in train the sequence of events which led to Grey and Lushington jumping ashore at Hanover Bay in north-western Australia, and nearly perishing of thirst just over a year later.[3]

Exploration as process

Explorations such as that undertaken by Grey and Lushington can be examined from several perspectives. The most common has been to cast the explorer in the role of romantic hero testing his character against adversity.[4] This approach is understandable, given the insatiable public demand, particularly pronounced in the nineteenth century, for accounts of outstanding personal courage, achievement and commitment, and the deeply felt urge among inhabitants of newly settled lands to create a panoply of home-grown heroes. But emphasis on the individual actor and his drama diverts attention away from exploration as a deliberate process by which peoples and nations attempted to increase their knowledge of the world and enhance their influence and power over it through trade, territorial acquisition and settlement, as well as through the diffusion of their ideas and values. When exploration is divorced from this broader context, important insights into the construction and application of geographical knowledge and the creation of empires are denied.

At a colonial level, for example, Overton[5] has shown that exploration and settlement exist in a close, symbiotic relationship, and that the pattern of settlement which evolves in an area can be understood only when we reject exploration as a series of unconnected events and view it as an 'interacting process closely linked to the development of a pioneer economy'. Major elements which interact include: the knowledge derived from all explorations – and not just the trail-blazing activities of the few; the motives for these explorations, successive subjective evaluations of the land surface; and the ways in which these are responded to by settlers. Regulating the level of interaction is the broad economic milieu, and particularly the demand for land.

The situation at an imperial level is simultaneously simpler and more complex. For imperial Britain, exploration was the spearhead of expanded trade and settlement, but, after the epic voyages of Anson and more especially those of Cook, exploration became a major vehicle through which Britain asserted its intellectual dominance, notably in the fields of geography and natural science. Cook's voyage of 1768–71, intended primarily to give astronomers a better vantage point from which to view the transit of Venus, changed forever the knowledge of the geography of the southern hemisphere, overthrew long-established notions of a southern equipoise counter-balancing the known land mass of the northern hemisphere, had a profound effect on conceptualisations of the plant and animal kingdoms, and led directly to the establishment of a penal colony at Botany Bay on Australia's eastern coastline in 1788. Its greatest significance, however, was to demonstrate the value of careful prior planning and the involvement of scientists and explorers in a partnership. Thereafter, few British expeditions were devoid of a scientific intent, typically determined by the involvement of the great learned societies in London with the Admiralty, the Colonial Office, and, more rarely, with the Board of Control, the body established in 1785 to supervise the affairs of the British East India Company.[6] Exploration became a three-phase activity of prior planning, the expedition itself, and a careful analysis and dissemination of its results. The net effect was that geography and empire became entwined and London became the control centre for Britain's exploratory endeavours.

It is ironic, perhaps, that while its distinctive flora and fauna continued to fascinate Europeans, Australia remained largely unknown in the fifty years after Cook's death. Its coastline had been examined by Flinders and Baudin between 1801 and 1803 and again by Phillip Parker King between 1818 and 1822, but it took settlers twenty-five years to break through the sandstone barrier to the west of Sydney and make tentative probes along the banks of the rivers flowing to the north west. Hume and Hovell did blaze a trail south west from Sydney to the southern coast, but it was 1829 before Sturt revealed that at least one of Australia's major river systems drained into the sea.[7]

Thus, our purpose here is to unravel some of the undercurrents in this imperial exploratory process, particularly as they relate to Australia, through a detailed reconstruction of the sequence of events which led to the Grey and Lushington expedition. No single case can reveal a complete picture but what this examination of the planning phase does achieve is to throw into bold relief the significance of the interplay of the personalities involved in the planning, their political and personal aspirations, and their perceptions of place in the determination of where, how, and by whom an exploration will be conducted and for what purpose. It also shows that the blending of contemporary geographical knowledge with personal and imperial ambition was central to all exploratory behaviour.

Agent and agency

For Grey and Lushington, Barrow's support was essential, but the inevitable consequence was that they became pawns in a bigger game, shaped, in part at least, by Barrow's own geographical knowledge and his views of empire. Barrow had established his reputation as an explorer through his travels in eastern Asia as a member of Lord Macartney's embassy to the Imperial Court at Beijing (1792–94), and, subsequently, in southern Africa during Macartney's governorship of the first British occupation of the Cape of Good Hope.[8] A close confidant of Sir Joseph

Banks, President of the Royal Society (1778–1820) and promoter of scientific discovery in all corners of the globe, Barrow had been involved in every significant government-sponsored exploration from his appointment as Second Secretary to the Admiralty in 1804, a position he held almost continuously for forty years. Following Banks's death in 1820, Barrow assumed his mantle as 'father of British exploration'.[9] Best known for his promotion of polar exploration over the thirty years until his death in 1848, Barrow was equally active earlier in central and northern Africa and sent a string of young men, Oudney, Clapperton and Denham, Ritchie and Lyon, and the Landers among them, to their glory or their death. On the Australian continent, he actively promoted the careers of Charles Sturt, Allan Cunningham, and Phillip Parker King, the son of a former governor of New South Wales and Banks's close family friend.

Barrow's writings had a wider influence for they helped to sustain an environment amenable to exploratory endeavour. His message to the nation in the eighteen books, fifteen entries in the *Encyclopaedia Britannica*, and more than two hundred anonymous articles in the *Quarterly Review* he wrote over a 50-year period was unambiguous. Apart from its intrinsic interest, exploration helped to maintain British pre-eminence in the field of science, particularly but not exclusively in geography. Equally, it tested the courage and honed the skills of the nation's future naval and military leaders and prepared them for any further foreign aggression. Moreover, it created opportunities for the expansion of trade.

As early as 1814 when planning the Tuckey expedition along the Niger, Barrow and Henry Goulburn, his Colonial Office counterpart, determined that the Colonial Office should take responsibility for land-based exploration while the Admiralty should sponsor the maritime component. Although most government-sponsored expeditions were jointly planned, this principle became firmly entrenched. The easy working relationship Barrow established with Goulburn (1812–21) was enjoyed with other under-secretaries in the Colonial Office, notably Wilmot Horton (1821–27) and particularly Robert Hay (1825–35), a close friend, former private secretary for thirteen years to the First Lord of the Admiralty, and co-founder of the RGS. His relationship with James Stephen, who replaced Hay in February 1836, was far less cordial, as Barrow's old-style conservatism and use of informal networks clashed with Stephen's liberalism and insistence on following established procedures, but by now Barrow had a range of prominent politicians to call upon for support. Many had served their apprenticeships as junior ministers within the Admiralty. By 1836, and recently knighted by William IV, his former Lord High Admiral, Barrow was at the height of his influence. His authority within the Admiralty was unquestioned and he was able to shape the direction of naval exploration through his protégé, Francis Beaufort, the naval hydrographer.

The creation of the RGS in 1830 strengthened his position. This grew out of the Raleigh Club, an exclusive dining club of distinguished travellers. In planning meetings chaired by Barrow, the society determined that one of its major functions would be to 'prepare instructions for such as are setting out on their travels . . . pointing out the research most essential to make . . . [and] to render such pecuniary assistance to such travellers as may require it'.[10] To achieve this and the greater purpose of collecting and disseminating geographical knowledge, the foundation Council of the RGS was drawn from senior members of the Foreign Office, the Colonial Office and the Admiralty as well as the Royal and Geological Societies, notable travellers, and, because they had a particular contribution to make to geographical knowledge, senior military and naval officers and officers of the East India Company's marine and military branches. This wide cross-representation was retained in subsequent years. Lord Goderich, the Colonial Secretary, agreed to become the foundation president. The king agreed to be patron.[11]

From the outset, the RGS actively pursued its objectives. In October 1834, through the intervention of Barrow and Hay, it gained government funding for expeditions in Guyana and inland from Delagoa Bay.[12] Then, in early 1836 and under Barrow's leadership, it set in train the planning of an expedition to complete the survey of North America's Arctic coastline.[13] The appearance now of Grey and Lushington provided the means to unravel the mysteries of inland Australia which had tantalised Barrow for two decades.

Australia: great river or inland sea?

Barrow first expressed public interest in the geography of Australia in 1814 in his major review of Flinders's circumnavigation of the continent.[14] At that time, Barrow concluded that Australia's river systems must flow into a great inland sea, a view he revised two years later when preparing sailing instructions for King's maritime survey. No doubt under Banks's influence and after a careful re-reading of Flinders's account and Dampier's *Voyages*, he expressed a strong hope that the mouth of a great inland river would be found near King Sound on the north-west coast where Dampier had experienced a tidal range in excess of 10 metres. King failed to find that outlet, although he had not explored the coast near King Sound in any detail. This, together with

John Oxley's encounter with a series of impenetrable marshes which impeded his progress down the Macquarie River north west from Sydney, led Barrow to conclude that 'the surface of this vast country somewhat resembles that of a shallow basin' in the centre of which were 'a succession of swamps and morasses, or perhaps a vast Mediterranean sea'. Throughout the 1820s his views fluctuated as each explorer's account became available, but he gradually moved towards a belief in a great river system. Sturt's discovery of the mouth of the Murray, previously overlooked by Flinders, encouraged Barrow into believing that King may have overlooked a similar river mouth on the north-west coast. He was confirmed in this view by Allan Cunningham's theorising in 1832 in an article commissioned for the RGS by Goderich.[15] Cunningham had established his credibility through a brilliant series of journeys through the plains and tablelands of eastern Australia which were notable for their botanical discoveries,[16] so that Sturt's similar conclusions, made two years later when he was in Britain, set the matter at rest.[17]

Thus, when Barrow met Grey and Lushington after the council meeting, he directed their attention to the north-west corner of the continent. He also advised them to see Stephen and seek Colonial Office support and financial assistance. This they did within a week. Stephen believed that support would be given provided that the RGS endorsed the geographical significance of the proposed expedition.[18] Negotiations now began in earnest.

Setting the agenda

Within two weeks of their initial approach to Barrow and assured of RGS endorsement, Grey and Lushington wrote to Lord Glenelg, the Colonial Secretary, seeking official endorsement. Their plan of attack had now begun to take shape. They proposed travelling by land north from the Swan River Colony until they encountered a major river flowing to the north west which they would follow to the sea before tracing its course as far inland as possible.[19]

Simultaneously, and following a council meeting called for that purpose, Barrow also prepared a memorandum for Glenelg. This reviewed in some detail the society's interest in the north-west coast and supported in glowing terms the initiative shown by Lushington and Grey, but outlined an alternative exploration plan. The young men should proceed direct to Swan River to seek the advice of the surveyor general John Septimus Roe, who had sailed with King and so was familiar with that coastline, as well as enlisting the support of the governor, Captain James Stirling, RN, to purchase a small boat from which to search for a major opening. If they discovered this, they should explore as far upriver as possible before returning overland to Swan River.[20]

Glenelg's response was to invite Grey and Lushington, and a deputation from the RGS to meet him separately.[21] At the first meeting on Monday 5 December, Grey and Lushington met a favourable reception for their views on the advantages of an overland expedition and left the meeting with instructions to prepare detailed costings. These were completed within a week.[22] Grey was already known to Glenelg through his friendship with William Grant, Glenelg's younger brother and private secretary.[23] Whether this influenced Glenelg's response can only be surmised, but it was to be of great value in coming months.

When the deputation from the RGS, consisting of Barrow, Beaufort, Washington and the geologist Roderick Impey Murchison, met Glenelg on the following afternoon, it came armed with a briefing paper prepared by Washington on Barrow's instructions.[24] Unlike Barrow's proposal of a week before, this outlined three options: an overland expedition similar to that proposed by Grey and Lushington; an exploration from the sea similar to that proposed by the RGS; or a joint venture in which Grey and Lushington would be attached to a naval survey vessel which would land them on the coast and provide support, while simultaneously surveying the adjacent coastline. Carefully constructed, the paper drew attention to the logistical difficulties and greater costs of the first two proposals to highlight the obvious advantages of the third.

These advantages were not fully enumerated in the paper but had been foreshadowed a week earlier in Barrow's memorandum to Glenelg.[25] In this, Barrow had observed that the object of the expedition was to explore a part of Australia 'which had never yet been visited', before he went on to say that not only was this object of 'real geographical interest' but that it may also be 'attended with important and advantageous results as regards the spread of civilization in this great country and at the same time conduce to the benefit of the commercial interests of Great Britain and India'. He was particularly gratified, therefore, that two well educated and appropriately qualified young officers were prepared to volunteer for this hazardous but extremely important service.

Barrow's imperial vision

But Barrow, in praising the worth of Grey and Lushington had a deeper, decidedly imperial, purpose. Once described as 'that great covetor of islands' by Lord Bathurst, then Colonial Secretary,[26] Barrow's views on empire had been shaped by his experiences in China and the Cape Colony under the tutelage of Macartney, a former governor of Grenada and the East India Company's presidency at Madras. At the Admiralty, he came under the direct influence of Henry Dundas, Lord Melville, the First Lord, who had been responsible when president of the Board of Control both for sending Macartney to China in an unsuccessful attempt to expand the China trade and for the occupation of the Cape of Good Hope in 1795. These acts were part of Dundas's larger intention of committing Britain to an eastern empire.[27] Barrow's wartime experiences as a senior member of the Admiralty added a further dimension.

Barrow's views, first expressed publicly in 1804 in an impassioned plea for the reoccupation of the Cape, written at Melville's request,[28] were refined over the next decade.[29] Like Macartney, Melville and many other strategists of the period,[30] Barrow concluded that an insular Britain, if dependent solely on its own resources, would always be at the mercy of continental powers like France whose considerable resource base could always be augmented from contiguous conquered territories. Britain's survival compelled it to gather its additional resources from an expanding maritime empire of trade. For trade to be secure, Britain had to control the sea lanes which, in turn, meant that not only did its navy have to be dominant but it also had to control key strategic points such as the Cape, Mauritius, Trincomalee, and Jakarta, upon which that trade depended and whose occupation was used to powerful effect during the wars against Napoleon (Figure 1.1). The social and economic distress following the wars led to further refinement and Barrow became convinced that some key points could be converted to colonies of settlement for Britain's redundant poor and for those people who owned a small amount of capital. As these colonies would become self-sufficient quickly, government expenditure for strategic purposes could be reduced dramatically.[31] He thus played an active role in the establishment of the Albany settlement in the eastern Cape in 1820 and the Swan River Colony in 1829 and became a strenuous advocate for emigration to southern Africa, the colonies on Australia's eastern coast, and upper Canada.[32]

To Barrow, as to many of his contemporaries, the post-war return to the Dutch of its East Indian possessions, made to cement stategic alliances within Europe, seemed to repeat the folly of the surrender of the Cape in 1802, particularly when it appeared that the East India Company would repudiate Thomas Stamford Raffles's occupation of Singapore which, like Jakarta, was one of the major control points for the heavily trafficked Sunda Straits used by East Indiamen heading for Canton and the lucrative tea trade. When a group of private traders approached Bathurst towards the end of 1823 with a proposal to neutralise any effect of this by establishing a base on Australia's

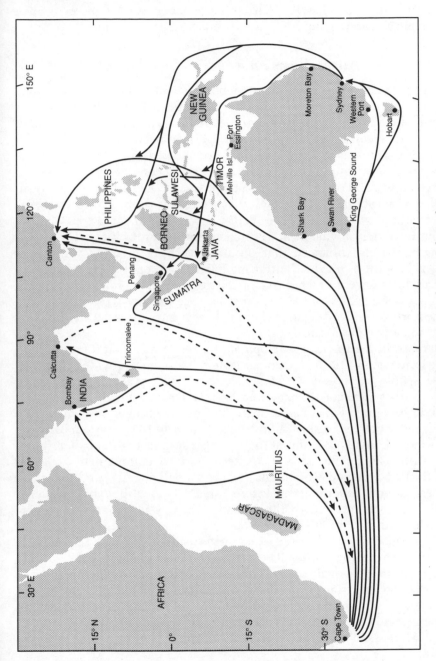

Figure 1.1 Early nineteenth-century sea-lanes in the Indian Ocean

northern or north-western coast to control the eastern passages through Lombok and Ombai straits, Barrow's assistance was sought to locate a suitable site and to draft the instructions for its occupation.[33]

Australia as an imperial beachhead

With the resultant occupation of Melville Island in 1824, Barrow turned his attention more closely to Australia, still a British penal outpost centred on Sydney with subsidiary settlements in Van Diemen's Land, but with great potential to become a bridge between the Indian and Pacific Oceans, as evolving trade patterns showed (Figure 1.2). Indeed, as Barrow knew from his time at the Cape, these convict settlements had been an integral part of the China trade since 1788 because shippers, in what was esentially a one-way trade, found it profitable to transport convicts to New South Wales before proceeding in ballast to Canton. The activities of whalers and sealers operating in Pacific and Southern Ocean waters enhanced that profitability by generating valuable if illegal additions for the Canton and Indian markets.[34] King's discovery of a safe passage through the Great Barrier Reef led a growing number of China-bound ships to use the inner passage through Torres Strait and the northern coastline instead of the longer outer passage around New Ireland and New Guinea.

Australia's economic development had been such that it promised to become one of Britain's most valuable possessions, provided claims to the whole continent could be enforced. The sealanes emphasised Australia's hexagonal shape and Barrow was confident that control could be gained with the occupation of its six corners, his 'stakes in a ring fence erected to keep out intruders'. In early 1826, Hay, on Barrow's advice,[35] convinced Bathurst to issue instructions to establish outposts at King George Sound and Western Port to complement those at Melville Island and Moreton Bay where a small convict settlement had been established the year before. Shark Bay, although considered initially, was rejected on the grounds that it was too arid to support settlement by any nation.[36]

Barrow was not consulted when the decision to abandon the Melville Island and nearby Raffles Bay settlements was made in 1829. Deeply disappointed, he made several subsequent attempts to generate interest in the reoccupation of northern Australia but could gain no official support.[37] The initiative taken by Grey and Lushington provided an unexpected opportunity to present the case to Glenelg. With Beaufort's and Murchison's help, he intended to exploit that opportunity to the full.

Should a major river be discovered on the north-west coast as he fully

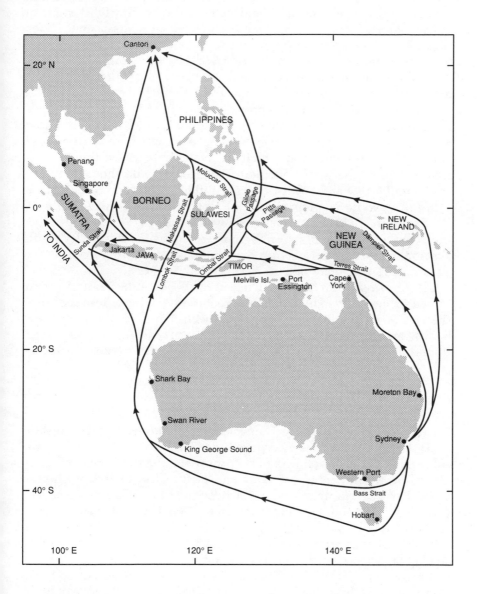

Figure 1.2 Preferred shipping routes from Sydney to Asia in the first half of the
nineteenth century

expected, the superior advantages of a settlement established at its mouth for the protection of shipping bound for India and China and 'from its power of extension inland by water communication' would be 'incalculable'. It could be another St Lawrence or a Mississippi. Referring to French and American preparations for similar scientific voyages to the area, Barrow felt compelled to stress the vulnerability of northern Australia to foreign occupation and the likely effect that this would have on the growing Australia–China trade which had expanded rapidly with the opening of Canton to private traders in 1834. This growth was such that the Admiralty was contemplating a detailed survey of Torres Strait which was part of the increasingly preferred route from New South Wales to China and which, in itself, emphasised the strategic significance of the whole northern coastline. Should this survey proceed, and Barrow and Beaufort were confident that it would, then the Admiralty would be prepared, in all probability, to join the Colonial Office and the RGS in a joint survey by land and sea. This would reduce the extent of the Colonial Office commitment as the Admiralty would absorb the costs of transporting Grey and Lushington to the north west and the RGS would provide the necessary expertise.[38]

Glenelg, fresh from his presidency of the Board of the Control, where he had successfully manoeuvred the East India Company into accepting the end of its trading monopoly, was well versed in the issues that Barrow and the deputation explored with him and, while shying away from the question of resettlement, gave his approval to the joint survey subject to his receiving a detailed plan.

Negotiating the details

Barrow now moved quickly. Using Beaufort's summary of their discussions with Glenelg, he convinced Lord Minto, First Lord of the Admiralty, of the desirability of re-commissioning the *Beagle* to survey Torres Strait and transport Grey and Lushington to the north-west coast. His own summary, now expanded into a detailed proposal, was endorsed by the council and forwarded to Glenelg within a week with a covering note to indicate that the Admiralty approved the proposal in full. Almost as an afterthought, for he hadn't discussed it with them, he added that, as the venture had now become a 'Naval Measure', Grey and Lushington may wish to proceed with their original intention to travel overland, in which case £1,000, managed by the RGS, would be sufficient to cover their costs.[39]

Over the next few weeks Stephen facilitated Grey's and Lushington's request for leave,[40] and arranged for the transfer of £1,000 from the Treasury to the society.[41] When Stephen notified him of Treasury approval,[42] Barrow was spurred into action once more and pointed out to Stephen that the Admiralty, while endorsing the desirability of surveying the unmapped portions of the Australian coastline, had postponed making a final decision until it knew what the Colonial Office intended. Official notification was now needed urgently as the Naval Estimates closed in ten days and the *Beagle* could not be commissioned until approval was given.[43] Stephen obliged with an immediate reply.[44]

Preparation of the *Beagle* was speeded up and George Windsor Earl, already known to the RGS for his travels in eastern Indonesia, was enlisted to assist the Hydrographic Office update its charts of the Arafura Sea and prepare sailing instructions.[45] It was therefore with some satisfaction that the council met on 13 February to endorse all that happened in the last five weeks before they joined with other members to hear Earl expand his already published views on the agricultural and commercial capabilities of Australia's northern coasts.[46] Unbeknown to them, Grey had seen Stephen two days before to complain bitterly about the society's most recent proposals.[47]

Grey and Lushington had met Barrow in mid January and were told that the *Beagle* would carry them to Swan River.[48] Grey subsequently had private discussions with Stephen and with Gordon Gairdner, the clerk in the Colonial Office responsible for Australian affairs. One of them had shown him Washington's briefing paper and indicated that the joint survey was the one most favoured by the RGS.

To this Grey took particular exception as it was quite unsatisfactory and 'not at all calculated to attain the desired ends'.[49] The coastal survey was really a task for the crew of the *Beagle*, would take several years to complete and would involve them as assistant surveyors only, a role for which they were not trained. More seriously, they would suffer 'the mortification of considering whatever celebrity or success might attend the expedition, would attach to the vessel and not to us'. The overland expedition, which Grey now wrongly and mischievously attributed to the RGS, was more acceptable but still suffered from serious drawbacks, not least of which was that they would have to travel 1,200 miles overland to get to the 'most desirable point' with their supplies 'far spent, the party fatigued in body, and wasted in spirit by months of incessant travel ... and the minds of all bent on how they were to return', so that they would not explore fully any major rivers they encountered.

His own proposal, actually the society's, was much more satisfactory but needed to be modified in the light of the information he had gathered over the preceding three months. He proposed that the *Beagle* land them at Swan River where they would buy stores and hire a small vessel before sailing to Kupang to buy ponies. From there they would head for Prince Regents River, 'and thus be landed on the point of greatest interest' from which they could make their way overland to Perth. By doing this:

> We shall be landed on the very spot which it is most necessary carefully to examine, we shall enter on this examination buoyed up by the hope, Masters of our time, and in full possession of all our energies both mental and bodily: the party will cheerfully trace each step that bears them closer to their friends and home, and the length of our journey will be at once diminished by half.[50]

While hopeful that the revised proposal would be accepted, even though it would cost more, Grey expressed his and Lushington's willingness 'to consider any other which may seem more advantageous'. Stephen's only formal response was to send the proposal to Glenelg with the comment 'As your Lordship has this subject almost exclusively under your personal cognizance, I transmit Mr Grey's letter at once to you.'[51]

When Grey pressed Glenelg for a response some five weeks later,[52] the Society's advice was sought once more. As before, Washington prepared another briefing paper which mounted a powerful attack on Grey's most recent scheme. Washington took the view that it may be possible to hire a small vessel at Swan River but any vessel hired there would be very expensive and would be quite unsuited to the detailed inshore survey which Grey's proposal required if it was to succeed. Neither Grey nor Lushington was suited to this task which was 'work requiring more than common attention by sailors, whose profession it is, and not effected by military men'. More seriously, however, they had 1,500 miles to traverse in their return to Swan River and absolutely no experience to draw upon.[53]

The joint survey also had serious drawbacks, not least of which was that the *Beagle* would commence its survey in Bass Strait before moving to the Barrier Reef. It could be two years before it was based off the north-west coast.

This led Washington to come down strongly in favour of an overland expedition which he was careful to point out was 'the original intention of Messrs Grey and Lushington', but which should now be heavily modified to minimise the possible effects of their obvious inexperience and apparent lack of judgement. Pointing out that Oxley, Cunningham, Hume, Sturt and Mitchell had many years of experience in Australia before they attempted to explore and yet all had encountered serious difficulties, he concluded:

> All circumstances considered, it seems expedient that the exploring party should make Swan River their headquarters – that they should by short

excursions endeavour to gain some experience of the mode of 'living in the bush' and of conciliating the natives – of making acquaintance with their manners and customs – in short to endeavour to feel their way, with the resources and still more the judgements and advice and experience of the colony to fall back upon – in this manner though much might not be done in the first season it would give every hope of double work [in] the second and that done in a superior manner to anything that could be expected from a party without experience dashing at once into danger – which probably the ardour and zeal of young Military Men not tempered with years might lead them to do.[54]

Barrow and Beaufort met Glenelg on 7 April, supposedly to resolve the concerns that Grey had raised. Instead, they resumed even more forcibly their claim for reoccupation of northern Australia as the means of asserting British control over the whole continent.[55] From the evidence they had gathered from the Admiralty records, it was clear that the earlier settlements had been sacrificed to the wilful misrepresentations of people who were ill suited or unwilling to do what was required to ensure success.[56] Yet, it was equally clear that the settlements had been successful and had been abandoned in a flourishing state. As a direct consequence, the whole northern coastline has been 'thrown open to any European or Asiatic power that might think fit to avail themselves [sic] and create another Singapore'.

When Glenelg replied that there was little likelihood of this happening because the earlier settlements gave Britain prior claim, they retaliated: 'if original discovery and the act of nominal possession constituted sovereignty, the Dutch might claim the larger portion of the coasts of New Holland and Van Diemen's Land'. Glenelg succumbed to the force of their argument and went so far as to agree that the new settlements at Port Essington and Cape York would be established by marines under the superintendence of a naval officer, for 'naval officers, from their experience of difficulties and practical habits or providing against them, have more resources at their command than most others'.

As for Grey's complaints, they were swept aside. Glenelg was persuaded to accept the original proposal of an overland survey, suitably modified to take account of the concerns Washington had raised. However, it was the end of the month before Grey and Lushington were informed of this decision and that the society had been requested to prepare detailed instructions.[57]

If Grey and Lushington were disappointed by this outcome, they gave no sign and continued preparations for their departure on the *Beagle*.[58] For its part, the council of the society met on 8 May and requested Washington to prepare detailed instructions for the Colonial Office to issue.[59] These were completed ten days later and, after receiving the required approval from Barrow and Beaufort, were forwarded on 23 May.[60]

Washington had further refined his views. This expedition was to be a reconnaissance only. Grey and Lushington, accompanied by at least two colonists familiar with exploration and two native guides, were to travel up to 700 miles (1,120 km) in a north-east direction from Perth before returning on a parallel route 100 miles (160 km) to the south. They were to record the precise location of major topographic features such as rivers, mountain ranges and lakes but were not to deviate from their line of march under any circumstance and, in particular, they were not to follow any rivers towards the sea. Depending upon their discoveries, these could be explored by a second expedition, the details of which were not specified.[61]

Appalled by the restrictions imposed on the expedition and apparently unaware that the tenor of Barrow's and Beaufort's discussions with Glenelg had led them to believe that these were appropriate, Stephen told Gairdner to meet Grey and Lushington and show them the proposed instructions.[62] Grey erupted. As soon as possible on the following Monday morning, he presented Stephen with an outline, not altogether accurate, of the steps the RGS had taken progressively to debase his original conception and diminish his role to the point where he was now a mere agent of the society 'in an expedition which I never contemplated undertaking and [which] does not hold out to me any prospect of such results, as those the hope of obtaining which has been my sole motive from the beginning of this affair'.[63]

Stephen arranged an immediate interview with Glenelg so that Grey could expand more fully upon the 'disappointments and mortifications' to which he had been subjected by the society's high-handed behaviour. Grey made the most of this opportunity:

> Engaged by your Lordship's great personal kindness to me, and also by my own ardent wishes, it would have been impossible for me to abandon a project on the success of which I had set my heart, but the real truth is that the project has abandoned me. I cannot see my way to the ends I have desired, and held up to Government and I therefore feel that it is a duty which I own to your Lordship in gratitude, and an obligation to myself as an honest man to decline.[64]

Faced with such a plea, and by way of compensation for Grey's obvious anguish, Glenelg requested Grey to prepare his own instructions. These were submitted the following day.[65] Approved immediately, they were issued to Grey and Lushington by the Colonial Office a week later,[66] at which time the RGS was informed that its services were no longer required.[67]

[28]

Thus, Grey emerged the victor five weeks before the *Beagle* sailed from Portsmouth on 5 July 1837 to take him to his destiny. He was empowered to hire a boat at either the Cape of Good Hope or Swan River and to proceed direct from there to Prince Regents River which he was to trace upstream as far as possible before returning overland to the Swan River Colony. He was to consult with Stirling and Roe on his 'first arrival at Swan River . . . and, according to the advice and information you may receive from them you will fill up the minor details of the plan', but, as he was to use his discretion in all matters, he was not required to consult with them before landing on the north-west coast.

Any disappointment Barrow may have felt in losing control of the expedition was assuaged by his conviction that northern Australia was soon to be reoccupied. Grey had served his purpose and Barrow, too, had emerged victorious. If he held any ill feeling towards Grey, it was not evident. Certainly, he did not oppose the putting forward of Grey's name for membership of the RGS in 1841, although Grey, perhaps showing an unusual sensitivity in view of the way he had exploited the society for his own ends, declined the honour.

Lushington was not so fortunate. As soon as he heard that Grey intended declining the society's proposed expedition, Lushington offered his services to the society on any terms that it determined.[68] His lapse of loyalty was repaid in many small and vindictive ways, the ultimate insult being to become the unknown partner in Grey's great adventure.

Reflection

In reflecting upon this sequence of events, Grey's own account, published five years later, provides a useful vantage point:

> We addressed a letter to Lord Glenelg, the Secretary of State for the Colonies, wherein we offered our services to conduct an exploration from the Swan River to the northward, having regard to the direction of the coast, so as to intersect any considerable body of water, connecting it with the interior, and in the event of such being discovered, to extend our examination of it, as far as circumstances may permit . . . The offer and suggestions were favourably entertained by Lord Glenelg, and further communications invited; and, the project being favoured by the support of the Royal Geographical Society, our services were finally accepted by the Government. More mature consideration, however, led to a material alteration in the first plan for . . . it was considered, for several reasons, more advisable that the exploration should commence from the vicinity of the Prince Regent's River, on the north-west coast, and be directed towards the Swan.[69]

While Grey points here to broad changes in planning, he makes no reference to fundamental disagreements between himself and the RGS, or to increasing tension with Washington, itself a reflection of larger tensions between military and naval officers, or to Barrow's negotiations with Glenelg, about which he was fully informed, or to the importance of Stephen's patronage. Such selective reporting from an ambitious young officer anxious to embark on a colonial career may be understandable. Yet, all of these elements were fundamental in determining the nature and direction of the proposed expedition and underline the need to reconstruct in detail the antecedents of exploratory activity if the significance of that activity is to be understood. Explorers' accounts cannot be taken at face value for they invariably portray a desired image of self.

Grey's statement also cloaks the deeper forces at work which provide the broader context for his own and other expeditions, two aspects of which are pertinent here. Firstly, exploration was a major vehicle for the acquisition of geographical knowledge, so much so that for most of the period of European expansion 'exploration' and 'discovery' have been interchangeable terms. The epic voyages of the nineteenth century such as Alexander von Humboldt's travels in South America and the numerous, and fruitless, searches for the North West Passage, while laden with heroic overtones, were essentially geographic or more broadly scientific in orientation.[70] Even minor voyages, such as Grey's, had this focus. This goes some way to explaining the essential tensions between Grey and the RGS, for Grey sought glory through discovery while the society sought information. And the society had authority on its side. After Bathurst's edict of April 1816, issued at the insistence of Banks and Barrow, all government-sponsored expeditions were required to maintain and then forward to London meticulous records of all that they saw and thought on their travels.[71]

The results were eagerly seized upon by armchair geographers, perhaps best exemplified in this period by Barrow himself, who were anxious to build up a comprehensive picture of their enlarging world. But it would be misleading to believe that this picture was constructed from facts alone, for no information could be separated from speculations and expectations as well as the theories and constructs held by its gatherers and interpreters. Mungo Park's two expeditions into central Africa, for example, spawned a string of abortive attempts spearheaded by Banks and then Barrow to prove that the Niger, whose route Park had part traced, either flowed into the Congo or swept across the continent to form the headwaters of the Nile – ideas first put forward by Herodotus nearly two millennia before.[72] Now and through Grey and Lushington, Barrow was trying to give shape to his speculations about the nature of the Australian land mass. For him, such speculations were essential: 'they are frequently the parents of exact geography and to them are owing some of its most brilliant and important discoveries – such for instance as those of Columbus, Vasco da Gama, Tasman, Cook, and Parry, which were all undertaken and effected on hypothetical grounds'.[73]

It is this dimension of information gathering that has led John L. Allen to asert that imagination and geographical exploration are inextricably intertwined.[74] As he concluded from a detailed examination of the Lewis and Clark expedition of 1804–06 across the United States, perceptions of place and existing systems and frameworks of geographical understanding, which together he designated as geographical lore, conditioned all aspects of exploratory behaviour. Allen argued that this behaviour should be seen as an interactive process, the major components of which were the articulation of certain goals gleaned from the geographical lore of the particular time and place, the implementation of a plan to achieve these goals, and the recording and interpretation of the results.

Secondly, and however important the gathering of information and the satisfying of curiosity were, they were not the sole or major determinants of exploratory activity. Indeed, as this example shows, it was rare that an expedition was free of other objectives and even more rarely was it free from being influenced by the motivations and personalities of those involved in determining and implementing its direction and purpose. That exploration was a major avenue for personal advancement is a well accepted feature of the exploration mystique but this was only possible because exploration has always been an important instrument of national policy, often with major commercial, political or strategic consequences. The Jeffersonian dream embodied in the Lewis and Clark expedition, for example, was not so much the discovery of an east–west version of the Mississippi but the opening of a riverine route to India.[75] Barrow was imbued with a similar vision so that his and Grey's near separate negotiations with Glenelg were complementary dimensions of the same entity. They constituted a meshing of geography and empire.

While Barrow's imperials ambitions were more aggressively territorial than either Banks's or his successor, Murchison's,[76] he was equally concerned with establishing the dominance of Britain's 'invisible' empire. As he asserted repeatedly through the pages of the *Quarterly Review*,[77] knowledge was power, and exploration produced knowledge, a view that was widely shared by his colleagues in the Royal Society and the Royal Geographical Society as well as within government circles. It was this desire to ensure intellectual as well as territorial supremacy that made exploration so attractive and ensured

[31]

that the planning of expeditions and the interpretation of their results were controlled from London. This, in turn, gave Barrow the basis of his influence. Recognised as the most knowledgeable geographer of his day, he had access through his personal, political and professional networks to the information and the infrastructure upon which successful execution depended.

As for the resultant expedition, it is a matter of record that Washington's fears about Grey's impetuosity and his lack of experience were borne out. Grey's two attempts to achieve the objectives he set himself were disasters. He retreated from the first to be rescued by the crew of the *Beagle* after penetrating a few miles inland and after being seriously wounded in an unnecessary confrontation with local Aborigines. In the second, following an avoidable boatwreck, he and his party trudged down the western Australian coastline, losing a man from starvation, and again having to be rescued, this time from Perth. His incompetence was recognised in the colonies and he was lampooned mercilessly in the colonial press as the inventor of the 'circumbendibus', a geometrical instrument which allowed its user to travel in circles.[78] Yet, the publication of his *Explorations in Western Australia* was a great success.[79] Modelled on the style of Fenimore Cooper, his favourite novelist, Grey cast himself in the role of an antipodean Deerslayer to emerge as a great explorer-hero. This made it easy for Stephen to recommend his appointment as governor of the troubled fledgling colony of South Australia[80] – the beginning of a colonial career spanning sixty years which encompassed Australia, New Zealand and South Africa and through which Grey established an enviable if ambiguous reputation as explorer, naturalist, linguist, anthropologist, colonial administrator, and politician.[81] It is apposite to note that the colonial-trained surveyor Augustus Gregory retraced Grey's path less than a decade later with no difficulty and no recognition.[82] But there was an essential difference. Gregory sought grass for expanding colonial flocks. Grey, ever with an eye on London, sought imperial fame.

Notes

1 Grey and Lushington to Barrow, 14 November 1836, *Journal MSS Australian 1836 (Grey and Lushington)* (hereafter *Grey and Lushington Papers*), Royal Geographical Society Archives.

2 *Council Minutes of the Royal Geographical Society* (hereafter *RGS Council Minutes*), 14 November 1836.

3 George Grey, *Explorations in Western Australia, 1837–1839*, London, 1841, I, pp. 67–80.

4 See, for example, Robert Sellick, 'The explorer as hero: Australian exploration and the literary imagination', *Proceedings, Royal Geographical Society of Australasia (South*

Australian Branch), LXXVIII, 1977, pp. 1–16.

5 J. D. Overton, 'A theory of exploration', *Journal of Historical Geography*, VII, 1981, pp. 53–70.

6 D. W. Mackay, *In the Wake of Cook: Exploration, Science and Empire, 1750–1801*, London, 1985, pp. 3–24.

7 See J. H. L. Cumpston, *The Great Sea and the Inland River*, Sydney, 1952. A succinct summary is contained in J. M. R. Cameron, 'The exploration of Australia', in Helen Delpar (ed.), *The Discovers: An Encyclopaedia of Explorers and Exploration*, New York, 1979, pp. 61–9. Recent assessments of Flinders, Baudin and Sturt include, respectively, K. A. Austin, *The Voyage of the Investigator 1801–1803. Commander Matthew Flinders, R.N.*, London, 1964; F. Horner, *The French Reconnaissance: Baudin in Australia 1801–1803*, Melbourne, 1987; E. Beale, *Sturt, the Chipped Idol*, Sydney, 1979.

8 See John Barrow, *An Account of Travels into the Interior of Southern Africa in the years 1797 and 1798*, London, 1801; John Barrow, *An Account of Travels into Southern Africa*, London, 1804; John Barrow, *Travels in China*, London, 1805; John Barrow, *A Voyage to Cochinchina in the years 1792 and 1793*, London, 1806. For a recent assessment of Barrow's role at the Cape of Good Hope, see M. Boucher and N. Penn (eds), *Britain at the Cape, 1795 to 1803*, Houghton, South Africa, 1992, esp. pp. 91–170.

9 Clements R. Markham, *The Fifty Years' Work of the Royal Geographical Society*, London, 1881, p. 14.

10 *Journal of the Royal Geographical Society*, I, pp. v-vi.

11 See *Minutes of the Raleigh Club*, RGS Archives, for 1830, especially the Geographical Society Prospectus of 4 August 1830. The king's acceptance of patronage, and Barrow's role in this, is recorded in *RGS Council Minutes*, 25 October 1830.

12 *RGS Council Minutes*, entries from 25 May 1833 to 18 October 1834.

13 *RGS Council Minutes*, 8 February 1836, 20 February 1836, 14 March 1836, 25 April 1836.

14 Barrow's interest in the question of the inland sea or great river may be traced from his anonymous writings in the *Quarterly Review*. See, in particular, XII, 1814, p. 15; XXII, 1820, p. 480; XXIV, 1820, pp. 70–2; XXXII, 1825, pp. 315–9; XXXVII, 1828, pp. 31–2. See also *Journal of the Royal Geographical Society*, I, 1830, pp. 2–4 and II, 1832, p. 318.

15 Allan Cunningham, 'Brief view of the progress of interior discovery in New South Wales', *Journal of the Royal Geographical Society*, II, 1832, espcially pp. 131–2.

16 W. G. McMinn, *Allan Cunningham. Botanist and Explorer*, Melbourne, 1970, esp. pp. 114–21.

17 Sturt to Hay, 24 February 1834, Colonial Office Records, Series 323, file 173, folios 390–1 (hereafter CO 323/173, 390–1).

18 *Grey and Lushington Papers*, (b), 21 November 1836.

19 Grey and Lushington to Glenelg, 28 November 1836, CO 201/258, 20–1.

20 Enclosed in Washington to Glenelg, 30 November 1836, CO 201/256, 479–82. See also *Grey and Lushington Papers*, (c).

21 Undersecretary Sir George Grey to Grey and Lushington, 3 December 1836, CO 202/35, 97–8; Grey to Washington, 3 December 1836, CO 202/35, 98–9.

22 Mr Grey's and Mr Dawson's estimates for Australian Expedition, 12 December 1836, *Grey and Lushington Papers*, (f).

23 William L. Rees and Lily Rees, *The Life and Times of Sir George Grey, K.C.B.*, Auckland, 1892, p. 16. See also James Macarthur to William Macarthur, 18 December 1836, *Macarthur Papers*, XXXV, 396, Mitchell Library, A2931.

24 Plan & estimate for expendition into Australia, 5 December 1836, *Grey and Lushington Papers*, (e).

25 *Grey and Lushington Papers*, (c).

26 Bathurst's undated minute at CO 324/75, 266; see also J. J. Eddy, *Britain and the Australian Colonies, 1818–1831*, Oxford, 1969, p. 235.

27 For Dundas's role in imperial expansion, see Holden Furber, *Henry Dundas, First Viscount Melville, 1742–1811*, London 1931; Vincent Harlow, *The Founding of the*

Second British Empire, London, 1952–1961; Michael Fry, *The Dundas Despotism*, Edinburgh, 1992.

28 John Barrow, *An Autobiographical Memoir*, London, 1847, pp. 251–3; Barrow, *Travels in Southern Africa*, II.

29 Barrow, *Cochin China*, pp. 334–8; *Quarterly Review*, VI, 1811, pp. 487–517; VIII, 1812, pp. 239–86; XII, 1815, pp. 444–6; XIV, 1815, pp. 1–38.

30 For example, Patrick Colquhoun, *A Treatise on the Wealth, Power, and Resources of the British Empire*, London, 1814; C. W. Pasley, *Essay on the Military Policy and Institutions of the British Empire*, London, 1810.

31 *Quarterly Review*, XII, 1814, p. 41; XXII, 1819, pp. 203–5.

32 *Quarterly Review*, XXII, 1819, pp. 203–24; XXIII, 1820, pp. 73–83; XXIV, 1820, pp. 55–72; XXVII, 1822, pp. 99–109; XXXII, 1825, pp. 311–42; XXXIII, 1826, pp. 410–29; XXXVII, 1828, pp. 1–32; XXXIX, 1829, pp. 315–44; LIV, 1835, pp. 413–29.

33 See J. M. R. Cameron, 'Traders, government officials, and the occupation of Melville Island', *Great Circle*, VII, 1985, pp. 88–99.

34 Australia's position in relation to the Asian trade is outlined in D. S. Macmillan, 'The rise of the British Australian shipping trade, 1810–1827: its problems, progress and promoters', in A. Birch and D. S. Macmillan (eds), *Wealth and Progress*, Sydney, 1967, pp. 1–20, and by F. J. A. Broeze, 'The cost of distance; shipping and the early Australian economy, 1788–1850', *Economic History Review*, 2nd series, XXVIII, 1975, pp. 582–97, and 'British intercontinental shipping and Australia, 1813–1850', *Journal of Transport History*, new series, IV, 1978, pp. 189–207.

35 Hay to Barrow, 12 February 1826, CO 324/85, 56–8; Barrow to Hay, 13 February 1826, CO 201/175, 11–12.

36 Bathurst to Darling, 1 March 1826, *Historical Records of Australia*, Series I, XII, pp. 192–3 (hereafter HRA I, xii, 192–3); Bathurst to Darling, 1 March 1826, HRA I, xii, 193–4; Bathurst to Darling, 11 March 1826, HRA I, xii, 218; Barrow to Hay, 7 March 1826, CO 201/175, 15–17.

37 See J. M. R. Cameron, 'The northern settlements: outposts of empire', in Pamela Statham (ed.), *The Origins of Australia's Capital Cities*, Melbourne, 1989, pp. 274–82.

38 Barrow to Glenelg, 13 December 1836, CO 201/256, 51–9. See also *Grey and Lushington Papers*, (f) and (g); *RGS Council Minutes*, 12 December 1836.

39 Barrow to Glenelg, 13 December 1836, CO 201/256, 51–9.

40 See CO 201/258, 25–6; CO 201/266, 629; CO 201/266, 631–2; CO 202/35, 139; CO 201/266, 634.

41 Stephen to Spearman, 3 January 1837, CO 202/35, 124–5; Spearman to Stephen, 6 January 1837, CO 201/265, 4.

42 Stephen to Barrow, 13 January 1837, CO 201/265, 293–5.

43 Barrow to Stephen, private, 14 January 1837, CO 201/64, 31–2.

44 Stephen to Barrow, 14 January 1837, CO 202/35, 129–30.

45 George Windsor Earl, 'Opinion on the best time for exploring the coast of Australia', *Journal MSS Australia 1837 Earl GW*, RGS Archives.

46 *RGS Council Minutes*, 13 February 1837. Earl's views were first published in May 1836 in *Observations on the Commerical and Agricultural Capabilities of the North Coast of Australia*. These were subsequently expanded less than a year later in *Appendices to the Eastern Seas or Voyages and Adventures in the Indian Archipelago in 1832–33–34*.

47 Grey to Stephen, 11 February 1837, CO 201/266, 636.

48 Grey and Lushington to Stephen, 19 January 1837, CO 201/266, 629.

49 Grey's arguments are outlined in CO 201/266, 638–43.

50 CO 201/266, 641.

51 Marginal note in Stephen to Glenelg, 15 February 1837, CO 201/266, 643.

52 Grey to Glenelg, 22 March 1837, CO 201/206, 644–5.

53 John Washington, 'Plans and estimates for exploring in Australia', 29 March 1837, *Grey and Lushington Papers*, (o).

54 Washington, 'Plans'.

55 Barrow's Memorandum of 5 April 1837, CO 201/264, 49.

56 Their argument is contained in Barrow and Beaufort to Glenelg, 10 April 1837, encl. 'Memorandum respecting the former settlements on the northern coast of Australia', 5 April 1837, CO 201/264, 37–55.
57 Undersecretary Sir George Grey to Grey and Lushington, 29 April 1837, CO 202/36, 31; see also Stephen to Barrow, 1 May 1837, CO 202/36, 31–2.
58 See Grey to (Undersecretary Sir George) Grey, 8 May 1837, CO 201/266, 646–7; Stephen to Grey, 13 May 1837, CO 201/266, 652; Stephen to Barrow, 16 May 1837, CO 202/36, 49; Barrow to Stephen, 20 May 1837, CO 201/264, 65; Stephen to Barrow, 22 May 1837, CO 202/36, 57.
59 *RGS Council Minutes*, 8 May 1837.
60 Washington to Stephen, 23 May 1837, Co 201/265, 297–303.
61 John Wasthington, 'Sketch of instructions for the expedition into Australia', *Grey and Lushington Papers*, (j).
62 Stephen's marginal note in Washington to Stephen, 23 May 1837, CO 201/265, 297.
63 Grey to Stephen, 29 May 1837, CO 201/266, 658.
64 Grey to Glenelg, 29 May 1837, CO 201/266, 654–5.
65 'Outline of a plan', 30 May 1837, CO 201/266, 664–5.
66 Undersecretary Sir George Grey to Grey and Lushington, 5 June 1837, CO 201/266, 663. See also Stephen's minute at CO 201/266, 667.
67 Stephen to Hamilton, 5 June 1837, CO 201/205, 306–7.
68 Lushington to Washington, 29 May 1837, *Grey and Lushington Papers*, (m); CO 201/266, 660.
69 Grey, *Explorations*, I, p. 2.
70 See G. M. Thomson, *The North West Passage*, London, 1975.
71 Bathurst to Macquarie, 18 April 1816, HRA I, ix, 114–16. See also instructions in Macquarie to Oxley, 24 March 1817, in John Oxley, *Journals of Two Expeditions into Interior of New South Wales*, London, 1820, pp. 335–61.
72 See E. W. Bovill, *The Niger Explored*, London, 1968.
73 *Quarterly Review*, XXXIII, 1826, p. 545. See also *Quarterly Review*, XXIX, 1823, p. 508.
74 John L. Allen, 'An analysis of the exploratory process. The Lewis and Clark expedition of 1804–1806', *Geographical Review*, LXII, 1972, pp. 13–36, and 'Lands of myth, waters of wonder', in D. Lowenthal and M. J. Bowden (eds), *Geographies of the Mind*, New York, 1975, pp. 41–61.
75 John L. Allen, *Passage Through the Garden: Lewis and Clark and the Image of the American Northwest*, Urbana, Illinois, 1975, pp. 59–67.
76 For acute assessments of Banks's and Murchison's views on exploration and the role they played in it see, respectively, H. B. Carter, *Sir Joseph Banks*, London, 1988, and R. A. Stafford, *Sir Roderick Murchison: Scientific Exploration and Victorian Imperialism*, Cambridge, 1989.
77 For example, *Quarterly Review*, XXX, 1823, p. 267; XXXVII, 1828, p. 539; XXXIX, 1829, p. 177.
78 *Sydney Herald*, 10 August 1838.
79 See, for example, *The Athenaeum*, 27 November 1841, pp. 907–9 and 11 December 1841, pp. 952–4; *The United Service Magazine and Naval and Military Journal*, January 1842, p. 20.
80 See Stephen's marginal comments in CO 18/23, 129; CO 18/23, 228; CO 201/292, 396.
81 J. Rutherford, *Sir George Grey, 1812–1898: A Study in Colonial Government*, London, 1961, esp. pp. 1–2.
82 A. C. Gregory, *Journals of Australian Explorations*, Brisbane, 1884, pp. 1–30.

CHAPTER TWO

Imperialism and the discourse of desiccation: the institutionalisation of global environmental concerns and the role of the Royal Geographical Society, 1860–1880

Richard H. Grove

Introduction

Concepts of artificially induced climatic change have a much longer history than one might imagine.[1] Nowadays, of course, they have become a part of our popular culture, and part of a widespread and possibly justifiable environmental neurosis. They are especially familiar and useful in providing the justification for new plans to prevent tropical deforestation and reduce outputs of carbon dioxide into the atmosphere.

It is, moreover, not well known that the fear of artificially caused climate change, and much of the modern conservation thinking which that anxiety stimulated, developed specifically in the tropical colonial context.[2] These fears and connections attained their most vigorous forms, in terms of deliberate state policy, during the heyday of immperialism. Geographers, and the Royal Geographical Society in particular, played a major role in formulating and then propagandising ideas about deforestation, desiccation and climate change, often as a basis for proposing large-scale forest conservation. However, the antecedents of this institutional role have to be sought very early on in the history of colonialism.

An awareness of the detrimental effects of colonial economic activity and, above all, of capitalist plantation agriculture (the potential profits from which had stimulated much early colonial settlement) developed initially on the small island colonies of the Portuguese and Spanish at the Canary Islands and Madeira.[3] It was on these islands that the ideas first developed by the Greek naturalist Theophrastus in his essays on deforestation and climate change were revived and gradually gathered strength, as his works were translated and widely published during the Renaissance.[4] For example Columbus, according to one of his biographers, feared, on the basis of his knowledge of what had happened after deforestation in the Canaries, that similar devastation in the West Indies would cause major rainfall decline.

Certainly these ideas were already fashionable by 1571 when Fernandez Oviedo, in Costa Rica, soon followed by Francis Bacon and Edmund Halley in England, began to theorise about the connections between rainfall, vegetation and hydrological cycle.[5] Edmund Halley's fieldwork on this subject, carried out on the island of St Helena during a summer vacation while he was a student at Queens College Oxford, showed remarkable insight. Furthermore, it was on St Helena that some of the earliest and best documented attempts were made to prevent deforestation and control soil erosion, both of which were serious by the end of the seventeenth century. These attempts were elaborately recorded by officials of the East India Company, which controlled the island. However the early conservation methods and local environmental thinking developed before 1750 on islands such as St Helena, and also in a similar fashion on Barbados and Montserrat, were purely empirically based, localised and often unsuccessful in application.[6] Indeed they were not based on any coherent body of climatic theory, despite the knowledge of Theophrastus's desiccationist hypotheses that already existed in some intellectual circles in Europe and South America.

The increasingly complex infrastructures of colonial rule under the British and French after the mid-eighteenth century provided the basis for the kinds of information networks needed to systematically collate environmental information on a global basis and to respond to perceived environmental crises with effective forms of environmental control based on unitary climate theories. These information networks were based primarily on the botanic gardens of Europe and the colonies and were a direct consequence of the rapid growth in interest in economic plant transfer and agricultural development which took place between 1750 and 1850.[7] But these networks were not sufficient on their own. As I shall argue in this chapter the development of conservationist ideas and early environmental concern was also critically dependent on the diffusion of desiccation concepts, or the formulation of a desiccationist discourse linking deforestation to rainfall reduction. Developing notions of species rarity, extinctions and endemism also played a significant although secondary part in early environmentalism.[8] To some extent it seems that the colonial networks of botanical exchange and the botanic gardens themselves acted as social institutions that encouraged the slow development of an environmental consciousness.[9]

Deforestation and climatic change

The linking of deforestation to climatic change and rainfall reduction (the essence of desiccationism) laid the basis for the initiation and proliferation of colonial forest protection systems after the Peace of Paris in 1763, particularly in the West Indies. The intense interest which developed during the eighteenth century, especially in France, in theories linking climate to theories of cultural 'degeneration' and human evolution assisted this process.[10] But after about 1760 empirical observations of deforestation and the impact of droughts in the colonies were now complemented by the widespread promulgation of desiccationist theories by metropolitan institutions in Britain and France, and especially by the Académie des Sciences in France and the Society of Arts in Britain. While deforestation in temperate countries, especially in North America, tended to be seen as beneficial, quite the opposite view pertained in many of the tropical colonies by the late eighteenth century.[11] Climatic change, it was believed, threatened not only the economic well-being of a colony but posed hazards to the integrity and health of the settler populations of the plantation colonies of the Caribbean and the Indian Ocean.[12]

The business of forest protection and tree planting had thus acquired, by the late eighteenth century, far more acute meanings in the colonial setting than it had in contemporary Europe.[13] The timing of the development of colonial forest protection actually depended both on the existence and complexity of institutions with an intellectual involvement in the colonies and on the pattern of diffusion of desiccationist ideas. Whilst the Royal Society had taken an early interest in forest preservation, colonial deforestation was not a concern of the society, even though it played a part in the development of the desiccationist discourse in the late seventeenth century. Instead the institutional connection between desiccation ideas and the colonial environment was made in the wake of the foundation of the Society of Arts in 1754.[14] Simultaneously the Académie des Sciences developed an interest in the matter, so that the intercourse between French and British intellectuals became of prime importance in the development of colonial environmentalism and indeed remained so until the mid-nineteenth century.

The elaboration of early desiccation ideas into complex physical and biological theories depended at first on the work of John Woodward at Gresham College in London (the founder of the first chair in geology at Cambridge University) in establishing the basic principles of transpiration.[15] In his *Vegetable Staticks* of 1726 Stephen Hales of Corpus Christi College, Cambridge, refined this work further in estimating the amount of moisture contributed by trees to the atmosphere.[16] Buffon's subsequent translation of Hales's work came to the attention of Duhamel du Monceau, the great French meteorologist and arboriculturist. In a popular work on tree planting, published in 1760, du Monceau developed, at length, the connections between trees and climate.[17] These ideas were then transferred across the Channel once again and were widely discussed at meetings of the new Society of Arts, which included the Marquis de Turbilly, Abeille and other members of the Académie des Sciences among its members.[18] However the whole matter might have remained academic had it not been for the fact that at least two members of the Society of Arts also served as members of the powerful Lords Commissioners for Trade in the Colonies, the body responsible for planning land use in the new West Indies possessions.

The most significant of these figures was Soame Jenyns, the MP for Cambridge.[19] It seems to have been due to his influence that the Lords Commissioners were appraised of desiccation ideas at some point between 1760 and 1763. With the signing of the Peace of Paris, the ceded isles of St Vincent, St Lucia, Grenada and Tobago all came under British rule. As part of their plans for the survey and division of lands on the isles, all of them still inhabited by substantial numbers of Carib Indians, the Lord Commissioners made provision for the gazetting of large areas of mountain land as forest reserve, for 'the protection of the rains'.[20] These were the first forest reserves ever to be established with a view to preventing climate change. The most extensive of them, on the highlands of central Tobago, is still in existence. In 1769 some very similar reserves, based on the same theory and with the same intellectual precedents, were established on Mauritius (then known as the Isle de France) by Pierre Poivre, the Commissaire-Intendant of the colony.[21] We know that Poivre had been an advocate of colonial forest protection for some time and that he had given, in 1763, a major speech in Lyons on the climatic dangers of deforestation.[22] This speech may go down in history as one of the first environmentalist texts to be based explicitly on a fear of widespread climate change.

Poivre's forest conservation programmes on Mauritius were encouraged by a government of physiocratic sympathies and by the botanists of the Jardin de Roi in Paris.[23] In the British Caribbean colonies, on the other hand, the more autonomous institutional influence of the Society of Arts remained critical to promoting tree planting and, to a lesser extent, forest conservation. The society had been instrumental in founding the botanic garden at Kingstown on St Vincent, the first such garden in the western hemisphere.[24] It was the existence of this garden and the activities of its superintendents, particularly those of Alexander Anderson, a Scottish Surgeon, that ensured institutional support for further forest protection measures in the

Caribbean, especially on St Vincent, and the other islands of the Grenada Governorate.[25] On St Vincent a comprehensive law was passed in 1791 to protect the Kings Hill Forest with the specific intention of preventing rainfall change.[26] This legislation was subsequently imitated on St Helena and subsequently in India.[27] Incidentally a major stimulus to forest protection activities in 1790 was the occurrence of droughts in tropical regions on a global scale. These events appear to have been caused by an unusually strong El Niño current in those years, which caused severe drought in southern India, St Helena and the West Indies as well as at locations in central America, especially in Mexico.[28]

Once colonial forest protection ideas, based on desiccation theories, were firmly installed, notably on Tobago, St Vincent, St Helena and Mauritius, they acquired a momentum of their own, assisted by emerging colonial botanic garden information networks, and particularly the lines of communication between the gardens at St Vincent, St Helena, Cape Town, Mauritius and Calcutta.[29] The influence of the metropolitan centres in these networks was actually relatively weak. This remained the case even after Sir Joseph Banks began his period of dominance of botanical science in Britain and Kew began to achieve pre-eminence. Although aware of the possibiities of environmental change, not least on St Helena, Banks cannot be counted among the major environmentalist pioneers. Kew only became a significant player in colonial conservation after Sir William Hooker became Superintendent.

Probably the first great environmental theorist in the colonial context, apart from Pierre Poivre and his colleague Bernardin de St Pierre on Mauritius, was Alexander Anderson. His *Geography and History of St Vincent* written in 1799 and his *Delugia*, an early geological history of the world, mark him out as a visionary environmental thinker and the pioneer of a generation of surgeon-conservationists and geographers.[30] The colonial expertise of men such as Anderson and Poivre meant that the role of metropolitan institutions in initiating 'centres of environmental calculation' (to adapt the terminology of Bruno Latour) remained relatively unimportant and derivative. Even much later, after 1800, the emergence of a school of environmentalists in India in the ranks of the East India Company Medical Service, long after desiccationist forest protection policies had emerged on the island colonies, was largely an internal and indigenous matter, drawing heavily on Indian environmental knowledge and tree-planting practice, and the desiccationist ideas put into practice on St Vincent and St Helena.[31] However, the influence of the Society of Arts remained important for a while. For example, William Roxburgh, the second superintendent of the Calcutta Botanic Garden, promoted extensive tree-planting policies in Bengal with the active encouragement of the Society of Arts.[32] Indian forest conservation practice and environmentalism after 1842 also drew on the climatic theories of Alexander von Humboldt and Joseph Boussingault, as well as on the forestry methods inspired by French physiocracy and its offshoot, German physiocratic cameralism.[33]

Initially local colonial scientific societies in India provided the impetus and professional authority necessary to establish the first forest conservation agencies in India to be based on desiccationist notions (specifically the Bombay Forest Department and the Madras Forest Department).[34] In the early 1850s, however, the proponents of forest conservation on an all-India basis found it necessary, in the face of state reluctance to finance such an establishment, to resort once more to a source of metropolitan scientific authority, namely the British Association for the Advancement of Science (BAAS). In 1851 the BAAS commissioned a full report on the 'physical and economic consequences' of tropical deforestation.[35] This helped to legitimate the theoretical and environmental basis for the subsequent development of an all-India forest administration. Seven years later the BAAS became the forum for discussions on 'the general and gradual desiccation of the earth and atmosphere' in the wake of a paper delivered by J. Spotswood Wilson.[36] This paper can be said to mark the onset of a truly global environmental debate in which processes operating at a global scale were being considered.

The Royal Geographical Society and conservationist discourse

In the ensuing two decades the BAAS continued to serve as a forum for advocates of forest and species preservation. Professor Alfred Newton and Alfred Russell Wallace both utilised the BAAS for launching their conservationist opinions and programmes and evoking discussions on extinction and deforestation.[37] To some extent, however, the BAAS proved unsatisfactory as an institutional setting for the airing of environmentalist views, partly because it only met for a limited time, once a year, and partly because its influence in the colonial context was relatively weak. As a result, the Royal Geographical Society displaced the BAAS, during the 1860s, as the most prominent institutional setting for the discussion of the desiccationist and conservationist discourses that were receiving so much attention from botanists and policy-makers in the colonies. At this period, it should be pointed out, the construction of environmental agendas and local land-use policies in the British colonies, in contrast to the French case, did not yet receive any backing or guidance at all from governmental institutions at the centre. In other words, there was no imperial environmental 'centre of calculation' sponsored by the state in Britain itself (and this remained the case until the establishment of the Imperial Forestry Institute at Oxford in 1924).[38] Instead any government centres of calculation were all situated at the imperial periphery, especially in Madras, Bombay, Port Louis, Cape Town and, more latterly, at Dehra Dun, the headquarters of the Indian Forest Service and training schools.

In France, on the other hand, the Imperial Forest School at Nancy had served as a centre of environmental ideas and training, much of it desiccationist in nature, since 1824.[39] The Royal Geographical Society, therefore, was effectively required to fulfil a centralising role and did so in a very important sense, particularly with respect to the transfer of forest conservation and desiccation ideas between India, where they had become well established, and the rest of the British colonies, above all those in Africa. In the course of being utilised in this way the RGS effectively played a role in the globalisation of desiccation concepts and hence a major part in the diffusion of a particularly exclusionist and hegemonic forest conservation ideology. In the course of acquiring this environmental role the RGS began to undergo a significant transition in terms of its own raison d'être and its terms of the influence which it exerted over the emerging agendas of academic geography.

The publication of Charles Darwin's *Origin of Species* in 1859 had set the scene for a decade of existential and religious crisis in which old assumptions about birth and death, time and chronology, religion and generation, already much fractured, were finally broken. These anxieties were mirrored, or coped with, in an unprecedented wave of environmental concern throughout the 1860s. Thus the decade saw the foundation of the all-India forest department, the founding of the Commons Preservation Society, the passing of the first British bird protection legislation and the publication of G. P. Marsh's *Man and Nature* and the publication of Dr Hugh Cleghorn's *Forest and Gardens of South India*.[40] The main focus of academic geography soon reflected this shift in the emergence of 'evolutionary physical geography' and in the birth of 'denudation chronology'. Indeed the RGS *Proceedings* of 2 May 1869 advertised Sopwith's geological models in wood, one of which was called 'Valleys of denudation'. And it was in the field of denudation and desiccation that the RGS and early environmentalists such as Hugh Cleghorn, John Croumbie Brown and George Bidie found much in common.

The intellectual ground had been well prepared by Livingstone's reports of what he believed to be evidence of chronic and irreversible desiccation in parts of the Kalahari and northern Bechuanaland. It was this data that first stimulated the writing in 1858 of a paper by J.

Spotswood Wilson on 'the general and gradual desiccation of the earth and atmosphere'.[41] This is one of the earliest papers on the 'greenhouse' effect and held out the stark promise of an early extinction of humanity as a result of atmospheric changes brought about by natural desiccation and augmented by the upheaval of the land, 'waste by irrigation', and the destruction of forests. Wilson quoted liberally from the works of Livingstone and other travellers, giving descriptions of desiccated landscapes in Australia, Africa, Mexico and Peru, all of which had 'formerly been inhabited by man', as Wilson put it. On 13 March 1865 a paper remarkably similar in theme, especially in its references to Southern Africa and to Livingstone's writings was given at the RGS.[42] Significantly an earlier version of the paper had first been given at the BAAS.[43] One may surmise that the fact that the same paper was then delivered at the RGS was due to the intervention of Colonel George Balfour of the Indian Army, a member of the RGS Council. George Balfour was a brother of Dr Edward Green Balfour, then deputy Inspector-General of Hospitals, Madras Presidency (and later Surgeon-General of India) and one of the earliest and strongest advocates of forest protection in India.[44]

At meetings of the RGS in 1865 and 1866, Balfour spoke at some length on developments in forest conservation that were then taking place in several different colonies. This display of his unrivalled knowledge was not a mere vanity. George Balfour clearly saw it as his task to propagandise what he saw as the merits of forest protection in stemming the threat of global desiccation. By 1865 the terms of a debate about the causes of desiccation in the tropics had been set far more clearly. While in 1858 it had remained acceptable to attribute the process to tectonic upheaval (which David Livingstone, for example, favoured) RGS discussions by March 1865 saw the appearance of an entirely new interpretation of global processes of degradation. Desiccation was not natural, James Wilson argued, 'but was entirely the consequence of human action'. Wilson felt that one could demonstrate this well in the case of South Africa. 'The human inhabitants [of the Orange river basin] are a prime cause of the disaster' he wrote, and 'the natives have for ages been accustomed to burn the plains and to destroy the timber and ancient forests . . . the more denuded of trees and brushwood, and the more arid the land becomes, the smaller the supply of water from the atmosphere'. Thus 'the evil advances', Wilson went on apocalytically, 'in an increasing ratio, and, unless checked, must advance, and will end in the depopulation and entire abandonment of many spots once thickly peopled, fertile and productive'.[45]

He followed this warning with a global survey of locations in which climatic changes had followed on deforestation. The lessons were clear, Wilson thought:

> In our own British colonies of Barbadoes, Jamaica, Penang, and the Mauritius, the felling of forests has also been attended by a diminution of rain. In the island of Penang, the removal of the jungle from the summits of hills by Chinese settlers speedily occasioned the springs to dry up, and, except during the monsoons, no moisture was left in the disforested districts. In the Mauritius it has been found necessary to retain all lands in the crests of hills and mountains in the hands of government to be devoted to forest, the fertility of the lower lands having been found by experience to depend upon clothing the hills with wood.[46]

Only draconian controls, it was implied, could stop a worldwide ruination of the forests of the British colonies and indeed the entire economic demise of large areas of country. Wilson was especially concerned with South Africa,

> it being a matter of notoriety . . . that the removal piecemeal of forests, and the burning off of jungle from the summits of hills has occasioned the uplands to become dry and the lowlands to lose their springs . . . it becomes of extreme importance to our South African fellow-subjects, that the destruction of the arboreal protectors of water should be regarded as a thing to be deplored, deprecated and prevented; and that public opinion on the matter should be educated . . . but we must not stop there. The evil is one of such magnitude and likely to bear so abundant a harvest of misery in the future, that the authority of law, wherever practicable, should be invoked in order to institute preventive measures. Not only should fuel be economized, but the real interests of the British colonies and Dutch republics, for many long years to come, would most certainly be represented by the passage of stringent enactments which should in the first place forbid, at any season or under any circumstances whatsoever, the firing of grass on field or mountain. The absolute necessity which exists for keeping as large a surface of the ground as possible covered with vegetation, in order to screen it from the solar rays, and thus to generate cold and humidity, that the radiation from the surface may not drive off the moisture of the rain-bearing clouds in their season, ought to compel the rigid enforcement of such a legal provision. Those colonial acts on this subject which are already in existence – for the Colonial Parliament at the Cape has found it necessary to pass restrictive measures – are not sufficiently stringent to be of much service, inasmuch as they are not entirely prohibitory, permitting the burning of the field at certain times of year.[47]

The main discussants of Wilson's seminal paper at the March 1865 RGS meeting were, on the one side doctors Livingstone and Kirk, who both contested social explanations of deforestation in favour of non-anthropogenic explanations of continental desiccation, which both believed to be taking place in Africa. Ranged on the other side of the argument were Francis Galton, the secretary of the RGS, (and a cousin of Charles Darwin), Colonel George Balfour and Lord Stratford de Redcliffe. Sir Roderick Murchison too, chairing the discussion, declared himself in favour of the interpretations offered by Wilson and in favour of his radical interventionalist solutions. Livingstone, for his part, pointed out that 'the author of the paper did not seem to know that many of his suggestions had already been adopted at the Cape, where immense quantities of Eucalypti were grown in the botanic garden for distribution among those who wished to plant trees. In four years the trees grew to a height of twenty feet'. Such exchanges of basic information serve to indicate the role which could be played by the RGS. However the discussion following the Wilson paper also exposed, in a somewhat embarassing fashion, the very slow nature of the diffusion of environmental information and ideas between colonies, even more, between colonies and the imperial centre.

This was particularly apparent to George Balfour who, in successive RGS debates, saw it as his duty to advertise the efforts made in particular colonies, above all in India, Mauritius and Trinidad, to protect forest and thereby forestall climatic change. Possibly as a result of Balfour's lobbying, two further papers given at the RGS, at meetings in June 1866 and in March 1869, dealt very specifically with the issue of state responses to deforestation and desiccation. The paper given by Clements Markham in 1866, ('On the effect of the destruction of forests in the western Ghauts of India on the water supply'), seems to have been intended to demonstrate and publicise the contemporary efforts being made to control deforestation in upland India.[48] The ensuing discussion took on much the same format as that of 1865, bringing together a whole variety of self-confessed experts and travellers from several different colonies. Initiating the discussion of Markham's paper Murchison commented that the subject of the destruction of forests 'was one of very great interest to all physical geographers'.[49] Murchison added that it was a subject upon which he had himself much reflected in reference to other countries, 'even our own country'. He was happy, he said, to see many gentlemen present connected with India; and he would, in the first instance, call upon Sir William Denison, late Governor of Madras, to make some observations upon the subject.

Under Denison's able administration, Murchison informed the gathering, some of those very forest protection operations had been undertaken to which Mr Markham had alluded. A three-cornered discussion then followed which fiercely debated the culpability or otherwise of the 'native' for deforestation. Denison believed that Indians 'cut down trees without hesitation, and no-one ever dreamt of planting a tree unless it were a fruit tree'. George Balfour, reflecting the anti-establishment attitudes of his conservationist brother, Edward Balfour, countered that it was 'the practice of rich Hindoos to sink wells and plant topes of trees'.[50] Another discussant, Mr J. Crawfurd, pronounced it his opinion that 'the presence of immense forests had proved one of the greatest obstacles to the early civilization of mankind' and made the assertion that Java, free from forest, was 'incomparably superior to all the other islands of the Indian Archipelago'. Balfour's unusual advocacy of the significance of indigenous knowledge reflected the beliefs of the first generation of (East India Company) colonial conservationists in India, in stark contrast to Denison's comments which typify the more racist, harsh and counter-productive exclusionism of much post-Company Indian forest policy after 1865.[51]

Sir Henry Rawlinson, in his contribution to the RGS debate, opined that 'it was a matter patent to every traveller, and it might be adopted as a principle in physical geography, that the desiccation of a country followed upon the disappearance of its forests'. It was this realisation that the emerging discipline of physical geography could be enlisted in the cause of global forest protection that seems to have persuaded the core of the Indian forest service establishment to patronise the meetings of the RGS in the late 1860s. Furthermore, in the absence of any other imperial institution, at least in London, showing any significant interest in the pressing issue of colonial deforestation, the RGS provided a sympathetic oasis in what was otherwise an institutional desert. Thus it was that on 25 January 1869 that Hugh Cleghorn, the inspector-general of the new Indian forest department, attended a meeting of the society addressed by Dr George Bidie entitled 'on the effects of forest destruction in Coorg'.[52] Murchison claimed on this occasion that 'it was highly gratifying to geographers to see various branches of natural history combined in illustration of a great subject in physical geography'.

Introducing Dr Cleghorn, Murchison suggested that 'we were more indebted than to any other gentleman in reference to this important question'.[53] Deforestation could best be understood, Cleghorn believed, in terms of an analysis of the amount of capital being invested in forest areas, principally by British planters. The native population in the Western Ghats, he pointed out, were almost universally of the opinion that the climate was drier on account of the changes that Europeans were gradually introducing. The Madras Forest Department, he added, was a new one, initiated only thirteen years before. It was gradually increasing in usefulness and it was now receiving the official attention that it deserved.[54]

Attempts to prevent deforestation in other colonies had, as late as the

1860s, received virtually no support from the imperial centre. Instead important propaganda for conservation was being created at the periphery, not only in India but, in particular, in South Africa. Thus some of the most strenuous extra-Indian efforts to promote forest protection and tree planting and restrict grass burning had been made by John Croumbie Brown, a missionary and the Colonial Botanist of the Cape Colony from 1862 to 1866.[55] However, local funding for these pioneer efforts had been removed, without protest from Whitehall, in 1866, and a resentful Brown had had to return to Scotland.[56] From there he proceeded to publish a stream of works on hydrology and forest conservation, many of which soon came to the attention of the colonial authorities in the Cape, Natal and elsewhere. The two most important of these works were *A Hydrology of South Africa* published in 1875, and *Forests and Moisture* published in 1877.[57] These works, far more influential in the colonial context than the writings of G. P. Marsh, drew heavily on the debates which had taken place at the RGS during the 1860s, and derived authority from them. In Brown's books the discourse of desiccationism was refined and made, in a sense, into an environmental article of faith. Furthermore his dicta on deforestation and climate were repeated and developed throughout the colonial context during the ensuing forty years.[58]

Brown's frequently expressed proposals for an Imperial School of Forestry (the idea itself was largely of his authorship) were ultimately developed at Coopers Hall in Surrey and Dehra Dun in India and eventually in the form of the Imperial Forestry Institute at Oxford in 1924, exactly a century, incidentally, after the foundation of the French Imperial Forestry School at Nancy.[59] In the long interregnum between the establishment of forest conservation in India and the establishment of forestry training in Britain the RGS had acted as a highly formative centre of debate and calculation and as a centre of academic authority of great practical use to such fervent early environmentalists as John Croumbie Brown. Above all, the society had served to legitimate a notion of global environmental crisis, articulated in a desiccationist discourse of remarkable political power, and the subject of a wide degree of consensus. The warnings against the 'evil' consequences of deforestation which were expressed at the RGS in the 1860s were closely connected with an emerging contemporary consciousness of the possibility of extinction which Darwin had sharply focused in 1859.[60] A sense of existential crisis and sense of impending loss was translated, through the RGS, into a highly empirical debate about deforestation and the possibilities of intervention and environmental institution building.

The hegemonic prescriptions for colonial forest control which the new consciousness stimulated and which the RGS encouraged can be interpreted, perhaps, as a desire to re-assert control over a new existential chaos and over environmental processes that might threaten the existence of humanity itself. Prescriptions for forest conservation, for grass-fire prevention and for irrigation can be seen in this sense as redemptive or in terms of atonement. Brown had originally been a Congregationalist missionary. He had found it logical and congenial to adapt the dire warnings of such desiccationists as Wilson and Bidie as a kind of environmental gospel. His otherwise 'scientific' accounts are sprinkled with references to Old Testament texts. The publication of *The Origin of Species* had simply helped to make the threat of desiccation more dire. Darwin made extinction a necessary part of natural selection annd evolution. This gave deforestation and desiccation a much strengthened meaning, necessitating human and conservationist intervention. New scientific theory could not however immediately displace religious meaning in the environment. Desiccation continued to be associated with expulsion from the Garden of Eden and with evil. If society failed to make conservationist amends for the evils of deforestation, extinction and ruin would follow. In the circumstances of this new thinking colonial conservation acquired the overtones of a kind of redemptive and confessional doctrine.[61]

For this purpose the evidence of desiccation needed, of course, to be global or cosmological, while its prescriptions needed to be universalist. It need hardly be said that practical policy prescriptions for counteracting desiccation, principally through forest reservation and soil conservation, would turn out to be highly palatable to the agendas of colonial rule, particularly when it came to controlling the landscape, and manipulating a 'chaotic' subject populations. Those who attended the RGS debates during the 1860s probably did not fully appreciate that. Instead, placed at the centre of such transitions, it is not surprising that geography itself should soon have been affected by a redemptive cosmology. The redemptive element was reflected particularly in the new dicipline of physical geography as it developed after 1870 and is best understood in the writings of Archibald Geikie. 'Evolutionary Geography', he wrote,

> traces how man alike unconsciously and knowingly has changed the face of nature . . . it must be owned that man in most of his struggle with the world had fought blindly for his own immediate interests. His contest, successful for the moment, has too often led to sure and sad disaster. Stripping forests from hill and mountain, he had gained his immediate object in the possession of abundant supplies of timber; but he has laid open the slopes to be parched by drought or to be swept down by rain. Countries once rich in beauty and plenteous in all that was needful for his

support are now burnt or barren or almost denuded of their soil. Gradually he had been taught by his own experience that while his aim still is to subdue earth he can attain it not by setting nature and her laws at defiance but by enlisting them in his service . . . he has learnt at last to be a minister and interpreter of nature and he finds in her a ready and uncomplaining slave.[62]

The final lines of Geikie's text indicate that, even while it assumed an environmentalist guise, geography continued to exhibit some of the attributes of a discourse of domination. However it was a discourse that was ultimately contradictory. Thus the efforts made by colonial conservationists and metropolitan geographers to understand the mechanisms of environmental degradation could hardly fail to touch the uncomfortable and dynamic connections between the kinds of economic development unleashed by imperial expansion and annexation and the alarming patterns of global environmental change that had become apparent to audiences at RGS debates after 1860.

Notes

1 See R. H. Grove, *Green Imperialism*, Cambridge, 1994. Also R. H. Grove, Conserving Eden', the East India companies and their environmental policies on St Helena, Mauritius and in Western India, 1660–1854', *Comparative Studies in Society and History*, 35 (1993), 318–51. I am indebted to Vinita Damodaran and Adrian Walford for their help in discussing some of the themes in this article, which was prepared while I was in receipt of a post-doctoral fellowship of the British Academy.
2 See R. H. Grove, 'The origins of environmentalism', *Nature*, 3 May 1990, pp. 5–11.
3 There is no single useful work on the environmental impact of colonial rule. However see the useful secondary summaries in Clive Ponting's *A Green History of the World*, London, 1991; and the useful regional study by David Watts, *The West Indies: patterns of development, culture and environmental change, since 1492*, Cambridge, 1987.
4 The most useful contextualisation of Theophrastus is in C. Clacken, *Traces on the Rhodian Shore*, Berkeley, 1967, pp. 49–51, 129–30.
5 See Edmund Halley, 'An account of the watry circulation of the sea, and of the cause of springs', *Philosophical Transactions of the Royal Society*, 142.7 (1694), 468–72.
6 Detailed information on the environmental history of the West Indies and on early conservation thinking on the islands can be found in D. Watts, *The West Indies*. For an excellent case study of Montserrat see Linda M. Pulsipher, *Seventeenth Century Montserrat: an environmental impact statement*, London, 1986, Institute of British Geographers.
7 Unfortunately there is as yet no coherent account of the growth of these networks on a global basis. For a sketchy and doctrinaire but still useful account of colonial botanic gardens see Lucille M. Brockway, *Science and Colonial Expansion: the role of the British Royal Botanic Garden*, New York, 1979. A forthcoming book by Richard Drayton, *Nature's Government*, New Haven, 1995, deals with the ideology and politics behind the rise of the Kew system. The origins of the Dutch colonial system of plant exchange are usefully covered for the seventeenth century by J. Heniger in *Hendrik Van Reeede tot Drakenstein and the Hortus Malabaricus: a contribution to the study of colonial botany*, Rotterdam, 1986.
8 My interpretation of the origins of western environmentalism clearly differs in its emphasis on the importance of the periphery from such orthodox explanations as

that given in David Pepper, *The Roots of Modern Environmentalism*, London, 1986. More recent work has been closer to the mark in its stress on the significance of Rousseau's circle in provoking a new environmental consciousness' e.g. see G. F. Lafrenière, 'Rousseau and the European roots of environmentalism', *Environmental History Review*, 14 (1990), 241–73. Even Lafrenière, however, does not recognise the direct impact of Rousseau's thinking on the beginnings of conservation in the French colonial context. Perhaps the best regional work to date on this is L. Ureaga, *La Tierra Esquilmada: las ideas sobre la conservacion de la naturaleza en la cultura espanola del siglo XVIII*, Barcelona, 1987.

9 The best study to date on the working and social influence of a single colonial scientific institution is probably J. E. Maclellan, *Colonialism and Science: Saint Dominique in the Old Regime*, Baltimore, 1992.

10 E. Spary, 'Climate, natural history and agriculture, the ideology of botanical networks in eighteenth century France and its colonies', unpublished paper presented at the International Conference on Environmental Institutions, St Vincent, West Indies, April 1991.

11 See K. Thompson, 'Forests and climatic change in America: some early views', *Climatic Change*, 3 (1983), 47–64.

12 For a typical contemporary expression of these views see Edward Long, *A History of Jamaica*, vol. 3, London, 1777, pp. i–iv.

13 For the significance of tree planting in Britain at this time see K. Thomas, *Man and the Natural World: changing attitudes in England 1500–1800*, Oxford 1983. J. Perlin, *A Forest Journey: the role of wood in the development of civilization*, Cambridge, Mass., 1991, and S. Daniels, 'The political iconography of woodland', in D. Cosgrove and S. Daniels, *The Iconography of Landscape*, Cambridge, 1988.

14 For the early programmes and ideology of the Royal Society of Arts see D. G. C. Adams, *William Shipley: founder of the Royal Society of Arts, a biography with documentation*, London, 1979; and Henry Trueman Wood, *A History of the Royal Society of Arts*, London, 1913.

15 J. Woodward, 'Some thoughts and experiments concerning vegetation', *Philosophical Transactions of the Royal Society*, 21 (1699), 196–227.

16 A. Clark-Kennedy, *Stephen Hales, DD FRS: an eighteenth century biography*, Cambridge, 1929. See Stephen Hales, *Vegetable Staticks*, London, 1727, p. 20. Hales owed a great deal to the pioneering chemical work of the Leiden establishment and cited Hermnn Boerhave's *New Method of Chemistry* translated into English in 1727 by P. Shaw and E. Chambers.

17 Duhamel du Monceau, *Des semis et plantations des arbres et de leur culture*, Paris, 1760.

18 Royal Society of Arts Archives, John Adam Street, London W1; Members' Files.

19 R. Rompkey, *Soame Jenyns*, Boston Mass., 1984, and 'Soame Jenyns, M.P., a curious case of membership', *Journal of the Royal Society of Arts*, 120, (1972), 532–42. See also PRO/CO/102/1 beginning 'representations of the commissioners' PRO Kew, Richmond, Surrey.

20 Public Record Office, Kew Richmond, Surrey. Ref. No. CO106/9 CO101/1 No 26.

21 See J. R. Brouard, *The Woods and Forests of Mauritius*, Government Printer, Port Louis, Mauritius, 1963.

22 Lecture to the Agricultural Society of Lyons, MS no. 575 folio 74, pp. 27–9; Archives of the Bibliothèque Centrale du Museum National d'Histoire Naturelle, Paris.

23 Louis Malleret, *Pierre Poivre*, Paris, 1974.

24 See Anon., *Premiums by the Society Established at London for the Encouragement of Arts, Manufactures and Commerce*, issue dated June 10th 1760. The society offered large cash prizes for tree planting and in this case for the establishment of a botanic garden, which it advocated in 1760. The garden was actually founded in 1763 by Robert Melville, the first Governor of the Grenada Governorate. See R. Dossie, *Memoirs of Agriculture*, 1789, vol. 3, p. 800.

25 There is as yet no useful or comprehensive biography of Alexander Anderson's extraordinary life. But see Lancelot Guilding, *An Account of the Botanic Garden in the*

Island of St Vincent, Glasgow, 1825, for some biographical details.

26 The act was proclaimed at St Vincent on 2 April 1791. The second reading of the Bill for the Act had taken place in the St Vincent Assembly on 13 November 1788. PRO CO263/21.

27 See Grove, 'Conserving Eden'.

28 For details of the unusual strength of the El Niño current in the years 1790–1792 (as well as for details of evidence of other major El Niño events, see W. H. Quinn and V. T. Neal, 'El Niño occurrences over the past four and a half centries', *Journal of Geophysical Research*, 92, C13 (1987 14449–61. The strength of the 1791 El Niño was recorded by J. H. Unanue in *El Clima de Lima*, Madrid, 1815.

29 This emerges in the correspondence of Sir Joseph Banks; see W. R. Dawson (ed), *The Banks Letters: a calendar of the manuscript correspondence of Sir Joseph Banks preserved in the British Museum, London*, London, 1958.

30 Alexander Anderson, 'Geography and history of St Vincent' and 'Deluga'; MS book manuscripts, Archives of the Linnaean Society, London.

31 See Grove, *Green Imperialism*.

32 British Library, India Office Library and Records (IOL) ref. no. F4/4/427. For details of Roxburgh's tree-planting experiments in Bengal and Bihar see Home Public Consultations letters, National Archives of India, New Delhi; especially letters dated 31 January 1798 and 23 May 1813, and 'Botanic garden letters 1816–1817'. His methods are detailed in HPC, NA1 Letter dt. 23rd Oct. 1812', E. Barrett (Acting Collector of Bauleah) to Richard Rocke, Acting President and member of the Board of Revenue, Fort William, Calcutta.

33 Alexander von Humboldt, 'Sur les lines isothermes et de la distribution de la chaleur sur le globe', Société d'Arcueil, *Memoires*, 3 (1817), 462–602.

34 See Grove, 'Conserving Eden'.

35 H. Cleghorn, F. Royle, R. Baird-Smith and R. Strachey, 'Report of the Committee appointed by the British Association to consider the probable effects in an economic and physical point of view of the destruction of tropical forests', *Report of the Proceedings of the British Association for the Advancement of Science* (1852), 22–45.

36 J. S. Wilson, 'On the general and gradual desiccation of the earth and atmosphere', *Proceedings of the British Association for the Advancement of Science, Transactions* (1858), 155–6.

37 Both Newton (the designer and drafter of Britain's first bird Protection legislation passed in 1868) and Wallace used their respective presidencies of sections of the BAAS meetings as platforms for propagandising their own conservationist agendas. Both men linked global deforestation and desiccation with species extinctions.

38 To some extent the forestry school founded at Cambridge in 1904 could be seen as a centre of environmental calculation and was certainly responsible for much innovation in colonial forest policy. It was more autonomous in some respects than the later Oxford Institute which replaced it in 1924. See Cambridge University Library Archives: files on the Forestry school.

39 For a period Indian foresters were trained at Nancy. See E. P. Stebbing, *The Forests of India*, vol's 1 and 2, Edinburgh, 1922.

40 There is no general work as yet on the 'environmental decade' of the 1860s. Separate works on and from the period are Lord Eversley (George Shaw-Lefèvre), *Forests, Commons and Footpaths*, London, 1912; George Perkins Marsh, *Man and Nature: or physical geography as transformed by human action*, New York, 1864; H. F. Cleghorn, *The Forests and Gardens of South India*, Edinburgh, 1861. For a very limited treatment, confined to Britain itself see J. Sheail, *Nature in Trust*, London, 1976.

41 J. S. Wilson, 'On the general and gradual desiccation'.

42 J. S. Wilson, 'On the progressing desiccation of the basin of the Orange river in Southern Africa', *Proceedings of the Royal Geographical Society* (1865), 106–9.

43 J. S. Wilson, 'On the increasing desiccation of inner Southern Africa', *Report of the British Association for the Advancement of Science, Transactions* (1864), 150.

44 See especially Edward Green Balfour, 'Notes of the influence exercised by trees in

inducing rain and preserving moisture', *Madras Journal of Literature and Science*, 25 (1849), 402–48.
45 J. S. Wilson, 'On the progressing desiccation'.
46 *Ibid.*
47 *Ibid.*
48 Clements Markham, 'On the effects of the destruction of forests in the western Ghauts of India on the water supply', *Proceedings of the Royal Geographical Society* (1869), 266–7.
49 Report of discussion, *Proceedings of the Royal Geographical Society* (1869) 267–9.
50 *Ibid.*, p. 268.
51 This is the later policy usefully characterised by R. Guha in 'Forestry and social protest in British Kumaon, 1893–1921', *Subaltern Studies*, 4 (1985), 54–101.
52 G. Bidie, 'On the effects of forest destruction in Coorg', *Proceedings of the Royal Geographical Society* (1869), 74–5.
53 Report of discussion, *Proceedings of the Royal Geographical Society* (1869), 75–8.
54 *Ibid.*
55 See R. H. Grove. 'Scottish missionaries, evangelical discourses and the origins of conservation thinking in Southern Africa', *Journal of Southern African Studies* (1989).
56 R. H. Grove, 'Early themes in African conservation; the Cape in the nineteenth century', in Anderson, D., and R. H. Grove (eds), *Conservation in Africa: people, policies and practice*, Cambridge, 1987, pp. 21–39.
57 J. C. Brown, *A Hydrology of South Africa*, Edinburgh, 1875; *Forests and Moisture*, Edinburgh, 1877.
58 See W. Beinart, 'Soil erosion, conservationism and ideas about development, a southern African exploration, 1900–1960', *Journal of Southern African Studies*, 11 (1984), 52–83.
59 E. P. Stebbing, *The Forests of India*, vol. 2–3, Edinburgh, 1922.
60 The background to this emerging consciousness of the possibility of extinctions is discussed in 'Hugh Edwin Strickland (1811–1853) on affinities and analogies or, the case of the missing key', *Ideas and Production*, 7 (1987), 35–50.
61 I discuss these issues at greater length in 'Conserving Eden'.
62 Archibald Geikie, 'On evolutionary geography', *Journal of the Royal Geographical Society*, 2 (1870), 232–45.

CHAPTER THREE

Imperial landscapes: photography, geography and British overseas exploration, 1858–1872

James R. Ryan

'With an empire that extends to every quarter of the globe ... the English have, perhaps, more to gain from the prosecution of geographical science than any other nation.'[1] So spoke the President of the Royal Geographical Society (RGS) in his anniversary address in 1860. Such sentiments found a common home in the RGS which, from its foundation in 1830, was the institutional hub of nineteenth-century British geography. Furthermore, geography's acclamations of national purpose hint at the deeply embedded relations between the science of geography and British imperial expansion. Through the influence of prominent Fellows such as Roderick Murchison, the RGS married its promotion of overseas exploration and survey to the needs and ambitions of an imperial nation.[2] It is indeed appropriate to describe Victorian geography as 'the science of imperialism *par excellence*'.[3]

However, discourses of geographical knowledge negotiated imperial ideologies in a variety of ways.[4] Though it has received little attention from historians of geography, photography comprises a highly significant mode of geographical knowledge. To the Victorian imagination, photography was a means of revealing the realities of far-away places as well as Britain's expanding presence in them. As research based on the photographic archives of the RGS shows, there are intimate and complex relations between practices of photography and geography in the context of British imperialism.[5] These relations worked in a number of ways, from the use of photography in the categorisation of racial 'types' to the use of lantern-slide lectures in the teaching of imperial geography to British school children. This chapter addresses a major facet of these relations: the uses of photography in the representation of landscape on British imperial expeditions in the second half of the nineteenth century.

To begin with, I consider how overseas exploration was central to geographical institutions, concepts and practices in the nineteenth century. I suggest that the fashioning of the scientific discipline of geography and the simultaneous development of photography from the late 1830s are related to the dramatic expansion of British imperial authority during this period. I then move on to elaborate my argument by discussing, in turn, three examples of overseas expeditions. These are: firstly, a scientific expedition (David Livingstone's Zambezi expedition of 1858–63); secondly, a military campaign (the Abyssinia expedition, 1867–68); and thirdly, the travels of a commercial photographer (John Thomson's travels in China, 1868–72).

I attempt to place the landscape photographs made on each of these three expeditions in the historical and cultural contexts within which their meanings were framed. I use the term 'landscape' to refer less to a particular physical environment than to a form of cultural representation; a way of seeing and knowing the world.[6] I thus conclude that, composed, reproduced, circulated and arranged for consumption within particular circles in Britain, these photographs reveal as much, if not more, about the imaginative landscapes of imperial culture as they do about the physical spaces pictured within their frame.

Exploration, geography and the development of photography

Although the science of geography was pursued through a number of channels, overseas exploration represented the focal point of the RGS and, as such, lay at the heart of the Victorian geographical enterprise. The promotion and practice of expeditions was the main means the RGS used to consolidate itself as the institutional centre of geographical science. Nevertheless, the scientific status of exploration, and its place in the emerging discipline of geography was always a matter of some debate amongst Fellows of the RGS in the nineteenth century.[7] As this essay attempts to show, the priorities of different expeditions were often reflected in the geographical roles assigned to photography.

From the initial proclamations of the parallel discoveries of Louis Daguerre in France and William Fox Talbot in England in 1839, photography was heralded by many as the ultimate means of capturing visual reality. Deploying the power of natural light, the 'sun picture' or 'pencil of nature' as photography was termed, seemingly allowed the natural world to reveal itself truthfully in pictures. However, photography's authority as an irrefutable means of depicting the world derives from well established pictorial traditions such as linear perspective. Thus the 'naturalness' of photography is thoroughly conventional. The cultural currency of the photograph as accurate visual evidence was further established through its use within nineteenth-century legal, artistic and scientific discourses.[8] The process of photography does not replicate that of human vision. Rather, photography conforms to a historically specific set of codes designed to represent visual reality in two dimensions.

Yet the myth of photography as an essentially truthful means of recording the world has proved to be highly durable. Today photographs are often assumed to be impartial witnesses to historical events; 'windows on the world' offering unmediated and objective visual access to different times and places. However, to use photographs in this way ignores the historical processes and cultural frameworks through which photography, as a form of representation, acquires and communicates meanings.[9] Far from being objective and neutral, photographs are highly selective in construction and frequently ambiguous in effect. The making of any photograph entails numerous factors, from the motivations and technical skill of photographers to the pictorial conventions of their cultural milieu. Photographic imagery itself takes a variety of forms, from captioned book illustrations to lantern-slides, all of which influence the subsequent meanings assigned to photographs as, for example, artistic commodities and/or scientific documents. Photographs need to be understood not simply as visual repositories of some frozen history but rather as complex moments in historical processes of representation.

The use of photography on European journeys of exploration had a number of historical precedents. From the mid-eighteenth century, overseas exploration became a highly significant site for the formulation and elaboration of cultural frameworks through which the dramatic, unequal encounters between Europe and the rest of the world were represented. These frameworks included travel writing, pictorial description and cartography. Whilst changing European expansionist ambitions had encouraged a movement from maritime to interior territorial exploration, the development of natural history from the mid-eighteenth century had elaborated a complex system through which the natural world was to be observed and catalogued. Widely varying natural phenomena could be categorised and placed within this grid of knowledge. This systematising project opened up new domains of visibility and was central to a new European 'planetary consciousness'.[10] Picturing or mapping the world as a means of understanding it was by no means a new practice or idea. Indeed, visual representation had occupied a central strand of geography since the seventeenth century.[11] However, in the 'Great Map of Mankind' of the late eighteenth century there was 'no state or gradation of barbarism, and no mode of refinement which we have not at the same instant under our view'.[12] Thus a picture of global geography was constructed within a

European frame of unity and order.[13]

This new way of seeing demanded forms of representation at once objective and comprehensive in their scope and dependent on the ordering eye of the scientific man. It is thus significant that interests in the possibilities of permanently securing images produced by the camera obscura, which might be thought of as a discursive desire for photography, may be traced to the mid-eighteenth century. Such interests were predominantly concerned with the appropriation of views of nature and of landscape.[14] Furthermore the development of a new, more empirical approach to the depiction of landscape during the same period was arguably greatly influenced by the work of artists on exploratory voyages.[15]

Almost immediately after the invention of photography in 1839, or rather the adoption of the word 'photography' and the various announcements in Europe of Daguerre's, Neipce's and Fox Talbot's techniques for obtaining permanent images from the camera obscura, attempts were made to employ the new technique on various European overseas explorations. Sir John Herschel, for example, tried unsuccessfully to have photographic apparatus included on the British Antarctic Expedition of 1839. In 1846, a writer in the *Art Union Journal* suggested that the camera should,

> be henceforth an indispensable accompaniment to all exploring expeditions . . . by taking sun pictures of striking natural objects the explorer will be able to define his route with such accuracy as greatly to abbreviate the toils and diminish the dangers of those who may follow in his track.[16]

However, despite such calls and the enthusiastic efforts of men such as Sir John Herschel and Sir David Brewster, photography's application on journeys of exploration was severely limited for more than a decade. The technical skill and bulky equipment necessary for early photography, together with prejudice for more established forms of depiction such as cartography, were the prime hindrances to its deployment on expeditions. Improvements in photographic apparatus, such as the development of Frederick Scott Archer's 'wet plate' collodion process (1851) and paper photography (1858) as well as the collapsing of Talbot's patent restrictions (1855), made expeditionary photography more practical, though no less technically demanding.

From the late 1850s, photographic apparatus was increasingly included in the baggage of many different British expeditions across the globe. Victorian exploration had a complex array of motives, including romantic adventure, commerical prospects, military conquest, as well as geographical discovery and the pursuit of scientific knowledge. Photography was deployed for a correspondingly wide range of purposes and effects. Commerical photographers, for example, keen to exploit the European demand for exotic scenes, set out to 'discover' foreign lands photographically.[17] Although professional travel photographers, scientific explorers, military men and missionaries deployed the camera for their own particular purposes, they frequently did so through a wider geographical discourse which was informed by European political and cultural hegemony. This is clearly illustrated by the three expeditions I have selected as case studies. Whilst they varied in size, motivation and effect, these expeditions demonstrate the different ways in which the pursuit of geographical knowledge and the practice of photography interact as imperial discourse.

Photography, as Paul Carter has pointed out, shared scientific geography's 'two-dimensional world view' and was an important means of legitimating geography's 'claim to reduce the world accurately to a uniform projection'.[18] Photography was initially noted in circles of geograhpical science for its potential as a technique to assist cartography. It was in this context, in 1841 that the President of the RGS praised photography's 'minuteness' and 'exactitude' as well as its seeming ability to act 'on the impulse of the moment, and with unerring certainty'.[19] Sir Roderick Murchison, at the time of his second term in office as President of the RGS (1856–58), encouraged photography's use as an accurate and economical means of reproducing maps.[20]

More importantly, the process of photography embodied the grammar of observation and depiction at the heart of geographical science's quest to expose the unknown. Indeed, Sir John Herschel, the man of science who coined the very term 'photography', had commented in 1861 that perfect descriptive geography should 'exhibit a true and faithful picture, a sort of daguerreotype'.[21] Since it was a technology based on the power of light, photography took on particular symbolic significance as part of geographical discourse informed by what has been aptly termed a 'providential theology of colonial praxis' whereby the mutual extension of Christian civilisation and scientific knowledge represented a transference of 'light' into the 'dark' recesses of the globe.[22]

Perhaps no greater significance was given to the inconography of light and dark than in the representation of the British imperial exploration of Africa, the 'Dark Continent'.[23] The RGS played a central part in the nineteenth-century exploration and colonial appropriation of Africa. The RGS derived much of its expeditionary drive from the African Association, founded in 1788, which it formerly absorbed in 1831. To mid-nineteenth-century explorers and fellows of the RGS, the interior of Africa offered a vast physical and conceptual space for adventure, geographical discovery and the pursuit of empire. It is thus interesting that possibly the earliest use of photography on an official British expedition with which the RGS was closely involved was David Livingstone's Zambezi Expedition of 1858–1863, to which I now turn.

The Zambezi expedition, 1858–63

The Zambezi expedition evolved as part of Sir Roderick Murchison's masterminding of national enthusiasm for the missionary-explorer David Livingstone and the RGS's expeditionary ventures in Africa. In his second presidency of the RGS (1856–58), Murchison recruited the official sponsorship (to the tune of £5,000) of an otherwise hesitant government and ensured that the RGS played a central and public role in the expedition's organisation and equipment.[24]

The official aim of the expedition, set out in David Livingstone's letters of instruction to his officers, was 'to extend the knowledge already attained of the geography and mineral and agricultural resources of eastern and central Africa, to improve our acquaintance with the inhabitants, and engage them to apply their energies to industrial pursuits, and to the cultivation of their lands with a view to the production of the raw material to be exported to England in return for British manufactures'.[25] The wider purposes envisaged by Livingstone included the extension of 'legitimate commerce' in place of the slave trade, and of Christianity, together marking the beginning of Africa's civilisation.

The expedition's scientific aims, projected by the RGS and Royal Society, demanded a range of expeditionary skills and techniques of representation, including photography. David Livingstone took his brother Charles on the expedition in the capacity of official photographer and cartographer. That photography was deemed important enough to merit an official place on the expedition is itself significant. The practice of photography was defined and controlled in relation the the aims of the expedition. In a letter to Charles, written in May 1858, David Livingstone suggested that from the early stages of the expedition Charles should get his photographic apparaus working in order to 'secure characteristic specimens of the different tribes', 'specimens of remarkable trees, plants, grain or fruits and animals' as well as 'scenery'.[26]

These official roles were not unique to photography. Thomas Baines, who was employed at Murchison's behest as the artist of the expedition, was also instructed by David Livingstone 'to make faithful representations of the general features of the country through which we shall pass'. As well as making sketches 'characteristic of the scenery', 'drawings of wild animals and birds', and delineations of 'specimens of useful and rare plants, fossils and reptiles', Baines was urged to 'draw average specimens of the different tribes we may meet with, for the purposes of Ethnology'.[27] In this latter instruction, David Livingstone was responding to contemporary interests in Ethnology which demanded the selection and classification of characteristic racial 'types'. As he instructed his brother on the selection of human subjects for photography: 'Do not choose the ugliest but, (as among ourselves) the better class of natives who are believed to be characteristic of the race.'[28]

Photographic equipment was also taken on the expedition by John Kirk, the botanist and medical officer of the party. Kirk was an experienced amateur photographer and it seems he was far more capable than Charles Livingstone, who was notoriously inept at all his expeditinary duties.[29] In November 1858, for example, Baines, Thornton and Charles Livingstone were left at Tete whilst the rest of the party travelled upstream in their expeditionary steamer, the *Ma-Robert*, to investigate the Zambezi's Kebrabasa rapids. Baines and Thornton were both ill and Charles chose to 'try and get on with photography', as so far John Kirk noted, 'he has made a mess of it'.[30] However, although the only surviving photographs are those made by John Kirk, both he and Charles Livingstone spent some time experimenting with photographic processes. Kirk had been a keen photographer since his undergraduate days and cetainly had greater knowledge and experience of the technique than anyone else on the expedition. He successfully made negatives using waxed paper rather than the collodion that Charles Livingstone tried to use. Kirk also worked with Charles in the early part of the expedition, helping him with his photographic processes.[31] Despite technical difficulties and environmental constraints, both successfully produced photographs which were sent back to London.[32]

Photographs were taken of principal geographical features, as defined by the aims of the expedition. For example, in November 1858 photography was one of the means of exploring and recording the Kebrabasa rapids. This set of rapids was the chief obstacle to David Livingstone's dreams of a navigable Zambezi which could open the interior of this part of Africa to civilisation and 'legitimate commerce'. David Livingstone's diary entry for 23 November 1858 records: 'On reaching the place we have as yet called Kebrabasa . . . Baines sketched whilst Mr. L. [Charles] photographed them.' A painting by Thomas Baines of the Kebrabasa rapids, made in November 1858, shows a camera and tripod on an outcrop of rock with Charles working under its black cloth, David standing next to him. Photographs were clearly used by Livingstone as witnesses to the expedition and as permanent records of the prospects of the country. Livingstone refused to admit that the Kebrabasa rapids could thwart his plans for a navigable Zambezi, even after he and Kirk (who knew better) had found even larger rapids further up stream, on

2 December 1858. Photographs and sketches were included with Livingstone's written report of 17 December 1858 to the new Foreign Secretary, Lord Malmesbury, extracts of which were read before the RGS at evening meetings soon after.[33] Describing the party's exploration and survey of the rapids Livingstone expressed his conviction that a steamer could pass over the rapids 'without difficulty' when the river was in flood. He noted that 'a careful sketch 1 and photograph 2 were made of the worst rapid we had then seen'.[34] However, far from showing what Kirk knew to be true and Livingstone refused to believe, this visual evidence was designed to reinforce Livingstone's assertion that a suitable steam vessel could navigate the entire length of the river. Informing the Foreign Office of the absolute obstacle to navigation posed by the Kebrabasa rapids would have thrown uncertainty not just over Livingstone's request for a new steam-ship, but over the entire future of the expedition.[35] Livingstone had included visual images with his dispatches, he explained, because 'I thought this the best way of conveying a clear idea of my meaning'. Livingstone noted in particular a photograph 'showing a dead hippopotamus while also exhibiting the rock in the river' and that 'another photograph exhibits the channel among the rocks'.[36] Livingstone thus used photographs as a form of geographical evidence; a means of proving the realism of his colonial vision.

Other images similarly reflected the cultural assumptions and political aspirations of the expedition. For example, John Kirk's photograph 'Lupata July 13th 1859' (see Figure 3.1), made on this expedition, represents a wall of impenetrable, twisted vegetation.[37] The photograph fits into a well established image of Africa as a place of barbarism and savagery that was both reinforced and complicated as more of the continent was explored and visualised by Europeans. The image of 'the Dark Continent', spun together from various impressions circulating in Europe since its earliest contacts with west Africa in the fifteenth century, began to solidify in the early nineteenth century as Europe, particularly Britain, began to see in Africa opportunities for 'legitimate' (i.e. non-slave based) commerce and, by extension, more permanent imperial prospects. By the 1850s explorers left Britain equipped with common images and assumptions about the nature of Africa and the Africans' place in nature.[38]

The dense foliage of Africa's tropical environments, for example, had been regarded by generations of Europeans as offering both the promise of abundant riches and the dangers of the unknown, of disease and death.[39] Richard Burton's book *The Lake Regions of Central Africa; a picture of exploration*, published in 1860 painted, as its subtitle suggested, yet another gloss on this image. The then President of the RGS,

Figure 3.1 'Lupata July 13th 1859.' By John Kirk

Earl Grey, remarked that the landscape Burton described had a 'repulsive aspect', being, 'a fever-stricken country that is skirted by a wide, low-lying belt of overwhelming vegetation, dank, monotonous, and gloomy, while it reeks with fetid miasma'.[40] Kirk's photograph represents this imaginary landscape: one that is not yet free of a state of barbaric, primeval geography.

Photography was one of a number of practices on the Zambezi expedition through which geographical images of eastern and central Africa were constructed. Although claiming to be scientific, representations of African landscape, whether in the form of maps, written reports or photographs, were by no means neutral since they were part of an expedition whose expressed goals envisioned the establishment of European settlement, commerce and civilisation. David Livingstone expressed this aptly in his introduction to the narrative of the Zambezi expedition, published in 1865:

> In our exploration the chief object in view was not to discover objects of nine days' wonder, to gaze and be gazed at by barbarians; but to note the climate, the natural productions, the local diseases, the natives and their relation to the rest of the world: all which were observed with that

peculiar interest, which, as regards the future, the first white man cannot but feel in a continent whose history is only just beginning.[41]

Photography fitted into this ethos since it was an additional means of representing, as visual 'truth', the perceived possibilities and problems of British imperialism in Africa. As botanist and medical officer to the expedition, Kirk was particulary interested in both 'natural productions' and 'local diseases' as Livingstone had termed them. Extensive medical debates in Britain and west Africa had long focused on establishing the causative connections between types of disease and 'miasmas' (pockets of poisonous air) associated with damp, luxuriant, and importantly, uncultivated, tropical environments. Kirk's choice of scene thus correlates with the concerns of mid-nineteenth century medical topography. It was framed by his interest in establishing the suitability of tropical Africa for development by the 'white races'.[42]

Kirk's photograph, like most of his surviving images, presents no signs either of the photographer or of the expedition of which he was part. By aspiring to scientific objectivity, photographs, like maps and indeed like the explorers themselves, could disguise the cultural assumptions and political visions which informed their operations. The choice of dense forest vegetation as a suitable scene for a photograph represents in part the exercise of that 'peculiar interest', as David Livingstone had put it, with which Europeans imagined Africa as a continent without history until colonisation. This imaginative stricture was also applied to the indigenous inhabitants of 'the Dark Continent'. Kirk's image presents an ostensibly empty landscape. Along with cartographic and literary representations, photography partriciped in a geographical discourse which emptied lived environments of their human presence and isolated indigenous peoples from their habitats.[43] In this sense, landscape views were the counterpart to representations of ethnographic 'types', where individuals were pictured against uniform backdrops and measuring screens to be placed within hierarchical frameworks of racial classification. In both cases the camera, like other technologies of representation, became a tool of *naturalisation*, of rendering the unfamiliar familiar and the unknown known; of converting complex environments into the constituent categories of European scientific knowledge.

Although absent from Kirk's photograph, the inhabitants of the areas visited by the Zambezi expedition were, in conceptual terms, central to the representation of savage landscapes. As Earl Grey noted, explorers like Burton projected 'the picture of one unbroken spread of vulgar, disunited, and drunken savagery over the entire land'.[44] Similarly, Francis Galton thought Burton's account conveyed 'a repulsive picture of a vulgar, boisterous and drunken savagery overspreading the land'. Furthermore, Galton shared Burton's 'unfavourable view' and poly-genist leanings.[45] The image of Africa's environment and inhabitants together constituting 'one rude chaos' was hardly new.[46] However from the late eighteenth century, Europeans had begun to consider the apparent fertility of the tropical environment as both cause and indicator of social idleness.[47] Whilst this was less significant for poly-genists such as Burton and Galton, monogenists such as Speke and Livingstone believed in the possibility of both environmental and human improvement. Images of vegetative pandemonium such as those of Kirk, signified not only the rich potential of the land, but also the absence of indigenous industry and labour. Thus, whilst photographs such as 'Lupata July 13th 1859' (Figure 3.1) re-inscribe an imaginative geography of the 'Dark Continent', presenting a disorderly, even threat-ening nature, they also offer a colonial prospect; on where wildness is taken for unruly fertility and barbarity is read as a blank space for improvement.

The official role of photography on the Zambezi expedition signified a wider recognition of the relevance of the technique to geographical expeditions. Indeed, Kirk's experience with photography on the Zambezi was used to furnish a section on photography in the well known publication by the RGS, *Hints to Travellers* (1865).[48] Reflecting the increasing emphasis placed on scientific exploration by the RGS, photography was promoted as a necessary adjunct to exploration. Despite the difficulties posed by harsh environmental conditions and restrictive apparatus, explorers expended much energy in the pursuit of a photographic record. In 1859–60 James Chapman accompanied Thomas Baines, after the latter had been dismissed from Livingstone's Zambezi expedition, to the Victoria Falls on the Zambezi. In a letter of January 1860 to Sir George Grey, the Governor of the Cape of Good Hope, Chapman explained how problems of illness, drought and poor sport had been compounded by a greater failure: 'Of all our little dis-appointments I regret none more deeply, and I am sure your Excellency will sympathise with me when I say that I come back without one good photograph.'[49] Thus the failure of securing a photographic record was increasingly seen to amount to a failure of exploration itself. This also applied to military campaigns.

The Abyssinia campaign, 1867–68

The Abyssinia Campaign of 1867–68, ostensibly to rescue a handful of Europeans, including the British Consul, imprisoned by the Emperor Theodore of Abyssinia after a series of diplomatic misunderstandings, allowed the practice and display of British imperial military and scientific might in a foreign field. The expedition involved a military force of nearly 13,000 men from Britain and India, under the supreme command of General Sir Robert Napier of the Royal Engineers. It was planned and executed with technical precision. Indeed, many histories of the war read like a grand construction project. Engineering feats (with the considerable help of Indian troop labour) included the building of piers at Annesley Bay and a coastal base at Zoola, a railway line ten and a half miles inland involving eight bridges, as well as roads, wells and pumps.[50] The expedition was equipped with advanced technology and armed with knowledge of Abyssinia's topography and climate procured, with the help of the RGS, from the written accounts, maps and sketches of earlier explorers. Well prepared, the imperial forces embarked on an expedition of around four hundred miles, from their coastal base at Zoola to Theodore's highland stronghold of Magdala.

As a mission to rescue European hostages, including some missionaries, from an apparently merciless African Emperor, this expedition, like other imperial campaigns of the period, morally sanctioned itself by adopting the tone of a holy crusade.[51] Furthermore, the campaign represented, to military and scientific establishments as well as to the audiences of popular journalism in Britain, an adventurous exploration into the cultural distance and the geographical unknown.

The detailed military planning, the considerable expense of the operations and the unprecedented publicity surrounding them all indicate major imperial investments in the campaign. As Freda Harcourt has argued, in the late 1860s this dramatic rescue mission was an effective means for Disraeli to focus a British public's attention on a distant imperial adventure and to provide a popular distraction from a range of domestic crises.[52] The significance of the campaign lies as much in its means, its modes of operation and legitimation, as in its predictable ends: the release of the hostages, the destruction of Magdala and the death of Theodore.[53]

Whilst it was considered by many to have been largely an 'engineer's war', this expedition demonstrates the significance of a whole range of technological and scientific skills in the exercise and legitimation of imperial military violence.[54] Indeed, the Abyssinia campaign demonstrates the extensive relationships between military force, geographical science and photography in the pursuit of imperial power. Photography played an important part in both the organisation and display of this exercise in imperial scientific might. Two sets of photographic stores and equipment (of which only one was actually used) were sent from England at the suggestion of the Director of the Royal Engineers Establishment, at Chatham. The equipment was supervised by a chief photographer, Sergeant John Harrold RE, and seven assistants, all attached to the 10th Company Royal Engineers.

The direct and indirect influence of the British army in the general adoption of photography throughout the British empire has often been under-estimated. Yet it was often officers and surgeons, such as Dr John McCosh in India, who were the first to take up photography. The Royal Engineers in particular, were quick to adapt photography to various tasks in their civil and military duties. From 1856, instruction in photography was included in the training courses of the Royal Engineers' establishment at Chatham. By 1860 the Royal Engineers were wielding cameras as far afield as British Columbia, India, Singapore and China.[55]

The use of photography by the Royal Engineers received much attention from journals of photography and science during and after the Abyssinia campaign.[56] Photography was regarded by many, such as the author of an article published in *Nature* in 1870 who praised the Abyssinia expedition as 'one of the most wonderful feats of engineering accomplished in modern times', as yet another application of science to modern warfare.[57] Indeed the labours of the Royal Engineers photographers in Abyssinia, according to another writer, formed 'no unimportant part of the cog-wheel of administrative machinery' which, he added 'worked so smoothly and surely'.[58] Photography was literally part of the engineering of the campaign. Indeed photography was most extensively employed as 'a printing-press in the field', accurately and rapidly compiling and reproducing maps, plans and route sketches.

Photography was also used to picture the progress and achievements of the imperial forces on this expedition. The photograph reproduced here, titled 'Focada camp and rock', (see Figure 3.2) is one of a series of landscape views, camp scenes and portaits made by the Royal Engineer photographers in Abyssinia. The photographs, along with some sketches, were subsequently assembled into albums and were presented to various institutions, including the RGS, by the Secretary of State for War in 1869. Images such as 'Focada camp and rock', are thus part of the official representation of the achievements of the campaign as seen from the perspective of a great imperial power overcoming harsh geographical conditions to vanquish a tyrannical ruler.

The application of photography to record scenes of the exploration received considerable attention during the campaign. Lieutenant-Colonel Pritchard, who commanded the 10th Company, was apparently 'indefatigable in his endeavours to obtain interesting subjects for the camera' and even General Napier had requested a photograph be taken of the dead King Theodore.[59] Carefully posed studies, which might just have easily been made in studios in London or Bombay, portray the main characters in the drama: from General Napier and his commanding

Figure 3.2 'Focada camp and rock (1868).' By the Royal Engineers

officers scanning maps, to groups of rescued Europeans conspicuously wearing their prison chains. Photographic images of the Abyssinian landscape are no less carefully staged. 'Focada camp and rock', like many of the photographic views of Abyssinia, presents a grand, imposing landscape. In contrast to John Kirk's photograph discussed above, the presence of the expeditionary force is not disguised. On the contrary, it dominates a landscape otherwise devoid of human presence. The scene is pervaded by a sense of order. The neat rows of white tents of the British camp, set against the dark, rugged backdrop of the massive rock, projects the discipline and organisational strength of the imperial forces against a disorderly and uninhabited landscape.

As well as having to operate photographic equipment in unfamiliar environmental conditions, the Royal Engineer photographers could not choose such subjects for themselves. They had to obey staff officer. One commentator regretted that much useless work was done because staff officers were ignorant of photography:

> sometimes the mules had to be halted and the boxes unpacked during a
> long march in a drizzling rain in order that a picture might be attempted of

some mountain or other, the top of which was enveloped in a dense fog simply because a staff officer had expressed himself to the effect that the whole would make a grand picture.[60]

For writers like John Ruskin, as John MacKenzie has noted, imperial warfare was intimately connected to the application of a way of seeing in which art and nature were inseparable from empire. Indeed, in 1865 Ruskin had elaborated many of his ideas in a series of lectures on war at the Royal Military Academy, Woolwich.[61] A true landscape, as imagined by Ruskin, was one in which an imperial civilising order was projected onto a chaotic nature. Thus although Ruskin found the artistic status of photography questionable, 'Focada camp and rock' represents, in part, a process through which the geography of Abyssinia was aestheticised and comprehended as landscape.

This process also involved the application of a scientific gaze. The Royal Engineers had for some time advocated photography as a scientific means of visually mapping the nature of a country. As one Captain Donnelly RE reported in the *British Journal of Photography* in 1860: 'Photographs of a country gave a most truthful and accurate idea of it. They would do more to give an accurate idea of any particular position than yards of description on foolscap.'[62] In 1863 the official photographer to the Royal Military Repository at Woolwich, John Spiller, reported in the *British Journal of Photography* how artillery officers' 'field days' with a photographic van on Woolwich Common gave 'proof of the value which attaches, to photography as a ready means of recording the geography and military features of a country'.[63] This faith in photography's capacity to reveal the geographical truth was linked to the global projects of exploration, surveying and mapping central to the RGS and to its associations, official and unofficial, with British military forces and the War Office.

The Abyssinia expedition was also an extensive campaign for geographical science. Largely through the considerable influence of its President, Sir Roderick Murchison, the RGS took significant roles in both the planning and successful completion of the campaign. The presentation of a set of photographs of the campaign to the RGS by the Secretary of State for War in 1869 was a recognition of its contributions. Prior to the expedition, the RGS went to some lengths to collect information on Abyssinia's topography, climate and diseases. This was made available to the War Office.[64] More importantly, Clements Markham, the RGS's secretary, accompanied the imperial forces as the official 'Geographer of the Expedition'.

For Murchison the campaign was a perfect means of advancing geographical science. In 1867, Murchison assured Fellows of the RGS that they could confidently rely on Markham to 'give us a masterly geographical sketch and a vivid description of the region he may traverse'.[65] Throughout the expedition extensive surveys and geographical observations were made. Markham's work was closely associated with the survey work of the Royal Engineers, particularly that of the Indian Trigonometric Survey. In his geographical reports Markham provided detailed descriptions of the meteorology, geology, scenery and resources of Abyssinia, as well as information on the geographical obstacles facing the forces. In their broad scope, their rhetoric of scientific accuracy, and in the way they incorporated a celebration of British scientific and military might, Markham's geographical reports correlate directly with many of the landscape photographs made by the Royal Engineers.

In his Presidential address to the RGS in May 1868, only a few weeks after the triumphal storming of Magdala, Murchison proudly claimed the Abyssinian expedition as a mission of imperial and scientific civilisation; as truly 'a fine moral lesson which we have read to the world'.[66] As Murchison pointed out, geographical knowledge was produced by the expedition but was also a means of its success. The expedition represented a double triumph for the RGS. It gave the Society a chance to play a public role within a dramatic imperial production whilst simultaneously reaping valuable geographical knowledge, including what may be read from photographs such as 'Focada camp and rock'.

Although they do not seem to have been employed as reconnaissance documents during the actual expedition, the photographs of Abyssinian landscapes provided a justification for extending the use of photography on future military campaigns. One commentator suggested that officers on reconnaissance should take small cameras to make stereoscopic photographs to accompany sketches and maps.[67] Furthermore, once added to the RGS's collections, the Abyssinian photographs, along with Markham's reports, surveys and maps, became part of an expanding, detailed archive of global geographical knowledge. By 1869 a place for photography in this archive was well established but by no means as secure as that of maps and written accounts. Photography's significance in the science of geography and in a more general Victorian geographical discourse was greatly extended by the activity of commercial photographers. It is to the work of one professional photographer in particular, John Thomson, that I now turn.

The journeys of John Thomson, 1871–77

Although the overseas travels of professional British photographers in the nineteenth century has received a good deal of attention, their work has been studied predominantly in relation to their technical skill and individual careers as photographic artists.[68] This is true of the studies of John Thomson, a professional photographer who travelled and photographed widely in China, Singapore and Cambodia between 1862 and 1872.[69]

Thomson had left Edinburgh in 1862 with the intention of settling and practising photography in Penang. He moved to Singapore in 1863 where he set up a studio in the city centre as a professional portrait photographer. From this work, Thomson developed the interest and finance to undertake his first photographic travels, visiting India via Ceylon (1864) followed by a longer journey to Siam (1865–66) from where he explored and photographed ancient ruins in Cambodia (1866). In 1868, Thomson settled in Hong Kong, where he established a studio. It was from here that, between 1868 and 1872, Thomson conducted extensive photographic expeditions in China. In 1872 he returned to England for good, opening a photographic studio in London. Thomson became well known in Britain from his subsequent publications which married reproductions of his photographs with written accounts of his travels.

Whilst Thomson's pioneering activities as a travel photographer and writer have attracted much attention, they are seldom related to his interest in geography.[70] However, Thomson regarded himself as much a geographer as a photographer. In 1866–67, when Thomson returned to Britain for fourteen months, he addressed the Geography and Ethnology section of the British Association Conference in Nottingham, exhibiting his Cambodian photographs.[71] More significantly, Thomson joined the Ethnological Society and was elected a fellow of the RGS. Through his photographic expeditions and subsequent involvement with the RGS as their instructor in photography from 1886, Thomson tirelessly promoted photography's place in the science of geography, particularly on expeditions.

In a lecture on 'Photography and Exploration' given to the Geographical Section of the British Association in Cardiff in 1891, John Thomson declared his faith in the usefulness of photography to British geographical exploration:

> Where truth and all that is abiding are concerned, photography is absolutely trustworthy, and the work now being done is a forecast of a future of great usefulness in every branch of science. What would one not give to have photographs of the Pharaohs or the Caesars, of the travellers, and

their observations, who supplied Ptolemy with his early record of the world, of Marco Polo, and the places and people he visited on his arduous journey? We are now making history and the sun picture supplies the means of passing down a record of what we are, and what we have achieved in this nineteenth century of our progress.[72]

Thomson consistently appealed to the veracity of photographic imagery, as both artistic truth and scientific fact. Indeed, Thomson regarded himself as both a professional photographic artist and a scientific explorer. His work integrates both aesthetic and scientific concerns.

The photograph reproduced here ('Trading boat': Figure 3.3) comes from Thomson's *Illustrations of China and Its People*, published in four large volumes in 1873 and 1874. This large and elaborate work aimed, as Thomson wrote in the introduction, to 'present a series of pictures of China and its people, such as shall convey an accurate impression of the country'. Thomson's intentions were thus to provide a complete and factual representation of China and the Chinese. The authority of his imaginative geography was sustained by linking his narrative to the visual images. Through large photographs and detailed written descriptions, Thomson narrates his journey of more than 4,000 miles through what he sees as a decaying empire, from Hong Kong, which he refers to as 'the birth place of a new era in eastern civilisation' to the Yangtsze-Kiang.[73]

The composition of this scene on the Yangtsze-Kiang River, titled 'Trading boat', carefully utilises the pictorial effects of water and its reflections. Indeed Thomson, like other skilled commercial landscape photographers catering for a European audience, worked within the parameters of a picturesque aesthetic. Like other commercial travel photographers, Thomson was also motivated by the romance of exploring new landscapes photographically. However he wanted his travels to contribute more directly to British geographical exploration; his photographs of foreign landscapes and people were to be regarded as accurate scientific records rather than simply attractive images.

The photograph shown here, 'Trading boat', represents a good deal more than simply a picturesque image, as is implied by the placing of the number 38 in the bottom left corner. This image is located towards the end of Volume III of *Illustrations of China and Its People*, following a narrative of Thomson's exploration of the upper Yangtsze, early in 1871. Although elements of a picturesque aesthetic are present, Thomson uses the image to support his record of the geography of the river, particularly in relation to its suitability for steam traffic. Throughout his written description of the river and surrounding country, Thomson

Figure 3.3 'Trading boat.' By J. Thomson (1871)

refers the reader to the numbered photographs. Through this combination, Thomson charts the river and possible obstructions to a steam vessel – the rocks in the channel, alluvial banks, and sections of rapids. Thomson suggests suitable points of deep water where steamers may anchor as well as 'eligible' sites for foreign settlements based particularly on their suitability for trading and the availability of coal. Thus Thomson's photograph incorporates aesthetic conventions and a rhetoric of scientific accuracy to support what he sees as the inevitable extension of western civilisation, through British commerce, into China.

Thomson's vision of China's landscape corresponded with his portrayal of its people. Thomson's portraits of, for example, footbinding, opium dens, punishments for criminals and street gambling as 'typical' scenes of Chinese life, contributed to a well established image of the Chinese as hopelessly addicted to a range of vices.[74] The vision of a country epitomised by such self-inflicted barbarity cast encouraging light on British imperial incursions within China. Foreign settlements and treaty ports were for Thomson the only signs of light in an otherwise general 'darkness that broods over the land'.[75]

John Thomson set out to explore parts of China with his camera at a time when Britain, as well as other western powers, was confidently asserting its commercial demands in the wake of the Second Opium War (1856–60). *Illustrations of China and Its People* both presumed and encouraged the expansion of western commerce and culture into China. The cost of such an elaborate work would have given it a relatively limited readership.[76] However many of its photographs were reproduced as woodcuts or half-tones in Thomson's other, less costly publications.[77] Thomson placed great emphasis on combining words and images to produce instructive and accurate geographical information. The photographs and written descriptions are combined to construct an image that had all the objectivity of seeing it for oneself. It was the photographic image that was the geographical truth since, as Thomson wrote: 'the faithfulness of such pictures affords the nearest approach that can be made towards placing the reader actually before the scene which is represented.'[78] The *British Journal of Photography*, in a review of *Illustrations of China and Its People*, identified Thomson as 'just the photographer of all others with whom we are acquainted best fitted to penetrate with his camera into unexplored regions . . . an artist, a photographer, a geographer, and a man of general scientific attainments'.[79] That Thomson's photographs were recognised as scientific documents as well as artistic scenes was further confirmed when *Illustrations of China and Its People*, together with four photographs of Chinese scenery, were exhibited in 1875 as part of the RGS collection at the International Geographical Congress in Paris where they won a second-class medal.[80]

The uses of photography within geographical exploration were intimately linked with the new social sciences, new practices of observation and new institutions emerging in Britain in the second half of the nineteenth century. By 1873 John Thomson had returned from his ten years of photographic work and travel in the far east and had set up a studio in London making, amongst other things, photographic portraits for the Royal Geographical Society. In 1877 Thomson turned his photographic lens on subjects at home when he set out 'armed with note-book and camera' to explore the 'highways and the byways' of London, 'bringing to bear the precision of photography' on the conditions of its 'humbler classes'. John Thomson's *Street Life in London* was published as a twelve-part series of photographic studies of both well and lesser known 'street characters' and scenes of urban topography. Thomson's photographs were each accompanied by written commentaries, either by Thomson or by the journalist and reformer Adolphe Smith. The work was subsequently published in one volume in 1878.[81]

Thomson and Smith were, as they put it, 'fully aware that we are not the first in the field'.[82] Greatly influenced by Henry Mayhew's *London Labour and the London Poor* (1861) Thomson's work draws on and adds to existing iconographies of urban characters and topography.[83] By claiming to document reality, Thomson's photographs and text naturalise the categorisation of individuals within various urban 'types' occupying particular urban spaces. *Street Life in London* presented its middle-class purchasers with well mapped social topography, from the flower girls of Covent Garden to the 'Crawlers' of St Giles. Furthermore, the work made explicit analogies between on the one hand non-European uncivilised 'types' and remote landscapes, and on the other, those unfortunate urban 'types' lost in the midst of civilisation, hidden in the dark spaces of the city.[84] Thomson's photographic work more generally demonstrates how explorations and representations of unknown foreign landscapes and peoples were closely tied, particularly in scientific discourses, to the exploration of the disruptive populations and urban landscapes of Britain.

Conclusions

The expeditions examined here varied in motivation, organisation and effect. However, each one was associated with geographical science in ways which interacted with the uses to which photography was put and the kinds of 'truths' it was used to create. The historical and contextual study of photography on such expeditions demonstrates the complex nature of different investments in photographic practices, including those of scientific exploration, military campaigning and commercial travel photography.

Whilst the meanings of photographs are never fixed, as modes of geographical knowledge these landscape images represent certain framings of the world.[85] Despite the common claims to objectivity and innocence, photography was practised on these expeditions from particular points of view. Whether photographs disguised the presence of the explorers altogether (Figure 3.1) or made them the main focus (Figure 3.2), such images obscure the fact that in both cases the expeditions served and relied upon wider networks of cultural and political power. The photographs made on these expeditions then are not visual witnesses to the discovery of unknown regions. Rather, they are a part of the projection of the imagined landscapes of an expansionist Britain. 'The struggle over geography', as Edward Said has reminded us, 'is not only about soldiers and cannons but also about ideas, about forms, about images and imaginings.'[86] The examples discussed in this chapter reveal how imperial power operates through aesthetic conventions and the framings of scientific knowledge as well as through, say, direct military force.

More particularly, these case studies testify to the significance of landscape, as a way of seeing, in the cultural representation of imperialism. Landscape, as produced by these expeditions, was framed in various ways by the cultural and material value projected onto the environments being explored by the explorers. Furthermore, in the process of visually mapping landscapes these expeditionary practices were simultaneously globalising a particular landscape vision. Indeed, the very idea of empire in part depended on an idea of landscape, as both controlled space and the means of representing such control, on a global scale.[87]

The adoption of photography on various forms of expedition was due to a myriad of initiatives taken by individuals and institutions working in circles of science, the military, and commerce. As the examples considered here show, the RGS played a central role, both formally and informally, in fostering the pursuit of geographical knowledge through overseas expeditions. It proved a significant force in promoting the adoption of photographic technology on varying forms of exploration.

Photography was of course only one medium used in the picturing of foreign landscapes as colonial prospects. As an 'art–science', as it was often termed, photography correlated with many other forms and conventions of representation, from travel writing to cartography. However, perceived to be both natural and scientific, photography incorporated a role as 'handmaid of the sciences' with an aestheticising project of global proportions. As one commentator put it in 1860, 'Our indefatigable countrymen are ascending the Nile, tracing the course of the Zambezi, navigating the Ganges, the Yang-tse-Kiang, the Mississippi, climbing the Alps, the Andes, the Himalayas, in fact wandering to every region of the habitable globe in search of the beautiful and the picturesque.' Indeed the 'pencil of the sun' seemed destined to reveal all the mysteries of the world to an expanding imperial gaze.[88] At the start of a highly successful career as a landscape photographer in India in 1863, Samuel Bourne assured his readers in the *British Journal of Photography*, 'there is now scarcely a nook or corner, a glen, a valley, or mountain, much less a country, on the face of the globe which the penetrating eye of the camera has not searched'.[89] Photography was invested with a truly imperial scope.

Notes

1 Earl Grey, 'Address to the Royal Geographical Society', *Proceedings of the Royal Geographical Society*, 4, 1860, pp. 207–8.
2 Robert A. Stafford, *Scientist of Empire: Sir Roderick Murchison, scientific exploration and Victorian imperialism*, Cambridge University Press, Cambridge, 1989.
3 David N. Livingstone, *The Geographical Tradition: episodes in the history of a contested enterprise*, Blackwell, Oxford, 1992, p. 160.
4 See Felix Driver, 'Geography's empire: histories of geographical knowledge', *Environment and Planning D: Society and Space*, 10, 1992, pp. 23–40.
5 This chapter is derived from part of a larger research project, based largely on the photographic archives of the Royal Geographical Society. See James Ryan, 'Photography, Geography and Empire, 1840–1914'. Ph.D. thesis, University of London, 1994.
6 See for example, Denis Cosgrove and Stephen Daniels (eds), *The Iconography of Landscape: essays on the symbolic representation, design and use of past environments*, Cambridge University Press, Cambridge, 1988; and Steven Daniels, *Fields of Vision: landscape imagery and national identity in England and the United States*, Polity, Cambridge, 1993.
7 See Felix Driver, 'Henry Morton Stanley and his critics: geography, exploration and empire', *Past and Present*, 33, 1991, pp. 134–66.
8 John Tagg, *The Burden of Representation: essays on photographies and histories*, Macmillan, London, 1988.
9 See Victor Burgin, *Thinking Photography*, Macmillan, London, 1982; Andrew Roberts, 'Photographs and African history', *Journal of African History*, 29, 2, 1988, pp. 301–11.
10 Mary Louise Pratt, *Imperial Eyes: travel writing and transculturation*, Routledge, London, 1992, pp. 15–37.
11 Livingstone, *The Geographical Tradition*, pp. 98–101.
12 Edmund Burke to William Robertson, 9 June 1777. Cited in P. J. Marshall and G. Williams, *The Great Map of Mankind: British perceptions of the world in the age of Enlightenment*, J. M. Dent & Sons, London, 1982.
13 Mary Louise Pratt, *Imperial Eyes*.
14 G. Batchen, 1991, 'Desiring production itself: notes on the invention of photography', in R. Diprose and R. Ferrell (eds), *Cartographies: poststructuralism and the mapping of bodies and spaces*, Allen & Unwin, London, 1991, pp. 13–26.
15 J. Hackforth-Jones, 'Imaging Australia and the South Pacific', in N. Alfrey and S. Daniels (eds), *Mapping the Landscape: essays on art and cartography*, University Art Gallery and Castle Museum, Nottingham, 1990, pp. 13–17.
16 *Art Union Journal*, 8, 7, 1846, p. 195. Quoted in John Falconer, *Commonwealth in Focus*, International Cultural Corporation of Australia Limited, Melbourne, 1982, p. 19.
17 Robert Herchkowitz, *The British Photographer Abroad: the first thirty years*, Herchkowitz, London, 1980; André Rouille, 'Exploring the world by photography in the nineteenth century', in André Rouille and Jean-Claude Lemagny, *A History of Photography: social and cultural perspectives*, Cambridge University Press, Cambridge, 1987, pp. 53–9.
18 Paul Carter, *Living in a New Country: history, travelling and language*, Faber and Faber, London, 1992, p. 33.
19 George Greenough, 1841, 'Presidential address' *Journal of the Royal Geographical Society*, 2, pp. 39–57.
20 Roderick Murchison, 'Presidential address', *Journal of the Royal Geographical Society*, 28, 1858, p. 155; 'Presidential address', *Journal of the Royal Geographical Society*, 29, 1859, p. 152.
21 John F. W. Herschel, *Physical Geography: from the Encyclopaedia Britannica*, Adam & Charles Black, Edinburgh, 1861, p. 2.
22 Livingstone, *The Geographical Tradition*, p. 171.

23 See Patrick Brantlinger, 'Victorians and Africans: the genealogy of the myth of the Dark Continent', *Critical Inquiry*, 12, 1985, pp. 166–203.

24 See Stafford, *Scientist of Empire*, pp. 172–82.

25 David Livingstone, Letter to Norman Bedingfeld, R.N., Screw Steamer, Pearl, 10th April 1858, in J. P. R. Wallis (ed.), *The Zambezi Expedition of David Livingstone 1858–63: the journal continued with letters and dispatches therefrom*, Chatto & Windus, London, 1956, p. 413. A similar summary of the official aims of the expedition is given in David and Charles Livingstone, *Narrative of an Expedition to the Zambezi & its Tributaries and of the Discovery of the Lakes Shirwa & Nyassa, 1858–1864*, John Murray, London, 1865, p. 9.

26 David Livingstone, Letter to Charles Livingstone, Screw Steamer, Pearl, 10th May, 1858, in Wallis, *The Zambezi Expedition*, p. 431.

27 David Livingstone, Letter to Thomas Baines, Screw Steamer, Pearl, 18th May, 1858, in Wallis, *The Zambezi Expedition*, p. 434. From the evidence of Baines's few portraits in the RGS, it seems he was little concerned with recording individuals for their general ethological significance as 'average specimens'. Indeed, whilst Baines records the individual names of those portrayed he makes no record of the dimensions of the heads of individuals pictured, as Livingstone had instructed.

28 From Letter to Charles Livingstone, Screw Steamer, Pearl, 10th May, 1858, in Wallis, *The Zambezi Expedition*, p. 431.

29 See Reginald Coupland, *Kirk on the Zambezi: a chapter of African history*, Clarendon Press, Oxford, 1928, p. 54. See also A. D. Bensusan, *Silver Images: the history of photography in Africa*, Howard Timmins, Cape Town, 1966, pp. 24–5. A number of photographs made by John Kirk on the Zambezi expedition survive in a private collection deposited in the National Library of Scotland, Acc. 9942/40 & 41.

30 Cited in Coupland, *Kirk on the Zambezi*, p. 129. For other comments by members of the expedition on Charles's lack of progress in photography see Bensusan, *Silver Images*, pp. 24–5.

31 In June/July 1858 Charles Livingstone, Kirk and Thomas Baines were left on Nyika Island when the rest of the expedition transported expedition equipment up river. Baines records in his diary for Wednesday 9 July (which should be the 7th): 'Mr Livingstone got out his dark tent for photography and we set it up . . . In the afternoon the stereoscopic camera was set up and we grouped ourselves and the Kroomen in (and) about the house and shed and took half a dozen views, some of which Kirk succeeded in developing at night.' Quoted in E. C. Tabler, E. Axelson and E. N. Katz (eds), *Baines on the Zambezi, 1858–1859*, The Brenthurst Press, Johannesburg, 1982, p. 120.

32 Oliver Ransford has noted that although Charles Livingstone's photographs are 'now almost entirely lost', they 'appear to have been reasonable and there is good evidence that his scientific work was satisfactory'. Oliver Ransford, *David Livingstone: the dark interior*, John Murray, London, 1978, p. 143. Although none appear to have survived, there exists, in the RGS archives, a reference card to '6 photographs taken by Charles Livingstone Esq. in South Central Africa'.

33 David Livingstone, 'Extracts from the despatches of Dr David Livingstone to the Right Honourable Lord Malmesbury (communicated by the Foreign Office)', *Journal of the Royal Geographical Society*, 31, 1861, pp. 256–96.

34 No. 1 Marginal note: 'No. 3 in Water Colour by Mr. Baines, separate.'; No. 2 marginal note: 'Photograph by Mr. C. Livingstone'.

35 Tim Jeal, *Livingstone*, G. P. Putnam's Sons, New York, 1973, pp. 202–14. See also Coupland, *Kirk on the Zambezi*, 1928, p. 136.

36 Wallis, *The Zambezi Expedition*.

37 Private collection deposited in National Library of Scotland, Acc. 9942/40. This photograph is also reproduced in Coupland, *Kirk on the Zambezi*, under the title 'Creepers in the bush'.

38 Philip D. Curtin, *The Image of Africa: British ideas and action, 1780–1850*, Macmillan, London, 1965; Brantlinger, 'Victorians and Africans'.

39 Curtin, *The Image of Africa*, pp. 58–87.

40 Earl Grey, 1860, 'Address to the Royal Geographical Society delivered at the anniversary meeting 28 May 1860', *Proceedings of the Royal Geographical Society*, 4, pp.
117–209.
41 David Livingstone and Charles Livingstone, *Narrative of an Expedition to the
Zambezi & Its Tributaries and of the Discovery of the Lakes Shirwa & Nyassa,
1858–1864*, John Murray, London, 1865, p. 6.
42 John Kirk, 'The extent to which tropical Africa is suited for development by the white
races, or under their superintendence', *Report of the Sixth International Geographical Congress, London 1895*, 1896, p. 526. See also Curtin, *The Image of Africa*,
pp. 185–7, 349–53; David N. Livingstone, 'The moral discourse of climate: historical
considerations on race, place and virtue', *Journal of Historical Geography*, 17, 1991, p.
413–34; and Morag Bell, ' "The pestilence that walketh in darkness": imperial health,
gender and images of South Africa c. 1880–1910', *Trans. Inst. Br. Geogr.*, N.S. 18,
1993, pp. 327–41.
43 For an example of this procedure in travel writing see Pratt, *Imperial Eyes*, p. 61.
44 Earl Grey, 'Address' pp. 117–209.
45 Francis Galton, 'Zanzibar', *The Mission Field*, 1861, 6, pp. 121–30, p. 128. Cited in
Raymond E. Fancher, 'Francis Galton's African ethnology and its role in the development of his psychology', *British Journal of History of Science*, 16, 1, 1983, pp. 67–79.
46 Marshall and Williams, *The Great Map of Mankind*, pp. 227–57.
47 Curtin, *The Image of Africa*, pp. 61–2.
48 'Extracts from a Letter from J. Kirk', *Journal of the Royal Geographical Society*, 39,
1865, pp. 290–2.
49 James Chapman, 1860, 'Notes on South Africa (from a letter; Otjimbinque, 30th Jan.
1860 to His Excellency Sir George Grey, Governor of the Cape of Good Hope)',
Proceedings of the Royal Geographical Society of London, 5, pp. 17–18.
50 For a brief account of the campaign see R. Whitworth Porter, *History of the Corps of
Royal Engineers*, The Institute of Royal Engineers, Chatham, 1889.
51 See John M. MacKenzie, 'Introduction: popular imperialism and the military', in John
M. MacKenzie (ed), *Popular Imperialism and the Military 1850–1950*, Manchester
University Press, Manchester, 1992, pp. 1–24.
52 Freda Harcourt, 'Disraeli's imperialism, 1866–1868: a question of timing', *History
Journal*, 23, pp. 87–109.
53 For a full and dramatic account of the campaign see Alan Moorehead, *The Blue Nile*,
Hamish Hamilton, London, 1962, pp. 211–80.
54 R. Whitworth Porter, *History of the Corps of Royal Engineers*, The Institute of Royal
Engineers, Chatham, 1889, vol. 1, p. 5, and vol. 2, p. 1.
55 For more details of the adoption of photography by the Royal Engineers, see John
Falconer, 'Photography and the Royal Engineers', *Photographic Collector*, 2, 3, 1981,
pp. 33–64.
56 Such journals were adding to a wealth of more general publicity in newspapers. See, for
example, 1867, 'The Abyssinian expedition', *The British Journal of Photography*, 14,
p. 389.
57 Anon, 'The application of photography to military purposes', *Nature*, 2, 1870, pp.
236–7.
58 H. B. Pritchard, 'Photography in connection with the Abyssinian expedition', *The
British Journal of Photography*, 15, 1868, pp. 601–3.
59 Pritchard, 'Photography in connection with the Abyssianian expedition', p. 603.
60 *Ibid.*, p. 602.
61 Cook and Wedderburn, *Works of John Ruskin*, 'War', 1865, Woolwich, Royal Military
Academy, in *The Crown of Wild Olive*, 1909, pp. 115–71. See MacKenzie, *Popular
Imperialism and the Military*, pp. 1–24, 4–5.
62 Captain J. Donnelly, RE, 'On photography and its application to military purposes',
British Journal of Photography, 7, 1860, pp. 178–9.
63 J. Spiller, 1863, 'Photography in its application to military purposes', *The British
Journal of Photography*, 10, pp. 485–7.
64 The Army Medical Department and the Topographical Department of the War Office

sent official letters of thanks to the Society's Librarian; 'Report of the Council', *Journal of the Royal Geographical Society*, 38, 1868, pp. 5–13.

65 Roderick Murchison, 'President's opening address', *Proceedings of the Royal Geographical Society*, 12, 1867, pp. 4–9.

66 Roderick Murchison, 'Sir Murchison's Address', *Proceedings of the Royal Geographical Society*, 12, 1868, p. 275.

67 Pritchard, 'Photography in connection with the Abyssinian expedition', p. 603.

68 See, for example, Robert Herchkowitz, *The British Photographer Abroad: the first thirty years*, Herchkowitz, London, 1980.

69 See Stephen White, *John Thomson; Life and Photographs; The Orient, Street Life in London, Through Cyprus with the Camera*, Thames & Hudson, London, 1985.

70 See, however, Elliott S. Parker, 'John Thomson, 1837–1921: RGS instructor in photography', *Geographical Journal*, 144, 1978, pp. 463–71.

71 A number of these were soon published in John Thomson, *The Antiquities of Cambodia: a series of photographs taken on the spot, with letterpress description*, Edmonston & Douglas, Edinburgh, 1867.

72 John Thomson, 'Photography and exploration', *Proceedings of the Royal Geographical Society*, N.S. 13, 1891, pp. 669–75, 673.

73 John Thomson, *Illustrations of China and Its People: a series of two hundred photographs with letterpress description of the places and people represented*, Sampson Low, Marston, Low, and Searle, London, 1873–74.

74 V. G. Kiernan, *The Lords of Human Kind: black man, yellow man, and white man in an age of empire*, Hutchinson, London, 1969, p. 162.

75 John Thomson, *The Straits of Malacca, Indo-China and China or Ten Years' Travels, Adventures and Residence Abroad*, Sampson Low, Marston, Low, & Searle, London, 1875.

76 Advertisements for the work in 1877 gave the price of three pounds, three shillings, per volume. It is unlikely that the print edition of *Illustrations of China* was greater than 1,000 copies. See White, *John Thomson: life and photographs*, p. 30.

77 These include: John Thomson, *The Straits of Malacca, Indo-China and China or Ten Years' Travels, Adventures And Residence Abroad*, Sampson Low, Marston, Low, & Searle, London, 1875; John Thomson, *Through China with a Camera*, A. Constable and Co., London, 1898; and John Thomson, *The Land and People of China: a short account of the geography, religion, social life, arts, industries, and government of China and its people*, Society for the Promotion of Christian Knowledge, London, 1876.

78 Thomson, *Illustrations of China and Its People*, introduction.

79 'Illustrations of China', *The British Journal of Photography*, 20, 1873, 569–70.

80 See John Thomson, Letter to H. W. Bates (Secretary, RGS) 12 June 1875, *MSS, RGS Archives*. See also White, *John Thomson: life and photographs*, p. 195.

81 John Thomson and Adolphe Smith, *Street Life in London*, Woodbury Permanent Photographic Printing Co., London, 1877, preface. Thomson's Woodburytypes and the accompanying text were issued in twelve monthly parts.

82 Thomson, *Street Life in London*, preface.

83 Henry Mayhew, *London Labour and the London Poor*, 1861. Issued in weekly parts, Mayhew's classic text was accompanied by wood engravings based on Richard Beard's daguerreotypes and sketches.

84 See, for example, Thomson's photograph, 'London Nomades', and accompanying text. Thomson, *Street Life in London*, plate no. 15.

85 The readings presented here are inevitably partial. I am aware that more might be said, for example, of the ways in which these landscape images inscribed a feminised 'nature'. See, for example, Gillian Rose, *Feminism and Geography: the limits of geographical knowledge*, Polity, Cambridge, 1993, pp. 86–112; and Jean Comaroff and John Comaroff, *Of Revelation and Revolution: Christianity, colonialism, and consciousness in South Africa*, University of Chicago Press, London, 1991, pp. 86–125.

86 E. W. Said, *Culture and Imperialism*, Chatto & Windus, London, 1993, p. 6.

87 See MacKenzie, *Popular Imperialism and the Military*, pp. 1–24, 5. See also Denis

Cosgrove, 'Prospect, perspective and the evolution of the landscape idea', *Transactions of the Institute of British Geographers*, 10, 1985, pp. 45–62.

88 F. F. Statham, On the application of photography to scientific pursuits' *The British Journal of Photography*, 6, 1860, pp. 192–3.

89 Samuel Bourne, 'Photography in the east', *The British Journal of Photography*, 10, 1863, pp. 268–70.

Acknowledgements

I wish to acknowledge the financial support provided by a Scouloudi History Research Fellowship at the Institute of Historical Research, which I held in 1993–94. The Royal Geographical Society generously granted permission to reproduce photographs from its collection. I also wish to thank Mrs Daphne Foskett for allowing me to reproduce Sir John Kirk's photograph from her private collection. I am grateful to Patricia Middleton and the staff of the National Library of Scotland for their help in tracing Kirk's photographs. Finally, I should like to thank Morag Bell, Robin Butlin, Felix Driver and Mike Heffernan for their insightful comments on an earlier draft of this essay. Needless to say, its final inadequacies are wholly my own responsibility.

CHAPTER FOUR

The Royal Dutch Geographical Society and the Dutch East Indies, 1873–1914: from colonial lobby to colonial hobby

Paul van der Velde

Introduction

With the exception of one article about the history of the Royal Dutch Geographical Society which was written on the occasion of its centennial in 1973, nothing has been written about the society from a historical point of view.[1] In my research into Dutch colonial policy during the age of modern imperialism (1870–1914), the Geographical Society has been an important agent. In this chapter I will try to clarify the attitude of the society towards Dutch colonial policy and also assess its influence on that policy against the background of modern imperialism. First, however, I will evaluate the present-day historical debate about modern imperialism and the place it occupied by Dutch imperialism. I will then discuss the nature of the society and finally provide a brief description of two scientific expeditions which it equipped: the first to Sumatra in the years 1877–79 and the second to New Guinea in the years 1904–05.

The Netherlands and modern imperialism

During the last quarter of the nineteenth century the appearance of the world changed dramatically under the influence of the industrial and technological revolutions which were then fast gaining momentum. Territories which had little or no contact with the world system, were indiscriminately incorporated into that system.[2] In most cases these territories were appended to the metropolitan cores as colonies. The imperialist process was initiated by several European countries in the 1870s, and Japan and the United States joined their ranks in the 1890s. Modern imperialism reached its zenith in the last two decades of the nineteenth century. For some European nations such as Great Britain and France this entailed the extension of already existing colonial possessions. For other countries such as Germany, Belgium, Japan and the United States, which did not yet have colonial possessions, it marked their emergence as imperial powers.

Notwithstanding the decline in its political power since the beginning of the eighteenth century, The Netherlands still possessed the major part of its seventeenth-century empire. It can be defined as a commercial empire, in contrast with the British empire, which was based on territorial power over vast stretches of land. In 1820 effective Dutch control of the Dutch East Indies was limited to a minor part of Java. The rest of the Dutch East Indies, refered to as the Outer Territories, was only nominally under Dutch control. In fact, the Dutch East Indies were at this stage an ambitious territorial claim which did not mirror global political realities. The Dutch government was perceptive enough not to join in the race for new territories in the 1870s. Nevertheless, from 1830 onwards it had marginally increased Dutch control in the Indies. According to the current historical debate two periods can be distinguished in this process of the extension of territorial control.

During the first or informal period, from 1830 to 1894, the marginal tightening of control may be attributed to the resident colonial officials, while the government officially observed a policy of non-intervention in the Outer Territories – the so-called 'abstention policy' (onthoudings-pollitiek). The second or formal period, from 1894 to 1914, was initiated by a successful military expedition to Lombok in 1894, which touched a responsive nationalistic chord in The Netherlands.[3] In 1896 the formulation of the pacification policy put an end to the abstentionism to which the government had tenaciously adhered for so long. From now on it was the central government which took the initiative in extending its control over the Outer Territories. The pacification policy was made palpable for all in 1901, when ethical arguments were adduced in support of a colonial policy or 'mission civilisatrice', and which cured the Dutch of their phobia of colonialism which they had nurtured for three-quarters of a century. In 1904 the Aceh war, which had lasted thirty years, was ended and at the close of the age of modern imperialism in 1914, the Pax Neerlandica prevailed in the whole archipelago.

This sweeping change in the relationship between mother country and colony was in part effected by the abolition in 1870 of the so-called 'Cultuurstelsel' of forced crop deliveries which paved the way for a liberalisation of this relationship. Private initiative, no longer hindered by official restrictions, was then free to take root in the archipelago, especially in Java. By contrast, the government was limited in its freedom of action by the Aceh War, which devoured a large part of the Dutch

Indies budget. Furthermore, it was impeded from extending its control over the Outer Territories by an international economic crisis, which lasted until 1895.

In spite of, or maybe because of this economic crisis, a strong growth trend can be traced in the number of private enterprises which founded subsidiaries in the Dutch East Indies. The increasing inter-meshing of the economies of The Netherlands and the Dutch East Indies and the increasing nationalistic involvement between them – 'Indies lost, disastrous cost' – brought about the collapse of the abstention policy in the mid-1890s. In short, national and economic motives were the cause of its collapse. At the same time imperialism was globalised when Japan and the United States started to compete in the imperialist race at the beginning of the 1890s. It was imperative that the Dutch government strengthen its position in the Dutch East Indies in order to remain a credible coloniser. In the current debate about modern imperialism The Netherlands are conspicuously absent.[4] On the face of it, this seems to be a paradox. Did not the Dutch possess an empire whose territorial extent was exceeded only by Britain, Russia and France? The fact that the concept of modern imperialism is strongly related to the expansion of the European powers in the last quarter of the nineteenth century, especially in Africa, perhaps helps to explain the absence of The Netherlands from the historical debate. In the Second Sumatra Treaty signed with Great Britain in 1872, it exchanged its last possession in Africa – Elmina on the coast of Ghana – for freedom of action in Sumatra. This was to plunge both countries into costly colonial wars: namely, the Ashanti war and the Aceh war.

As a basis of a reassessement of The Netherlands within the framework of modern imperialism, I propose to divide modern imperialism into two phases: the appropriation phase, 1870–95, during which the great powers occupied vast territories; and the perpetuation phase, 1895–1914, during which the grip on these newly acquired territories was strengthened. The Dutch Geographical Society, which was founded in 1873, has a role in such a reassessment, and it is also possible to view the foundation of the society within the framework of the increase of private initiative in the Dutch East Indies.

The nature of the Royal Dutch Geographical Society

In February 1873 four geography teachers from grammar schools founded a geographical society based on principles in line with the broad geographical movement in Europe at that time. In their view, the foundation of such a society could stimulate and channel the nascent interest in geography. This awakening interest in geography was not surprising in an age in which 'La géographie est devenue la philosophie de la terre [in which geography had become the philosophy of the earth]'.[5] The increase in geographical knowledge resulting from these explorations would benefit trade, industry, shipping and plans for colonisation. The founders informed like-minded people of their intentions and this resulted in a meeting in the rooms of the Diligentia club in Amsterdam on 2 March 1873.

The meeting was presided over by Professor P. J. Veth, who, with the first secretary of the society, C. M. Kan, can be considered to be the founders of modern scientific geography in The Netherlands. Veth was already well known for his standard works in the field of colonial geography, which at that time included anthropology and ethnography of the Dutch East Indies.[6] The discussions about the goals of the society during that first meeting centred on the question of whether the society should restrict itself to increasing geographical knowledge or whether it should also engage in disseminating that knowledge. In other words: should the organisation be of a practical or of a scientific nature? The practical path was chosen unanimously: 'Even the semblance of a learned society should be avoided.'[7] In this manner the society followed the pragmatic tendency which had manifested itself from 1870 onwards in other European geographical societies.

The meeting also decided that education should be a goal of the society but this should not be explicitly stressed. Nevertheless a lobby launched by the Geographical Society for the founding of the first chair of geography in The Netherlands bore fruit in 1877 when this was established at the University of Amsterdam, which was founded the same year. Until the beginning of the twentieth century, Kan, the first professor of geography, devoted his courses to colonial geography.[8] Examination of a collection of eighty letters written by Kan to Veth, reveals that all attempts by Kan to theorise were stifled by Veth, who stressed the practical nature which he believed geography should have.[9] This is why there was not much theorising about geography in The Netherlands until the beginning of the twentieth century. Dutch geographers leaned heavily on theories developed in France and Germany. It may be interesting to note that ten years later, in 1887, the Royal Geographical Society succeeded after much lobbying in establishing a Readership in geography at the University of Oxford. In a report of the society drawn up in 1886 it came to the conclusion that:

> there is no country that can less afford to dispense with geographical knowledge than England . . . [yet] there are few countries in which a high order of geographical teaching is so little encouraged. The interests of England are as wide as the world. Her colonies, her commerce, her

emigrations, her wars, her missionaries, and her scientific explorers bring her into contact with all parts of the globe, and it is therefore a matter of imperial importance that no reasonable means should be neglected of training her youth in sound geographical knowledge.[10]

The same strictures could have been applied to the Dutch situation.

One of the other questions addressed by the Dutch Geographical Society concerned the encouragement of emigration. Only a slight majority was in favour of this policy. The opponents argued that by encouraging emigration the society would be considered to be one of colonisers. As will be seen, this strong anti-colonisation lobby would decline in less than a year. The foundation meeting of the Geographical Society on 3 June 1873 was attended by forty-four people. In his inaugural address the president, Veth, reviewed the two main goals of the society, which followed naturally from each other: the nationalistic goal of striving for and maintaining ancestral pride; and the dual scientific-economic goal of increasing and propagating geographical knowledge. Drawing his inspiration from Darwinism, Veth rejected the idea that The Netherlands would not be able to survive competition with the big powers. He focused the attention of his audience on the vast colonial heritage of The Netherlands, which, according to him, was respected by every civilised nation. This colonial heritage – the Dutch East Indies, in particular – offered excellent opportunities for adding to the sum of geographical knowledge. The geographical knowledge thus acquired should then be disseminated in order to increase the possibilities for opening up new trade routes.[11]

After the inaugural address the board members were chosen. In a commemorative lecture on the occasion of its tenth anniversary, Veth commented on the composition of the board: 'that representatives of trade, industry, the army and navy were represented on the board together with teachers of geography ... guarantees the interests of science and pragmatism'.[12] The closing address at the foundation meeting, again given by Veth, dealt with the Aceh war, which had broken out at the end of March 1873. His talk was based on his booklet 'Acehnese–Dutch relations' which had been published in May. In this booklet Veth spoke his mind freely: 'It is my conviction that in our struggle with the Acehnese, we represent civilisation and humanity in the face of barbarism and cruelty ... let there be no doubt that we must pursue this just war.'[13]

During the first months of its existence the society concentrated on Aceh as a future field for research. It petitioned the colonial secretary very energetically and after the first expedition had turned out to be a complete failure, it pressed strongly for a second military expedition to

Aceh. The society blamed the failure of the expedition, and previous expeditions to other territories, on a lack of geographical knowledge: 'If only we had possessed geographical knowledge of these territories blood and treasures could have been spared.'[14] In this very argument lay a basis for the growing popularity of geographical science, because in coming to terms with their defeat at Sedan at the hands of the Germans, for example, France had also blamed its defeat on the lack of geographical knowledge of its officers.

The interests of the society were by no means limited to the Dutch East Indies but extended to other parts of the world.[15] The Polar region, the Congo River and southern Africa also enjoyed the attention of the society. A predominantly nationalist motive accounted for its interest in the North Pole. The society was under the impression that Spitsbergen was still a Dutch possession. This is understandable in view of the fact that before the Berlin Conference of 1885, actual occupation of territory was not a precondition of ownership. An acknowledged claim sufficed. The economic motive concentrated on reviving the once flourishing whaling industry. The society's interest in the Congo River was governed by an economic motive. The Rotterdam-based Afrikaansche Handelsvereeniging (African Trade Society) had a large network of trading posts along the Congo River which comprised a sort of informal empire. The society's interest can also be explained from a strategic point of view, since the territory bordered on the colonial possessions of other countries in southern Africa.

The society's special interest in southern Africa was dictated by national sentiments due to kinship with the southern African Boers. The society viewed the Boers as the torchbearers of national Dutch vitality, in which their brethren in The Netherlands were completely deficient. In the view of the society, southern Africa with its enormous economic potential was the ideal region for Dutch emigration. Would this be the opportunity to create a Dutch-speaking cultural and economic empire, an opportunity which it had missed in the seventeenth century?

Leaving aside the pan-Dutch dream which continued to linger in the background, it transpired that the society's other ideas met with wide response in upper-class bourgeois circles, which is borne out by the number of such persons who became members of the society. Prince Hendrik, nicknamed the Seafarer, gave his support by becoming patron of the society. He expected good results for trade and industry from the society. This royal support was made evident when in 1888 the society was granted the right to call itself the Royal Dutch Geographical Society. In the space of three years the Geographical Society had become an effective colonial lobby, which bore overt witness to its imperialist nature. Equipping a scientific expedition to the land of economic promise, Sumatra, was a concrete expression of this nature.

The Sumatra expedition (1877–79)

The economic goal of this expedition – the discovery of a transportation route for the coal of the Ombilin Field, which had been discovered in the 1860s – was inextricably interwoven with the underlying political goal: expansion of effective control in that part of Sumatra through which the society expected to find a transportation route. This area was controlled by Sultan Taha, who had been driven out of Palembang in 1858 and had sworn to kill every Dutchman who tried to set foot on his territory. Everybody with some knowledge of the political conditions prevailing in Sumatra should have been aware of the risks the expedition members were running. Furthermore, in the past it had frequently been the case that the murder of a Dutchman by natives had given impetus for military expeditions which had sometimes led to expansion of effective control.

Since the majority of the Dutch in the Dutch East Indies were opposed to these kinds of provocation, they also opposed the scientific expedition of the society because they viewed it as a veiled attempt to extend control over Sumatra in an amoral way. 'It is evident that this private expedition . . . will have to be backed up by a military expedition . . . but in extending our control moral means should prevail over military violence.'[16] Although under pressure from Governor-General J. W. van Lansberge, the Council of the Dutch East Indies had decided to fund the expedition, but it was opposed to the extension of effective control in the Outer Territories, as were the government of the interior and the army. Their fear of creating a second Aceh was deep rooted because of its possible military and financial consequences.

Notwithstanding the support of the Governor-General, the provincial government and the navy, their room for manoeuvre was limited by the non-cooperation of the aforementioned branches of government. Van Lansberge was quite clear about what was at stake: 'The expedition will not only enhance our scientific knowledge but will also increase our political influence in the heartland of Sumatra. In the course of time it can eliminate the distrust of the natives for the Europeans and give them the correct view.'[17]

In The Netherlands the expedition received much support from influential members of the society. Also by partly funding the expedition, the Dutch parliament backed the society's initiative. More significant still was the support of the two colonial secretaries involved in the planning and execution stages of the expedition. Nevertheless, both secretaries, W. van Goltstein and F. Alting Mees, continued to reiterate in a somewhat hypocritical way that the private character of the expedition should be maintained at all times. However, they were fully aware that the so-called private character of the expedition meant little in an environment in which every white man was looked upon as a representative of Dutch authority. Therefore one can draw the con- clusion that Dutch colonial secretaries were prepared to run the risk of an unnecessary military expedition.

Disregarding staunch opposition in the Dutch East Indies, Van Lansberge planned to create an incident on the basis of a tried and tested recipe. After the first attempt by the expedition members to penetrate into Jambi had failed because one of the vessals of Sultan Taha, the Roja of Singuntur, had barred the way, the Governor-General gratefully accepted the 'proposal' of the Governor of Palembang, A. Pruys van der Hoeven. The latter proposed sending a gunboat up the Batang Hari River to a rendezvous where he would await the arrival of the expedition members. Citing one of the provincial officials, Van Canne: 'If Pruys van der Hoeven succeeds in steaming up river, I think the malevolent Roja of Siguntur will back down.'[18] This hope remained unfulfilled because the Roja managed to stop the expedition members for a second time. They failed to reach the rendezvous and the Governor of Palembang was forced to return to Palembang due to a fall in the water level of the river. Neither did the showing of the flag have the desired intimidating effect, since the gunboat was spotted by only a few natives. In the absence of a direct confrontation between the Governor and the Sultan, no incident occurred which would have warranted the launching of a military expedition. An attempt later that year by expedition members to penetrate the Limun territory was doomed to failure due to a lack of military support which, it turned out, was a precondition to any successful attempt to penetrate the heartland of Sumatra. The telegram from the leader of the expedition to the president of the organising committee in The Netherlands, Veth, bears testimony to the failure of the expedition. 'Armed people–stop–forced us–stop–leave Limoen– stop–travel Djambi impossible–stop–ask instructions–stop.'[19]

The unrest created by the expedition in Sumatra (and Java) inspired the third colonial secretary involved – Van Bosse (a wealthy sugar plantation owner) – to let the expedition die a natural death. In his opinion the extension of effective control should be accomplished gradually by provincial officials. 'I think it would be an advantage if our provincial officials are able to act when they think the time is ripe and that their actions should no longer be complicated by the peregrinations of the scientific expedition of the Geographical Society.'[20] If Van Bosse had lived longer – he died at the beginning of 1879 – the political aim and its concomitant economic goal might eventually have been realised, albeit in a roundabout way.

But Van Bosse's successor, O. van Rees, a repatriated member of the Council of Dutch East Indies, was a staunch supporter of the abstention policy. Van Lansberge, whose tenure was drawing to a close, could not but concede to the point of view of the new colonial secretary. The same was true for the society, whose president attributed the failure of the expedition to the fact that the political conditions in Sumatra had been completely misrepresented by the colonial government. According to him the most important outcome of the expedition was that it had disclosed the true nature of the prevailing political conditions in Sumatra, which he and the society would like to have imagined otherwise.[21]

The society in the 1880s and 1890s

The attempts by the society to organise similar expeditions to other parts of the Outer Territories at the beginning of the 1880s, for example to New Guinea at the height of the Berlin Conference, failed due to lack of support from parliament and the Colonial Ministry. Nevertheless the society funded a one-man expedition by the son of its president to Benguela in southern Africa. It was a complete failure and resulted in the death of Veth's son. In contrast to government circles, the trade and industry sector continued to support the society's expedition plans. At the end of the 1880s, the society succeeded in obtaining support from the government for a one-man expedition to Flores.

A. Wichmann, a geologist, was sent to Flores by the society to verify the rumours about rich tin deposits on that island. The expedition did not bear fruit but Wichmann's remark that the Kokka tribe had shown itself very friendly towards the Dutch nation began to take on a life of its own. On the basis of Wichmann's remarks, both the Resident of Timor, G. G. de Villeneuve, and the director of the Department of Education and Religion, W. P. Groeneveldt, a future president of the society, were in favour of sending a government expedition to the tribal area of the Kokkas. The mining engineer, R. van Schelle, who was sent to Flores at the end of 1889, barely escaped with his life after his expedition was attacked by the 'Dutch-loving' Kokka tribe. This incident provoked a military expedition, which was sent to Flores in May 1890 and resulted in the pacification of the Kokkas. Again the aspirations of the society had been served in a roundabout manner.[22]

The society's pioneering role in promoting an imperialist policy was taken over by the Society for the Advancement of Scientific Research in the Dutch Colonies, founded in 1890. It operated from The Netherlands as well as from the Dutch East Indies, which had obvious advantages. Its many expeditions to the sensitive Borneo region played a crucial role in extending effective control there.[23] In the meanwhile the decline in membership of the society which had set in in the middle of the 1880s continued. In these years the society gradually embraced a more scientific course, thus narrowing its support base. At the end of the century the membership had dropped by 30 per cent in comparison to the 1885 figure of nearly one thousand members (see Figure 4.1).

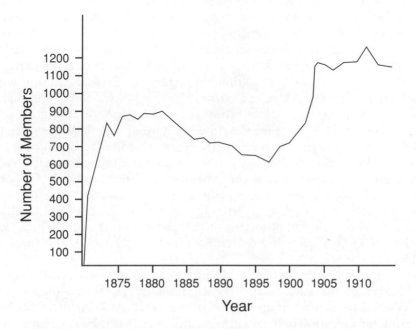

Figure 4.1 Membership of the Royal Dutch Geographical Society, 1873–1914

As mentioned earlier, economic and social interrelations between The Netherlands and the Dutch East Indies were multiplying. Thus, pressure on the government to abolish the abstention policy continued to mount. This internal pressure was stepped up by external pressure arising from the participation of Japan and the United States in the imperial race, which forced the Dutch to strengthen their position in the Dutch East Indies: an archipelago that bordered the spheres of influence of both newcomers. Therefore the abstention policy was definitively replaced by the pacification policy. Aceh served as guinea pig for the implementation of this policy which was vigorously pursued by General J. B. van Heutsz. The strong nationalist upsurge resulting from the outcry of indignation about the imperialism of Great Britain at the outbreak of the Boer War paradoxically cleared the way for an active imperialist policy, which, in 1901, was given the misleading label of 'ethical'.

The south-west New Guinea expedition (1904–05)[24]

The Society thrived on the surge of nationalism. The new president, J. W. Ijzerman, and the new secretary, A. L. van Hasselt, capitalised on this trend and the membership surpassed the 1885 level, reaching 1,200 members. Due to a financial reorganisation by the Amsterdam banker, A. W. van Eeghen, the Society re-emerged ready for the fray. Its attention now was directed both to east and west. Between 1900 and 1914 it co-sponsored seven expeditions to Surinam. Furthermore it was a ready tool in the hands of the Colonial Secretary, A. F. W. Idenburg, who within the framework of the 'ethical' policy, strove to fill the many blank spots of the map of New Guinea. In choosing between the expedition plans of the Society for the Advancement of Scientific Research in the Colonies and the Geographical Society, the colonial secretary opted for the plan of the latter because it was aimed at a topographical survey of the island. Idenburg gave the expedition military support. Only one socialist protested about the expedition plans when these were brought before parliament, but, apart from this, there was no opposition whatsoever. This was symptomatic of the change in attitude which had taken place in Dutch and Dutch East Indies societies in the space of thirty years.

The expedition was a failure. There were conflicts between the representative of the society and the commander of the troops and the accessibility of the terrain had been over-estimated. Furthermore the protracted digestion of the meagre results lasted until 1908 when a book about the expedition was published.[25] This made it clear that the society could not be used as an instrument of policy. On the advice of H. Colijn, a future prime minister, a government team was established which, by the time it was disbanded in 1915, had mapped 80 per cent of New Guinea. Thereafter, the society resigned itself to purely scientific pursuits.

Conclusion

Due to the emphasis in the debate on modern imperialism in its first appropriation phase, historians are still of the opinion that Dutch imperialism did not exist. However, when we take the perpetuation phase into consideration, it becomes clear that the Dutch with their active imperial policy were second to no other imperial power. The pinnacle of Dutch imperialism can be located in the period 1900–1914. The subjection of the whole archipelago, 'From Sabang to Merauke', the Dutch equivalent of the British influence from 'the Cape to Cairo', meant that the Dutch had become the fourth largest imperial power.

The foundation of the society in 1873 occurred in the initial years of the appropriation period. Within a few years the society became the rallying place for pro-imperialist forces in Dutch society which were nationalistically, culturally and economically motivated. Thus the society metamorphosed into a colonial lobby which denounced the conservative colonial policy of the government. Under the veil of a scientific expedition to Sumatra, it tried to torpedo the government's policy. The conservatives held the upper hand until 1895, when under external and internal pressure which had been building up during the appropriation period, their policy was replaced by an agressively imperialistic one.

During the latter period, the society helped to lay the foundation for the new imperial policy – the pacification policy which the society wholeheartedly supported during the perpetuation stage. However, its expedition to New Guinea proved to be a political failure. It turned out that the society could not be used as an instrument of policy. The society would resign itself during the last years of imperialism (1914–40) to purely scientific pursuits in an empire which it had helped to found.

Notes

1 R. Schrader, 'Honderd jaar Koninklijk Nederlands Aardrijkskundig Genootschap, 1873–1973', Geografisch Tijdschrift, 8, 1974, pp. 1–164.

2 I. Wallerstein, The Modern World System: Capitalist Agriculture and the Origins of the European World Economy in the Sixteenth Century, New York, 1974.

3 C. Fasseur, 'Een koloniale paradox. Nederlandse expansie in de Indonesische archipel in het midden van de negentiende eeuw, 1830–1870', Tijdschrift voor Geschiedenis, 92, 1979, pp. 162–86. See also, M. Kuitenbrower, The Netherlands and the Rise of Modern Imperialism, New York, 1991.

4 H. L. Wesseling, 'Bestond er een Nederlands imperialisme?' Tijdschrift voor Geschiedenis, 99, 1986, pp. 214–25.

5 H. Brunschwig, Mythes et réalitiés de l'imperialisme colonial Francais, Paris, 1960, p. 24.

6 P. G. E. I. J. van der Velde, 'De projectie van een Groter Nederland. P. J. Veth en de popularisering van Nederlands-Indie, 1848–1895', Tijdschrift voor Geschiedenis, 105, pp. 367–82. P. G. E. I. J. van der Velde, 'P. J. Veth (1814–1895) as empire-builder', in R. Kirstner (ed.), The Low Countries and Beyond, Los Angeles, 1993, pp. 13–27. It is interesting to learn that Veth had just finished his translation of A. R. Wallace's, The Malay Archipelago, which would inspire him to write his three-volume standard work on Java.

7 Tijdschrift van het Aardrijkskundig Genootschap (TAG), 1, 1876, p. 5.

8 J. van Beurden, C. M. Kan (1837–1919). Zijn opvattingen over de geografie in relatie tot die van zijn tijdgenoten in het buitenland', Nijmegen, 1988 (MA thesis).

9 Universiteitsbibliotheek Leiden. Collectie Westerse handschriften, BPL 1756.

10 D. R. Stoddart, *On Geography and Its History*, Oxford, 1986, p. 87.
11 *TAG*, 1, 1876, p. 6.
12 *TAG*, 2 (second series), 1885, p. 144.
13 P. J. Veth, *Atchin's betrekkingen tot Nederland*, Leiden, 1873, p. 132.
14 *TAG*, 1, 1876, pp. 11–12.
15 The following is based on my MA thesis: P. G. E. I. J. van der Velde, 'Een vergeten koloniale lobby. Het Koninklijk Nederlands Aardrijkskundig Genootschap en Sumatra, 1874–1879', University of Leiden, 1986.
16 *Java Bode*, 16 augustus 1877.
17 Dutch State Archives, The Hague. Archives of the Ministry of Colonies, 1850–1900. Letter of Van Lansberge to Van Goltstein of 24 June 1876.
18 State Archive, Utrecht. Archives of the Royal Dutch Geographical Society. Letter of Van Canne to Van Hasselt of 2 March 1878, in a map on the Sumatra expedition.
19 *Ibid.*, Telegram of Van Hasselt to Veth, 22 July 1878.
20 General State Archive, The Hague. Archives of the Ministry of Colonies, 1850–1900. Letter of Van Bosse to Van Lansberge of 2 November 1878.
21 *Midden Sumatra. Reizen en onderzoekingen der Sumatra expeditie, uitgerust door het Aardrijkskundig Genootschap*, 1877–79, Leiden, 1881, p. 8.
22 P. Jobse, 'De tin-expedities naar Flores, 1887–1891', University of Utrecht, 1980 (MA thesis).
23 A. A. Pulle, 'Overzicht van de lotgevallen en werkzaamheden van de Maatschappij ter bevordering van het Natuurkundig Onderzoek der Nederlandse Koloniën', *Bulletin ter bevordering van het Natuurkundig Onderzoek der Nederlandse Koloniën*, 99, 1940, pp. 1–84, 22.
24 P. G. E. I. J. van der Velde, *De Zuidwest Nieuw-Guinea expeditie van het Koninklijk Nederlands Aardrijkskundig Genootschap*, 1904–05, Leiden, 1983.
25 *De Zuidwest Nieuw-Guinea-Expeditie 1904/5 van het Koninklijk Nederlandsch Aardrijkskundig Genootschap*, Leiden, 1908.

CHAPTER FIVE

The provincial geographical societies in Britain, 1884–1914

John M. MacKenzie

In the history of the nineteenth-century geographical movement, Lyon and Bordeaux have always featured more prominently than Liverpool and Manchester. There are a number of reasons for this. The Lyon and Bordeaux societies were founded respectively in 1873 and 1874. They represented a French search for a utilitarian commercial geography, a conscious reaction to the supposedly 'pure science' approach of the Paris 'Société de Geographie which led to the foundation of the schismatic Paris Commercial Geographical Society in 1876. These societies emerged in a decade of growing recession in the European economy when the French were searching for an imperial alternative to failure in Europe at the hands of the Prussians. Moreover, Lyon and Bordeaux seemed to be key transmitters of the geographical contagion which swept mainland Europe, north Africa, and points as far afield as the Dutch East Indies. Quebec, Japan and Argentina. Almost forty societies were founded in the 1870s (at least ten of them in France) and many played a prominent role in the public and commercial agitations associated with the 'new imperialism' of the last quarter of the century.[1]

By contrast, the British provincial societies – the Scottish (with four branches), Manchester, Liverpool, Tyneside, Southampton and Hull – made a late appearance in the 1880s and 1890s. They seemed to succeed rather than precede the developing scramble for Africa. If they represented a strategic or intellectual break with the dominant Royal Geographical Society of London, founded in 1830, it was a split that took more muted and restrained forms than their Gallic counterparts. Only the Scottish made any substantial contribution to exploration and at first sight it seems hard to chart any influence exerted by them upon government or commercial policy. Nor did they once seem to have a major effect on the development of geography as a discipline. They appeared to reflect a passing public mood, a temporary excitement that was more symptomatic than instrumental. In short, with 'geographical fever' as with European monetary and economic union, the British seemed late and half hearted.

Moreover, historians of British geography have found it difficult to escape the metropolitan focus. Of the six identifiable British provincial societies, only two, the Manchester and the Royal Scottish have survived. Southampton (1897) and Hull (briefly flourishing, if flourishing is the word, at the beginning of the twentieth century) have almost literally vanished without trace. The Liverpool, founded 1891, expired in 1933. The Tyneside (1887) was moribund in the 1920s and for much of the 1930s, was refounded in 1937, only to disappear in World War Two. Even the two survivors have not kept archives, and as a result the sole sources for the history of these societies are their published journals, annual reports and transactions, together with the local press and fragmentary references in personal biographical material. The centennials of the Manchester and Scottish societies in 1884 produced some secondary material, but most of this failed to establish a full historical contextualisation.[2] And none of it attempted a comparative study of the provincial geographical movement as a whole.

In fact, a study of the British provincial geographical societies has the capacity to shed much light on the operations of 'municipal imperialism' in Britain, on the popularity and influence of travellers and propagandists in the imperial cause, on the ways in which geography was harnessed to the mobilisation of public opinion, and on the hopes, often ultimately unrealised, of the captains of commerce, shipping and industry in alerting their employees and workers in general to the utility of geographical knowledge in competing with and confounding competitors. Moreover, membership lists help to illustrate the ways in which the political, commercial and intellectual objectives of the societies were reflected in the social and economic backgrounds of both leaders and led within them. Their shifting composition also reflects both disillusionment and change, the dissipation of the initial energy of foundation and the need to move geography in fresh directions.

There can be no doubt that the sudden rush to the geographical colours in Britain between 1885 and 1891 was related to the economic alarms of the time. And these economic discontents arose from a recession which had placed free trade on the defensive, stoked up the final scramble for command of the earth's resources, and convinced the British that they had entered upon a new and highly competitive age. New foreign tariffs, a naval scare, crises in the Sudan and the North West Frontier, and a conference (at Berlin) which appeared to move the diplomatic centre of gravity towards central Europe all rendered the middle years of the 1880s a time of national anxiety. Imperial commentators, academics, journalists and other members of what are now called the 'chattering classes' became prone to two convictions in the period: first that British governments, as exemplified by failure in the Sudan and concessions to Germany in Africa, lacked both patriotic conviction and essential knowledge; second, that in the new European struggle the fitness of the state to survive and prosper could only be achieved if the public were trained in an understanding of the world's resources.

This may seem like a bold formulation of the conditions which produced the new geographical societies, but an extensive reading of the early meetings and rhetoric of these bodies confirms that they were founded in a mood of defensive reaction. The provincial societies were created as a response to the conviction that the Royal Geographical Society was neither utilitarian nor imperial enough, that the metropolis could no longer be trusted to act as a true guardian of the national interest.[3] None of the societies restricted itself by using the label of 'commercial geography', but there can be no doubt that, in varying degrees, this was their initial thrust. In the late 1870s there had been an attempt to create a society for commercial geography in London, an exploit in which Verney Lovett Cameron, the African explorer, had been implicated. This had its provincial counterpart in the effort to establish such a society (provisionally entitled 'The Society of Commercial Geography') in Manchester in 1879. Both failed, but when the Manchester society was at last founded in 1885 Cameron gave it his blessing and became one of its first honorary members. But, just as the word 'commercial' was omitted, there were no overt schisms in the British geographical movement (at least until the Geographical Association was founded in 1893). Equally, there was to be no federation of societies and the RGS both patronised and was suspicious of the provincial upstarts.[4]

Geography and municipal imperialism

The English societies were founded to meet a combination of particular civic needs and a pressing national interest. In the Scottish case, the national interest operated at two levels. British and north British; civic concerns were plural and complex; and, as Elspeth Lochhead has argued, there were also considerable intellectual and scholarly impulses in its foundation. The objectives of the societies made these emphases abundantly clear. Manchester devoted itself to the promotion of all branches of geographical knowledge, 'especially in its relations to commerce and civilisation'.[5] Its third objective fleshed this out:

> To examine the possibility of opening new markets to commerce and to collect information as to the number, character, needs, natural products

and resources of such populations as have not yet been brought into relation with British commerce and industry.

The society also dedicated itself to enquire into all questions relating to colonisation and emigration as well as to the dissemination of geographical information. In an objective reminiscent of the contremporary Imperial Institute (of which all these societies became local agents) the society also pledged itself to make collections of specimens of raw materials and commercial products.

The Scottish society produced a balanced list of utilitarian and scientific objectives. The first prospectus, approved at a meeting of its council on 28 October 1884, dedicated its members to the pursuit of knowledge and its practical application which were 'daily becoming more important for the reputation and prosperity of a people sharing in international commerce and in colonisation as do the people of Scotland'.[6] Its objects included demonstrating the utility of geography, support for exploration, publication of explorers' accounts and material useful for those 'engaged in founding British settlements and trading stations in different parts of the world', together with statistical accounts of both climatology and ethnology. To these were added particular concerns with Scottish topography and the development of the discipline of geography within higher and other branches of education.

Liverpool followed the objects of Manchester sufficiently closely as to suggest that the framers had the Manchester list before them, though the Liverpudlians expressed them in a rather more terse form. Key phrases such as 'commerce and civilisation', 'new markets for commerce, their natural products, resources and needs', 'questions bearing upon colonisation and emigration' reappear, but Liverpool stressed its desire to establish a 'library and museum' to contain a 'collection of maps and charts, gazetteers, works of travel and geographical science, specimens of commercial products and ethnological examples'.[7] Indeed, the Liverpool society subsequently devoted such financial resources as it could muster to the library and museum, failing as a result to establish a journal. This priority neatly illustrates its desire to offer a commercial membership utilitarian returns. The Tyneside society had slightly different emphases: its first object was 'the acquisition and diffusion of geographical knowledge throughout the district in which the society's sphere and influence extend'. It placed a stronger stress on 'uniting the efforts of all who are interested in the science of geography' and on pressing for the recognition of geography in education. Its prime utilitarian end was to offer help and encouragement to emigrants.[8] But the history of the origins of all the societies reflects their attempt to fulfil civic needs in order to advance the national interest.

The original proposal for a Manchester society had been initiated – according to later accounts which may have had respectability in mind – not by businessmen, but by clerics, in conjunction with professors at Owens College.[9] In 1879 the Bishop of Salford, the future Cardinal Vaughan, and the Bishop of Manchester had together called a meeting at the Chamber of Commerce for the 'promotion of trade with Africa' and urged the founding of a geographical society. The Lord Mayor made the intentions of such a society abundantly clear in his address to this meeting. The society would 'inaugurate a movement in the face of our declining commerce, to see whether some other large sectors of the human race could not be brought into commercial intercourse with this manufacturing country'. All those in public positions and with capital invested in manufacturing concerns should set about the revival of trade by all means possible, and a geographical society could go beyond the 'ordinary operations of Chambers of Commerce' by pursuing 'scientific and geographical researches . . . with the view of discovering fresh fields for the commerce of this country'.[10] This development was, indeed, closely bound up with a scheme by Manchester merchants to find new markets in east Africa by securing a concession from the Sultan of Zanzibar (such a concession had already been granted to William Mackinnon) and, following the suggestion of Consul Holmwood, build a railway into the interior. It proved a chimerical notion, but it was clearly a poor relation of the grander French railway-building schemes.

Indeed, the French model was not far from their minds. Vaughan later referred to the fact that many valuable geographical societies had been founded in France and that Manchester should follow this example. Subsequently, he presided over a meeting at the Chamber of Commerce which set about framing rules, appointing leading men to the council and forming a provisional committee. The response of businessmen was so slight (130 replies to 2,000 letters) that the attempt was abandoned, only to be resurrected, as the society's journal put it, when 'a great change for the worse had come over the commercial world'.[11] By 1884 the recession was deeper and considerable distress was being felt throughout the cotton industry. It was in this severe economic climate that Manchester politicians and businessmen were fiercely active in opposition to any extension of influence in Africa by the protectionist Portuguese. To them the government appeared to be colluding in such an extension through the proposed Anglo-Portuguese Congo treaty. Equally they were alarmed at the prospects of the growth of French power, particularly in the Niger regions. A powerful coalition of Manchester figures, including the MPs Jacob Bright (brother of John) and John Slagg (an influential cloth merchant), J. F. Hutton (president of the chamber) and Elijah Helm (secretary of the chamber) were active in leading opposition in the Commons, in establishing connections with Scottish figures like Mackinnon, and in delegations to the Foreign Office.[12]

It comes as no surprise to find that it was this group who were instrumental in bringing H. M. Stanley to Manchester to issue propaganda both for the free-trading and supposedly philanthropic International Association of King Leopold of the Belgians (one of the first honorary members of the Manchester society) and for the founding of geographical societies. The visit of Stanley on 21 October 1884 occurred under the auspices of the Manchester Athenaeum, but it was presided over by John Slagg, as part of a conscious effort to recreate a geographical society. Stanley's address at the Free Trade Hall once more made a direct connection between Africa and geography: 'Central Africa and the Congo Basin; or, the Importance of the Scientific Study of Geography'. Stanley offered his audience what they wanted to hear by suggesting that since Britain was dependent upon abroad, and since all questions of markets, resources, ports, shipping, and technological developments relied upon geographical knowledge, geographical societies were vital for survival. Businesses could only prosper if the 'merchant's clerk and book-keeper ... manufacturer's assistants, employees, clerks, and packers ... even down to the smallest boy of the factory' had some understanding of geography. Hence, it should be studied by 'every resident, male or female, in the country'.[13]

At the first meeting of the society on 27 January 1885, Vaughan revealed why he was concerned to promote the study of geography for the interests of trade and commerce. On the one hand, he was anxious for the welfare of the 'hundreds of thousands of hands upon whose wages the welfare, comfort, and happiness of millions of men, women, and children depended', while on the other the society 'would quicken the human sympathies in our hearts towards our neglected and degraded brothers and sisters in far off lands, and lead us to reach out a hand to lift them out of the dark night in which their existence had been shrouded for so many centuries'. Thus he feared for a dark night of social unrest if new markets were not found for the products of workers' hands, linking this to the traditional and redemptive cry of Christianity, commerce and civilisation. Moreover, the Manchester society would be able to do that which the RGS could not do 'for the special interests of Lancashire and the manufacturing districts'. Jacob Bright similarly emphasised the specific needs of Manchester and linked the necessity of an understanding of geography to 'the universal scramble of the nations for the possession of distant islands and barren coasts and unhealthy regions' of the globe, to know more 'about many of the coveted places'. While Lord

Aberdare, the visiting president of the RGS and chairman of the anti-French United African Company, forerunner of the Royal Niger, considered that 'many an irretrievable blunder' had been made in the Foreign and Colonial Offices for want of geographical knowledge, Professor Boyd Dawkins considered that Britain 'would be heavily weighted in the race for commercial and political supremacy by the neglect of the study of geography'.[14]

One of the prime movers of the early days of the Manchester society was J. F. Hutton, west African merchant, influential member of the African committee and later president of the Chamber of Commerce and member of parliament for North Manchester. Hutton led delegations from the city to the Colonial and Foreign Offices on several occasions, arguing in 1876 against the cession of the Gambia to France and against the Anglo-Portuguese treaty in 1884 as but two examples. He was active in Brussels in 1878 in the establishment of the international association for maintaining free trade in the Congo regions under the personal suzerainty of King Leopold; he was a founder and director of both the United African Company (later the Royal Niger Company) and (from 1888) Mackinnon's Imperial British East Africa Company. Stanley was a personal friend who stayed at Hutton's home on several occasions. Hutton, in short, was the embodiment of 'municipal imperialism'.[15] At one of the society's meetings in 1886, Hutton inveighed against the government's capitulation to Germany in Africa and asserted that:[16]

> the interests of British merchants in Africa had hitherto been lamentably neglected by the Government, and the permanent officials of the Foreign Office had displayed surprising ignorance with regard to African matters. It was sincerely to be hoped that our trade with that Continent would in future receive more attention at the hands of the Government than it had hitherto done.

On his death in 1890, the society's obituary underlined the point:[17]

> If our Government were to consult men like Mr. Hutton who are intimately acquainted with all the details of trade, and also with the geography (this latter the Government official is always ignorant of) of the foreign countries they have to deal with, we should see fewer of those blunders made which so often act ruinously to British interests.

Even if, as Driver has pointed out, the ideology of a society is not formed by the remarks of its prominent members,[18] there can be little doubt that in its founding and early years the Manchester society was primarily interested in geography as an aid to statecraft, with the protection of the interests of Manchester manufacturers and merchants, even if dressed up in the language of philanthropy, and above all with the extension both of markets and of sources of raw materials.

This is well illustrated by repeated references in the 1880s to the advance of German interests in Africa and the south Pacific, by a heavy concentration on Africa in the early years, and by a rapid abandonment of support for Leopold's supposedly internationalised Congo in favour of pressure for the protection of British positions (for example, on the Zambezi and in the Nyasa (Malawi) region). Analysis both of the contents of the meetings, the resulting publications, and of the materials added to the map collection in the first few years reveals a predominant fascination with Africa. By the 1890s interests were already broadening, but it was Africa that had the power to bring together the coalition of noblemen, senior clerics, academics, businessmen and politicians that went into the founding of the society. And it was Africa which produced the liveliest debates on the Congo, on liquor traffic, on steamers and the development of plantations, on the need for a charter for the North West Africa Company, and on whether there was any conflict between missionary and commercial motives, debates that were extensively noted in the *Manchester Guardian*, whose celebrated and influential editor, C. P. Scott, was a member.[19]

It is a striking illustration of civic imperialism that so many of the leading citizens of the region were active rather than ornamental members. The Lord Mayor of Manchester and the mayors of sourrounding burghs hosted meetings in their parlours and town halls. Peers led delegations to London. Support was mobilised for the entertainment and civic honouring of distinguished visiting speakers. Senior clerics continued to be active for the first few years and, as the analysis of membership (below) indicates, the society maintained within its membership lists a strikingly high proportion of civic dignitaries. The Dean of Manchester may have had a real point when he indulged in a Darwinian joke at the third meeting of the society, held in Cheetham town hall. The geographical society, he remarked, supplied 'the missing link' in the social, educational, and commercial life of 'this great community'.[20]

Elspeth Lochhead has drawn a distinction betweeen the Scottish society and those of the English provincial cities. While the latter were commercial associations, the Scottish was, in her view, a genuinely intellectual and learned body, benefiting from the status of Edinburgh as a great centre of scientific endeavour.[21] There can be no doubt that the Scottish society had a more learned membership than its English counterparts and that it developed more rapidly and convincingly as a scientific body making a significant contribution to the development of the discipline of geography. However, in its origins there are more similarities than differences. The Scottish was also inaugurated by an address by H. M. Stanley – in Edinburgh on 3 December 1884. It brought together a similar coalition of social, political and civic forces, pursued the fashionable fascination with Africa, and pressed for government protection for British missionary, commercial and settler interests on the continent. After all, the exploits of Scottish explorers, missionaries, consuls and traders had made Africa a distinctively Scottish concern. It inevitably appealed to the memory of Livingstone, whom many saw as the greatest Scot of the century, and founded the Livingstone gold medal in his name. His daughter and son-in-law, the Livingstone Bruces, were closely involved with its establishment and were instrumental in bringing Stanley to Scotland. Its founders were particularly anxious to curb Portuguese and German influence, since Scottish concerns such as the missions and philanthropic companies in the Nyasaland (Malawi) region, which it described with some justification as a Scottish colony, were distinctly insecure in the 1880s, caught between the Germans in Tanganyika and the Portuguese in Mozambique.[22] There was little confidence that the British government would protect them.

The founding of the society, as with that of Manchester, was set in a European context. The Edinburgh cartographer, J. G. Bartholomew 'had been impressed by the number of geographical societies in Europe and the way many of them were associated with political and economic projects'.[23] It was Bartholomew in discussion with A. Livingstone Bruce who brought about the preliminary meeting of the society on 28 October 1884 which in turn led to the framing of its objectives. Stanley (delivering the same address he had used in Manchester) duly appeared not only in Edinburgh, but also in Glasgow and Dundee (he visited Aberdeen in 1890), providing the crowds from whom the membership of the society would be drawn. The fashionable controversy about Africa, together with the status of geography and exploration soon brought into the society a galaxy of leading noble, political, civic, commercial, religious and educational figures in Scotland. Shipowners and imperialists were represented by (Sir) William Mackinnon of the British India Line (later founder of the Imperial British East Africa Company), Sir Donald Currie of the Castle Line, and Sir Charles Aitchison, Lieutenant-Governor of the Punjab.[24]

A great deal more work remains to be done on the distinctive characteristics of the four branches of the Scottish society, but Roy Bridges has indicated the extent to which the Aberdeen branch reflected the civic, academic and religious interests of that city. The Earl of Aberdeen, a Vice-President of the Edinburgh society, took the chair at its first meeting in February 1885 and (Sir) George Adam Smith, son of the Free Church's foreign mission secretary, was active in arranging lectures in the various cities to develop the strength of the branches. In Aberdeen, his father, Dr George Smith, minister of the Queen's Cross Free Church, was an influential figure, together with (Sir) David Stewart, Lord Provost of the city, and prominent businessmen including a storekeeper, a banker and shipowner, a railway magnate and a granite merchant. Among the academics were both the principal and the professor of humanity (latin) at the university. The latter, James Donaldson, expressed anxiety about stemming the decline of Britain and the need for Scotland to honour its own explorers. For Donaldson, Aberdonians could expect to reap utilitarian rewards from membership of the society and an understanding of geography: 'our manufacturers would get hints, young men would have an opportunity of knowing what land they ought to go to and what kind of work they might accomplish, and our commercial men would see openings for their energy, and equally our philanthropic men would have opportunities of knowing where they could do good'.[25]

Bridges has calculated that of thirty-one major meetings in Aberdeen between 1885 and 1895, no fewer than eleven were devoted to Africa. The contents of the society's journal, the *Scottish Geographical Magazine*, similarly reflect this concentration, as do the identities of recipients of the society's gold medal and Livingstone gold medal (endowed by Mrs Livingstone Bruce). Stanley received the first gold medal and the initial Livingstone award went to Sir Harry Johnston. Other recipients included Alfred Milner and George Taubman Goldie, although Arctic and Antarctic explorers, contemporary heroes in what were seen to be the last unexplored regions, became more prominent in Edwardian times. While the first expedition to be funded by the society was that of H. O. Forbes to New Guinea in 1885 (where there were, of course, growing anxieties about the power of Germany, reflected in the founding of geographical societies in Australia in the same year), the second was Daniel Rankin's of 1888 to carry out explorations of the Zambezi basin. He succeeded in discovering the Chinde mouth of the river, which was more open to navigation and less liable to silting than other navigable channels. It was subsequently entered by Consul Harry Johnston on HMS *Stork*.

The Tyneside society, like that of Manchester and at least the Aberdeen branch of the Scottish, had its origins in the interest of a clergyman. The Revd F. O. Sutton, curate at All Saints' Church, Newcastle, had been the secretary of a geographical society when a student at Cambridge and thought that it would be a good idea to found such an institution in Newcastle.[26] He interested the mayor, Sir Benjamin Browne, but it was agreed that the project should be postponed until after the jubilee celebrations in 1887. The suggestion was duly discussed by merchants at the Newcastle 'Change, and a young merchant called G. E. T. Smithson, who subsequently became the secretary of the society for more than ten years until his early death in 1898, was drawn in. A meeting was held in a Quayside restaurant towards the end of 1887 and at a further meeting at the Literary and Philosophical Society the decision was taken to found the society. Senior academics and Earl Percy became involved. The first committee declared that:[27]

> Without a knowledge of the methods of trade adopted by countries beyond the sea, of the quality and description of goods they have to offer, and wish to receive, of the means of ingress and egress to and from these countries, and the transport necessary for the distribution of such goods, home trade would by-and-by find itself confined to old channels worn by competitors, and antiquated by the birth of successive generations of healthy, vigorous trade pioneers.

The language is contorted and the metaphors forced, but the utilitarian anxieties are transparent.

In the annual report of 1889–90, the society underlined its commitment to these practical objectives, proclaiming that it was now generally acknowledged throughout the district that 'the Tyneside Geographical Society is supplying a want long felt by a community which has ever been renowned for its commercial, no less than for its scientific activity'. In that same season Eli Sowerbutts, the secretary of the Manchester society, lectured a Newcastle meeting, under the chairmanship of the mayor, on 'John Bull's Estate, how he got it and what he is going to do with it'.[28] This commitment to empire and conviction in the efficacy of the society was expressed anew in 1894:[29]

> Once our buildings and appliances are complete, it is hoped they may form a centre of attraction and source of information, not only to the surrounding community, but also to visitors from every part of the Empire . . . Too long has Newcastle upon Tyne been considered too unimportant, by Colonists for instance, to make it a stopping place on their trade journeys. They arrive in London, go to Edinburgh (they all go to see Edinburgh), thence to Glasgow, Liverpool, Manchester, and back to London, distributing orders, or obtaining them, en route.

Newcastle, the report went on, has for too long been associated with only coals and grindstones, even if this reflection no longer fits economic reality, and the society's task was to publicise the wider economic attractions of the city.

H. M. Stanley was induced to stop off and pay a visit only by the offer both of the freedom of the city and of an honorary DCL (Doctor of Civil Law) at the University of Durham. The society expressed its appreciation both to the mayor for the former and to various academic members for arranging the latter. Well might they be grateful: it was a meeting which set the society on its feet in terms of membership and temporary financial health. A visit by Lord Roberts in 1894 produced a fresh outburst of imperial enthusiasm for the military hero. It was on the strength of the Stanley visit that the society was able to move in 1891 to its own premises, a vacated Presbyterian Church at Barras Bridge, which became known as the Geographical Institute. It offered accommodation for a large lecture hall, a minor hall, a reading room, a map room and library, a gentlemen's smoking room, a lady members' private room, a council room and attendant offices. In this building, the society attempted to turn itself into a club which members could visit between 9.00 a.m. and 10.00 p.m. Afternoon tea and light refreshments were supplied by a caretaker whose quarters were connected to the members' rooms by speaking tube and a small lift.[30]

It is surprising in many ways that, given the significance of Liverpool as a great port with major shipping connections with the Americas and West Africa, the geographical fever should have reached it so late. It was not until 1891 that the usual combination of civic, commercial, religious, educational and political interests set about creating a geographical society. This may be partly because the African Trade Section of the Liverpool Incorporated Chamber of Commerce had not only acted as a pressure group for Liverpool intersts, but had also been accustomed to discussing geographical issues and had even brought notable explorers to speak to businessmen in the city. Moreover, Liverpool shipping concerns, as represented by Alfred Jones and John Holt, were implacably opposed to the Royal Niger Company and its monopoly on the Niger which Manchester supported because the company astutely allowed in cottons without duties.[31] It is no accident that the years 1891–92, when the dispute raged at its most intense, mark exactly the time when Liverpool turns to the solace of a geographical society.

It is indeed apparent from the early days of the society that Liverpool's pride was at stake. In the annual report of 1893 the council proclaimed that it should at least be as successful as Edinburgh, Manchester, and Newcastle, 'not one of these being so favourably placed as Liverpool, which from its commerce, population, and unrivalled opportunities of obtaining the experiences of travellers returning to England, should hold a position second only to the parent society'.[32] Yet the initial difficulties which prompted this observation were exacerbated by the major trade recession of 1894.[33] In a continuing appeal for members, the council pointed out that 'much naturally is expected from the Liverpool Geographical Society, having in view the unrivalled position of the city, its wealth and population, and last but not least, the fact that its trade ramifications bring it into actual contact with every portion of the globe'.

Yet the society was never fully clear about its purpose. It found that lectures were its most successful events, yet it sought to stress that its objects were esentially utilitarian, that it should satisfy the commercial needs of the city. In 1894 it was at pains to point out that 'Your Council desires earnestly to impress upon all members that the Geographical Society was not founded simply to supply the general public with the latest Geographical discoveries in the form of popular Lectures, but to render assistance to the commercial interests of the city, and thus benefit the whole community'.[34] It was for the same reason that the society decided that it could not afford to publish a journal, despite some embarrassment that it could not reciprocate the donations of journals from the other societies. Moreover, it felt that a journal could not succeed because of the difficulties of securing original material that had not been published elsewhere. Lectures before the society were published in brief summary form with the reports and accounts.

The LGS decided instead to use its relatively meagre resources to establish a library, map room and 'museum' of specimens to be consulted by its business and commercial members, who could become acquainted with the 'conditions and Trade of all countries'. Yet this attachment to commercial utility did not seem to produce the desired results. In 1902 the society was complaining that 'in a City like Liverpool, whose existence is dependent on its commerce, a Society whose object is to increase the knowledge of the commercial wants of the world generally, should receive far more encouragement from merchants, shipowners and the community at large, than it does'.[35] It was a querulous note that was seldom absent from the annual reports. By 1911 it was lamenting that 'the number of members making use of the Library is not very large', despite the fact that it was open daily, 'is well suplied with geographical literature, maps and atlases, and is maintained for their accommodation and convenience'.[36]

It may be that this stress on utility came from the influence exerted by shipowners, sugar and other commercial interests in its membership. Among these (Sir) Alfred Jones of the Elder Dempster shipping company was the most important. He donated considerable sums to the society and acted as its chairman of council for several years. He was instrumental in founding the Liverpool School of Tropical Medicine in 1899, was invited by the Colonial Office to produce schemes for the regeneration of the West Indies after he had helped to found the banana industry of the Canaries. Yet he also represented a great split. In 1893 he had resigned as Chairman of the African trade section when he dropped his opposition to the Royal Niger Company on receiving a long-term contract for his firm from Goldie.[37] As will become apparent below in the survey of membership, he helped to maintain his power in the society by enrolling large numbers of employees and also sought to contribute to its social cachet.[38]

It is perhaps not surprising that given the fractiousness of Liverpool interests, it was unquestionably the lectures and its educational activities which were the most successful aspects of the work of the LGS. (The educational projects and propaganda of the various societies, a highly significant part of their objects, will not be examined in this chapter.) The LGS failed to secure a visit from H. M. Stanley, since it was founded too late to catch either his tour in late 1884 or his round of lectures in 1890 after his return from the Emin Pasha Relief expedition. Since these 'Stanley spectaculars' had had considerable effect upon both the growth of membership and the finances of the other three societies, this was clearly a major loss. Still, attendance at the Liverpool lectures rose from 90–100 in 1892 to 600–700 in 1893. It was claimed that they were being maintained at a similar level in 1903. Nansen filled the Philharmonic Hall in 1897 and visits from Captain Scott, Sven Hedin, Sir Ernest Shackleton and Commander Evans also drew the crowds.

Jones, together with the society's president, the Earl of Derby and local academics, were able to draw on the goodwill of the Chamber of Commerce, the civic establishment, and the university. Regular meetings took place in a lecture theatre at the university, at the Royal Institution and at the small concert room of the St George's Hall. The annual general meetings were held in the town hall under the chairmanship of the Lord Mayor and were invariably 'largely and influentially attended'. Conversaziones were held at the Free Library and Museum or at the Walker Art Gallery, at which the members were presented to the Earl and Countess of Derby. Derby was president until his death in 1908 when he was succeeded by his son, a sense of family obligation which had been exemplified in 1899 when two vice-presidents, Sir Henry Tate, the sugar magnate (who had endowed the University College in 1881 and was the benefactor of the Tate Gallery) and Thomas Ismay, shipowner, were succeeded on their deaths by their sons William Tate and J. Bruce Ismay. In 1897 Alfred Jones held a garden party at his residence, Oaklands in Aigburth, which was attended by no fewer than 1,600 people. J. L. Bowes, the wealthy consul for Japan in the city and another vice-president, frequently threw open his magnificent collection of Japanese ceramics to visiting members and distinguished visitors to the city. If nothing else, the society clearly offered a significant entrée to the highest echelons of Liverpool society.[39]

Without a journal, with very little scholarly base, with utilitarian purposes that seemed to fall well short of their hopes, and incidental social and civic advantages that were limited by the mortality of its founding fathers, the LGS never seemed to be wholly secure. That lack of security comes through in the barely concealed recriminations and plaints that are a feature of its annual reports. Its dependence on lectures by visiting celebrities was an acknowledged source of weakness. In 1906 the annual report lamented that 'As year by year the world's surface becomes better known and more perfectly mapped out it is a matter of increasing difficulty to light on anything absolutely new either in travel or research'.[40] A society dedicated to such a restricted view of geography was clearly doomed to fail. In 1911 it illustrated its continuing search for influential patrons by electing the powerful shipowner Sir Owen Philipps (founder of the Royal Mail group and the future Lord Kylsant) as one of its vice-presidents. By 1914 it was showing cinematograph films and sadly noted that 'many of our younger subscribers have, unfortunately, been called away owing to this terrible war'. The LGS missed the principal African passions of the 1880s (though it had many speakers and sessions on Africa in the 1890s) and flickered fully into life only when stimulated by the visits of Arctic and Antarctic explorers in the late Victorian and Edwardian years.

Membership

The provincial geographical societies all had mushroom growths, but their memberships fluctuated considerably before World War One and an analysis of membership lists indicates that they also changed in character in interesting ways. The data has been collected from the early journals of the societies and varies in quantity and quality. The Scottish society maintains full membership figures in its annual reports to council, though actual membership lists are not printed. The Liverpool society published membership numbers consistently even when its fortunes were clearly flagging. The Tyneside, however, became more reticent as numbers declined, while the Manchester journal included relatively little member information. Tyneside published lists of members only in 1889, 1890, 1892 and 1894, so it is difficult to establish changes in composition over any considerable period of time. However, even these four years indicate interesting developments. Both Liverpool and Manchester consistently published their membership lists, though the Liverpool listing (like those of Tyneside) offers the great advantage of also printing addresses, which Manchester supplied only in its first two years. These addresses provide the opportunity to judge the regional spread of the membership, together with the family connections and business affiliations of members.

The Royal Scottish Geographical Society maintained its membership

(Table 5.1) remarkably well in the 30 year period between 1885 and
World War One. Indeed, as a national society it compares well with the
RGS which had a membership of 2,387 in 1872 and 4,031 in 1900. By the

Table 5.1 *Membership of the RSGS, 1885–1914, including breakdown into
membership in Edinburgh, Glasgow, Dundee, Aberdeen, English and foreign*

	Total	Edin.	Glas.	Dun.	Ab.	Eng. and for.
1885	990					
1886	1,088	585	253	104	64	82
1887	1,121[a]	611	247	94	66	103
1888	1,144	632	240	89	60	123
1889	1,160	646	239	83	54	138
1890	1,574	897	354	100	76	147
1891	1,495	859	327	89	71	149
1892	1,475	815	313	81	68	198
1893	1,417[b]					
1894	1,417	792	280	85	52	208
1895	1,325	749	264	80	43	189
1896	1,414	773	283	95	78	185
1897	1,598	896	313	109	107	173
1898	1,613	911	314	113	104	171
1899	1,592	908	300	115	101	168
1900	1,546	900	288	110	93	155
1901	1,453	852	261	105	84	151
1902	1,316	752	242	97	79	146
1903	1,372	789	250	108	83	142
1904	1,464	850	263	105	92	154
1905	1,790	1,045	369	148	92	136
1906	1,852	1,086	368	149	100	149
1907	1,773	1,024	373	133	96	147
1908	1,709	968	370	135	93	143
1909	1,746	977	372	139	107	151
1910	2,127	1,115	559	164	140	149
1911	1,994[c]	1,047	498	163	132	143
1912	1,898	999	471	158	123	147
1913	1,727	920	422	149	113	123
1914	1,805	921	450	183	120	131

Notes:
The Scottish society had a high proportion of life members, reaching 251 in 1899
and a peak of 270 in 1907.
[a] This figure included 19 members entered in more than one branch.
[b] No branch breakdown was provided for this year.
[c] This error in the addition is in the original. The figures in fact add up to 1983.
Source: Annual reports published in *Scott. Geogr. Mag.*

end of its first year, it had achieved almost 1,000 members and passed the 2,000 mark in 1910. After a relatively slight decline in the period at the end of and immediately after the Boer War, the numbers rose strongly towards World War One. The teacher associate scheme, which started modestly in 1907, was obviously a considerable success, multiplying almost ten-fold in the space of seven years. All the branches experienced blips of membership apparently associated with the Boer War. The Edinburgh branch fluctuated between a pre-Boer War peak of 911 in 1898, a low of 752 in 1902 (a decline of 17.5 per cent) and a high of 1,115 in 1910. The Glasgow branch had a pre-war peak of 314 in 1898, a low of 242 in 1902 (representing a 22.9 per cent drop) and then more than doubled to a high of 559 in 1910. The Dundee and Aberdeen branches also declined in membership during the Boer War period (Dundee 14.2 per cent; Aberdeen 24 per cent) and grew strongly towards World War One. The teacher associate members are, incidentally, wholly separate from these figures (see Table 5.2).

How are these fluctuations to be explained? It may be that the societies do offer a sort of fever chart of imperial activity and international tension. But the Boer War drop may have resulted not so much from any sence of imperial disillusion as from the fact that members were caught up in the war or were too busy in war-time conditions to be actively involved. Moreover, notable speakers were less available and major events less likely to be organised. Equally the strong growth up to World War One may be explained by the society's capacity to become academically respectable (as well as attract celebrated speakers) as by the growing anxieties of the Edwardian age. The success of the teacher associate scheme may help to confirm this.

Table 5.2   *RSGS, teacher associate members, 1907–14*

1907	22
1908	34
1909	54
1910	155
1911	154
1912	166
1913	163
1914	213

Source: *Scott. Geogr. Mag.*

However, the Scottish society was as susceptible to the 'spectactular' as other societies. The second visit of H. M. Stanley to Edinburgh in 1890 produced a considerable jump in membership. The large growth in 1905 can be explained by the visits of Captain Scott who spoke in large halls in all four centres in 1904. The peak of 1910 was achieved because of the visits of Sir Ernest Shackleton. Many new members joined, as for similar 'spectaculars' at the other societies, in order to be sure of having seats for the meetings he addressed. Such growth had a tendency to be 'soft' as the succeeding two to three years indicate. Recognising this, the Scottish society introduced in 1890 a membership entrance fee of one guinea in addition to the annual subscription of the same amount. No doubt its officers hoped that a flood of uncommitted members would be deterred by having to pay two guineas for one celebrity meeting. This entrance fee lapsed after the turn of the century, but was reintroduced in 1910. Thus the Scottish society was much the most expensive to join, yet this had little effect upon the strength of its membership. It also ensured its considerable financial health. In both these respects, the Scottish case is strikingly different from those of the English provincial societies as will become apparent. In key years its membership was almost as great as the other three put together; it maintained a high proportion of life members; and its finances were secure.

There seems little doubt that the Scottish society had a higher quota of adherents who held membership of other learned societies or were actively engaged in education than the English provincial societies (Table 5.3). However, the composition of the council in 1885 represented the range of interests that promoted geography as a utilitarian

Table 5.3   *Composition of membership of the RSGS, 1885 and 1895*

	1885	%	1895	%
Total	798[a]		1,325	
Of which:				
Learned societies	105	13.2	124	9.4
University staff	18	2.3	31	2.3
Titled	34	4.3	71	5.4
Army/naval officers	15	1.9	63	4.8
Clergy	36	4.5	48	3.6
MPs	17	2.1	13	1.0

Note:
[a] This figure does not cohere with the RSGS published figure of 990 for 1885. However, snapshots of membership may have been taken at different times in the first year. Alternatively, the figure of 798 may represent that of the main Edinburgh branch.
Source: Adapted from Elspeth N. Lochhead, 'The Royal Scottish Geographical Society: the setting and sources of its success', *Scott. Geogr. Mag.*, September, 1984.

discipline vital to the continuing power and imperial success of the United Kingdom. It contained a striking list of Scottish luminaries: no fewer than 16 peers, representing both main parties, 3 MPs, 7 senior academics, 4 leading clerics, 1 admiral of the fleet and 1 imperial governor, as well as shipowners as celebrated as Sir Donald Currie and Sir William Mackinnon, and publishers like William Chambers, Adam Black and J. G. Bartholomew. Thus the undoubted academic respectability of the Scottish society, its position in the intellectual life of the 'stateless nation' of Scotland, and its significant role in the development of the discipline of geography does not detract from the fact that it was founded for essentially national and imperial reasons. What is different about the Scottish society is that when the imperial propagandist purpose fell away, it had the strength to survive as a significant intellectual body.

The journals of the Manchester society did not publish the membership totals, so they have to be calculated for selected years from the lists of members. Table 5.4 gives the membership of the Manchester society in 1885, 1886, 1890, 1892, 1899, 1903, 1910 and 1914. The peak year for Manchester was 1890, a mere five years after its founding. Thereafter, there is a steady decline, marked at the end of the Boer War, but with some recovery before World War One. In 1914, however, it was still 150 below the 1890 figure.

Table 5.4   *Membership of the Manchester Geographical Society, 1885, 1886, 1890, 1892, 1899, 1903, 1910, 1914*

1885	392[a]
1886	444
1890	873
1892	803
1899	741
1903	648
1910	703
1914	719

Note:
[a] Brown gives 383, but the list, which includes 19 honorary, corresponding and life members, adds up to 392.
Source: *J. Manchr. Geogr. Soc.*

Table 5.5 breaks down the membership into the social categories which can be divined from titles and qualifications. Professional people are indicated by degrees, professional associations, and titles such as Reverend, Doctor or Professor. Membership of learned bodies, such as the Royal Geographical Society, the Linnean Society, The Royal

Geological Society, the Imperial Institute and the Royal
Anthropological Institute are not included in this list (unless combined
with degrees or titles), for they do not necessarily indicate professional
status. Justices of the Peace are listed separately, since (together with
civic leaders) there is a remarkably large contingent of them. In the late
nineteenth century, membership of the bench would indicate primarily
an involvement in local government and, in some cases, land
ownership.

Table 5.5 *Membership of the Manchester Geographical Society, 1885–1914*

	Professional	Women	Business addresses	JPs	Councillors, aldermen, MPs, peers	Military
1885	53	9	49		38[a]	2
1886	55	14	38		39[a]	4
1890	78	23		61	60	5
1892	72	28		72	56	8
1899	83	34		79	44	12
1903	58	31		70	46	12
1910	67	90		78	27	10
1914	75	110		69	25	9

Note:
[a] Inclusive of JPs.
Source: *J. Manchr. Geogr. Soc.*

Since addresses were only supplied in 1885 and 1886, it is difficult to
assess the rise and fall of business affiliation, family membership or
regional spread. While the names of some members are specially
attached to firms in 1885 and 1886, in many cases business addresses are
less easy to identify than those of Liverpool. Many addresses in the
centre of Manchester may be business rather than residential, so the
figures for manufacturing, commercial and other company connections
may be under-estimates. Among the companies represented were
cotton spinners, a clothier, a packer, an iron works, and the Hadfield
Steel Foundry in Sheffield. Inevitably, several mills and sundry
merchants appear in the list together with, more unexpectedly, a hat
manufacturer.

Among the more striking characteristics of the Manchester membership is the fact that no fewer than 38 clergymen were members in 1890, a
figure which dropped to 13 in 1914. In 1890 23 members claimed
membership of other learned societies; this rose steadily to 48 in 1914,
but remains a very small proportion (2.6 per cent and 6.7 per cent). In

1900 there were 11 consuls (mainly of foreign countries in Manchester) among the membership, but this had dropped to 5 in 1914. Women constituted a mere 2.2 per cent of the membership in 1885, 2.6 per cent in 1890, but by 1910 and 1914 the figure had reached 12.8 per cent and 15.3 per cent respectively. Manchester had a category of associate membership, for half a guinea, which offered the opportunity to attend lectures, but not to participate in the business of the society or introduce guests. In 1885 there were 44 associates (11.2 per cent) and by 1914, 100 (13.9 per cent). The weakness of associate membership is well illustrated by the fact that in 1899 there were 114 of them (15.4 per cent), while in 1903 there were 69 (9.6 per cent). Moreover, the majority of women were associates, denying them full involvement or voting powers in the society.

The Tyneside society reveals a dramatically different profile from the others. Perhaps wisely, the Tyneside did not create the category of associate member and encouraged the membership of complete families. It did this by a change in its rules in 1889. Its full membership fee was ten shillings (all the other societies charged a guinea), while 'ladies and youths under 21' (often, but not exclusively additional family members) were charged five shillings, as were those who lived outside Newcastle and whose businesses were not in the city. This attempt to broaden the membership base succeeded dramatically, as the shift from 1889 to 1890 indicates. The visit of H. M. Stanley in 1890 brought in a rush of members and prompted the secretary of the society, G. E. T. Smithson, to regret that there could not be Stanley spectaculars every year – 'if we cannot command such phenomenal success in the future, we must see to it that we hold the ground already gained'. Cheap membership flourished in such conditions, but Smithson rightly predicted that it was unlikely that there would 'ever again be the same enthusiasm over the exploits of any traveller'.[41] The Tyneside policy contrasts strikingly with that of Scotland and ultimately it operated to the detriment of academic respectability or financial health.

Table 5.6 lays out the membership of the Tyneside society between 1888 and 1898 when the figures cease to be published, indicating that after the peak of 1897, the following year may well represent the start of a relatively steep decline. By the beginning of the century, membership must have dropped by several hundred, for throughout the Edwardian period the annual reports (in publications that became increasingly sparse) lamented that a society of its importance should have a membership below 1000 and repeatedly suggested that financial viability could only be secured if this magic figure were passed. Attempts to found branches to ensure a greater geographical spread in the north east failed. In 1889 efforts were made to establish branches in Hexham, Morpeth and South Shields. The Revd John Mackenzie, a noted missionary imperialist speaker at all the geographical societies, was induced to go to Hexham, but only the South Shields branch flourished intermittently between 1895 and 1902. Branches in North Shields and Durham also occasionally secured speakers in the middle 1890s, but they were discontinued in 1896 and 1897 respectively.

Table 5.6   *Membership of the Tyneside Geographical Society, 1888–98*

	Newcastle	South Shields branch
1888	213	
1889	400	
1890	768	
1891	not available	
1892	1,134	
1893	1,011	
1894	985	
1895	1,054	
1896	1,277	
1897	1,327	327
1898	1,259	362

Source: *J. Tyneside Geogr. Soc.*

Table 5.7 provides an analysis of members for 1889, 1890, 1892 and 1894. In 1892 there were 157 family relationships expressed in the membership lists, some involving as many as five or six members of the same family. In 1894 the figure had grown slightly to 157. The strength of the female membership so close to the founding of the society is remarkable and unique among all the societies. In 1892 the business affiliations included several shipping companies, collieries, banks, a brewery, a gas works and a paper works. In that year no fewer than 24 clergymen were members, as well as 10 MPs. Moreover, the addresses of members indicates that, more than any other society, the Tyneside, reflecting its name and its subscription policy, included people from the entire north-eastern region. By 1894 the women's membership was growing strongly, but professional and business memberships were already in decline. The number of clergymen had dropped to 17 and only 8 members gave school addresses. Only 3 members declared affiliation to other learned societies.

The Liverpool membership (Table 5.8) reached a peak of 748 in 1898 and maintained a high point, with just a slight decline, during the Boer War. From then on, however, steady decline set in, with a low of 459 in 1912. As usual, the society was prone to the effect of the 'celebrity

Table 5.7   *Membership of the Tyneside Geographical Society, 1889, 1890, 1892, 1894*

	Professional	Women	Business addresses	MPs, JPs and others in public life (incl. peers)	Military
1889	30	27	11	26	7
1890	52	138	35	39	16
1892	71	235	49	40	10
1894	51	277	38	38	10

Source: *J. Tyneside Geogr. Soc.*

spectacular', which was invariably held in the Philharmonic Hall. As noted above, Liverpool never succeeded in enticing Stanley, unlike Manchester, Edinburgh (and other Scottish centres) and Newcastle, but the visit of Fridtjof Nansen in 1897 caused a sharp rise in membership from 611 to 746. The lectures of Captain Scott in 1904 and of Sir Ernest Shackleton in 1909 only stemmed the speed of decline, but the appearance of Commander Evans in 1913 produced a jump in membership from 459 to 570. The annual report enthused that Evans had 'delivered a most thrilling account of the Capt. Scott Expedition to the South Pole in 1911–1912. A very large audience attended, and his most brilliant address proved deeply interesting'.[42] However, whereas these celebrity visits emphasised the strength of the Scottish society, they reflected the weakness of that of Liverpool. Associate membership, instituted to broaden the membership base, was in reality a major source of decline. While there were 42 associate members (who paid half the subscription fee and were not permitted to vote or to introduce visitors to the meetings) in 1893 (10.3 per cent of the total), there were 167 (29.3 per cent) in 1913. Even in 1914, when associate membership dropped dramatically (clearly a group vulnerable to disassociation, taking out membership only for the significant occasion) to 114, they constituted 22.9 per cent of the total.

These years also saw a shift in the basis of the membership. Table 5.9 indicates membership for the years 1893, 1895, 1898, 1903 and 1913 and 1932 divided into occupational and social categories. These figures demonstrate the close involvement of the business community of Liverpool in the society in the 1890s and their rapid disengagement in the Edwardian era. There seem to be two main reasons for this. The first is the fact that the society numbered among its early members prominent businessmen who appear to have insisted that employees of their companies took out membership or alternatively paid for them as a

Table 5.8 *Membership of the Liverpool Geographical Society, 1892–1914*

1892	316
1893	406
1894	486
1895	518
1896	611
1897	746
1898	748
1899	742
1900	730
1901	725
1902	692
1903	668
1904	650
1905	640
1906	610
1907	588
1908	553
1909	543
1910	504
1911	464
1912	459
1913	570
1914	498

Source: Reports, Liverpool Geogr. Soc.

means of heavily subsidising the society. The most notable example of this is Sir Alfred Jones of the Elder Dempster shipping company. Table 5.10 indicates the numbers of Elder Dempster employees in the lists and their rapid disappearance after the death of Jones in 1909, shortly after he had presented the gold medal of the society to the Swedish explorer of central Asia Sven Hedin. Many of these members were associates. The second explanation is presumably the realisation that the society had less to offer in terms of commercial utility than had originally been hoped for. Jones and others believed that membership of the society and access to lectures as well as to books and maps in its collections would be a vital aid to the education of his office employees and ships' officers. The continuing weakness of the Liverpool body and the existence of other sources of such materials removed such confidence in the efficacy of society membership.

Some other companies were represented on a smaller scale while in addition there were always two to three company corporate memberships which did not reveal the names of individuals. Inevitably, company membership was primarily associated with shipping, although

Table 5.9   *Membership of the Liverpool Geographical Society 1893, 1895, 1898, 1903, 1913, 1932*

	Professional	Women	Business addresses	MPs, JPs and others in public life	Military
1893	35	21	98	28	25
1895	45	44 (23 assoc.)	128	40	25
1898	45	119 (72 assoc.)	163	44	25
1903	37	128 (78 assoc.)	120	51	20
1913	28	141 (80 assoc.)	47	38	10
1932	23	94 (48 assoc.)	37	36	12

Source: Reports, Liverpool Geogr. Soc.

Table 5.10   *Employees and directors of Elder Dempster as members of the Liverpool Geographical Society, 1895, 1898, 1903, 1913*

	Full members	Associate members
1895	4	28
1898	5	30
1903	3	35
1913	2	00

Source: Reports, Liverpool Geogr. Soc.

there were also a number of merchants, insurance firms, engineers, the Tate sugar concern and the United Alkali company represented among the members. A notable absentee was Sir William Lever of Lever Brothers who did not become a member until just before World War One by which time many other companies were departing.

The other dramatic change is the extraordinary growth of women's membership. Women constituted 8.5 per cent of membership in 1895, 16 per cent in 1898, 19 per cent in 1903, 24.7 per cent in 1913. This is a particularly striking increase considering that men with full membership were able to take their wives or other female relatives to meetings. The Liverpool Geographical Society clearly became a significant arena for women's involvement in a form of public and intellectual life in the city which was open to them (Women also addressed the provincial societies, although Mary Kingsley insisted on having her paper at Liverpool read for her).[43] It is apparent that without women's membership the society would have collapsed at an earlier date, given the decline of all other forms of membership. It is also worth noting that the professional and military membership (a category which included those with military, naval and merchant naval titles) also declined, as did the membership of clergymen. In 1895 13 clerics were listed; in 1903 only 4; and by 1913 just 1, the bishop. The society simply seems to have failed to fulfil expectations for these social groups, while professionals had new outlets, including, of course, the Geographical Association. Only those in local and national politics and public life seemed to have kept on their membership as an expression – or an obligation – of civic pride.

These figures indicate that of all the societies, Liverpool's was by far the most strongly business orientated. This is reflected not only in the high numbers of business addressess and employees of single companies like Elder Dempster, but also in the relatively low levels of professional membership, falling from 8.6 per cent in 1895 to 4.9 per cent in 1913. Professional membership at Tyneside also dropped proportionately, from 7.5 per cent in 1889 to 6.2 per cent in 1892 and 5.2 per cent in 1894, while that of Manchester increased from 8.9 per cent to 10.4 per cent between 1890 and 1914. These figures have to be treated with caution since they are dependent on the printing of degrees and professional qualifications, not all of which may have been declared by the members. However, Victorians and Edwardians were punctilious about such matters and most such qualifications are probably listed. We can be confident about the figures for clergymen and these unquestionably show a rapid decline in Liverpool, Tyneside and Manchester after the first heady days. In Manchester they were clearly replaced by other professionals.

To sum up the English societies, then, it would appear that the Liverpool was founded with strongly business and commercial motives, which receded rapidly after the turn of the century. Women's membership grew to compensate, but it never developed a solid academic base which would have been represented in degrees, professional qualifications and membership of learned societies. By contrast, although it has always been assumed that the Manchester society had a strongly commercial motivation, the most striking characteristic of its membership is the heavy representation of civic leaders, politicians and magistrates. It starts with a very low female membership, but in common with the other English societies, it is the adherence of women which keeps membership figures up when they would otherwise have experienced a catastrophic decline. The Tyneside society, with its remarkable female membership from the start, its extensive family membership and its regional spread indicates that it offered educational interest and entertainment which transcended its commercial or business interests.

[118]

However, the Tyneside society encouraged full membership by avoiding associateship, while at Liverpool and Manchester wives and other female relatives could have attended meetings on the basis of a male membership. Although the Manchester society succeeded in making the transition to a learned body, all exhibited decline before World War One.

The Scottish society uniquely maintained its membership and it is notable that the branch in Glasgow, where the main shipping, commercial and industrial interests were of course to be found in Scotland, was never any larger than one-third of its Edinburgh counterpart until the five years before World War One when it more closely approximated to one-half. Thus the membership would appear to be greater in the scholarly and intellectual centre of Edinburgh. However, many leading businessmen (for example, Sir William Mackinnon in the early days of the society) were affiliated to the Edinburgh branch, although they had estates in the west of Scotland. This was also true of many aristocratic and political members. Moreover, the importance of Leith and other ports on the Forth should not be discounted. However, the strength and growing success of the Royal Scottish Society contrasts strikingly with its provincial English counterparts, a contrast that was re-emphasised by relative financial prosperity. The Scottish society gives the impression of being metropolitan rather than provincial. It reflected the growing intellectual and cultural self-confidence of Scotland, re-asserting itself in the face of English dominance.

Transformation and decline

One index of the relative strengths of the societies can be found in their financial reports. There is no space to examine these in any detail, but just a brief glance indicates a wide desparity in financial health. The Scottish society was able to invest £1,000 in Glasgow and South Western Railway funded debt at 4 per cent as early as December, 1885. By 1892 it had added to this £400 in North British consolidated 4 per cent stock, while investments in New Zealand government (£300) and Dominion of Canada stock (£400) followed in 1894 and 1911. In 1898 Great Central Railway stock at 4½ per cent was added to the portfolio, while the low-yielding New Zealand stock was abandoned in 1903 for more North British railway stock at 4 per cent. From 1898 the society maintained a steady investment portfolio in excess of £2,200 which climbed to £2,600 in 1911. It was able to do this because its income consistently exceeded expenditure. In 1890, its subscription income was £1,765 and its receipts from lecture tickets amounted to £463. In 1897 the latter reached £738. In 1901 Mrs Livingstone Bruce endowed the medal in the name of her father with the considerable sum of £1,000, which was promptly placed in the society's favourite North British railway stock and accrued a steady income. The Livingstone account was kept above the £1,000 mark thereafter. In 1902 the society had a separate Antarctic exploration fund with £285 in hand. In the years before World War One, the society's balance sheet topped the £5,000 mark with an income and expenditure account in excess of £2,000.

As with membership, the RSGS was in many respects more financially healthy than the three English societies put together. Its considerable income enabled it to invest in railways and colonial governments in a manner which interestingly mirrored its scholarly interests. By contrast, the Tyneside society with its low subscription rates had a membership income which seldom topped £200. The society was strikingly dependent on lecture income. In 1890 Stanley's visit raised £405 against £254 in expenses. This helped to pay for the move to the Barras Bridge church, the purchase of which was funded by a loan of £500 each (at 4 per cent) from seven sympathetic gentlemen. Specific appeals paid for furniture, while the debt on the alterations and embellishment of the building was paid off by a memorial fund in memory of Smithson, the first secretary, amounting to £304 in 1899. In 1903, visits of Sven Hedin and Baden-Powell actually cost more than was covered by income from lectures, presumably because of the lavishness of the fees or entertainment offered these gentlemen. By 1904 the society's main income came from hall rents, amounting to £427 on a total income of £755. This was a precarious source, for when the use of the halls dropped dramatically before World War One the society found itself in real financial difficulties. From 1903 onwards it ran a bank overdraft. Apart from the building, presumably secured against the loans, there were no investments and no exploration funds.

Manchester and Liverpool similarly had a relatively low subscription income: in 1885, the total income of the MGS was just over £300, reaching £484 in 1886. Nevertheless, the MGS was able to secure the freehold of its premises at St Mary's Parsonage in 1894 and after the turn of the century raised £3,000 in £10 shares to redevelop the site, ensuring its financial health.[44] Liverpool invested £400 in the Mersey Docks and Harbour Board in 1898, presumably partly on the strength of the proceeds of the visit of Nansen. It still had modest investments as late as 1914, but never had permanent premises of its own.

But the relative strength of the societies can be gauged in other ways too. The power and influence of the RSGS was illustrated by the granting of the prefix 'royal' as early as 1887 to rival its London counterpart. In the winter of 1891–92 it was audacious enough to open a London branch and at least two meetings took place. As the Scottish society, in common with all the provincial examples, was open to women members, this could have posed a real threat. However, the RGS swiftly granted reciprocal membership and the London branch was wound up. Moreover, the society, recognising the dangers of being too dependent upon the large meeting addressed by distinguished figures, also founded in 1888 a Geographical Club which it described as being suitable for 'more intricate discussions'. In addition, its Scottish placenames committee, using correspondents throughout the country, provided significant information over many years to the Ordnance Survey. And as Freeman has illustrated, the *Scottish Geographical Magazine* became a significant international publication which concentrated on Scottish geographical studies while still finding space for wider international interests.[45] By 1900 and certainly by 1914, it was apparent that the RSGS had transcended its imperial origins and had become a fully fledged learned society.

Manchester also survived by maintaining its professional interests, developing a scientific membership, publishing a significant journal which carried some major articles and also concentrated on its region. In 1887 it founded a group called the Victorians who were charged with developing the practical work of geography. Neither Liverpool nor Tyneside were able to develop either the scholarly or 'practical' work. Liverpool attempted to create a businessman's commercial resource and failed when abandoned by the very businesmen it sought to help. It never published a proper journal and its complaints about the difficulties of finding something new in the way of travellers' tales and celebrated speakers were symptomatic of its failure to adapt. Tyneside also offered more entertainment than scholarship and when the entertainment value – and its hall rents – began to pall, it had no scholarly resources to draw on.

All pressure groups enjoy an initial burst of energy which is followed either by adaptation to a new purpose, or decline. These four societies were originally founded – in a period notable for the appearance of imperial associations – as propagandist bodies for economic concerns that were closely bound up with imperialism.[46] They were dedicated to the promotion of national survival by making the metropolis more aware of municipal and regional interests. In some respects they represented (particularly in Manchester's resentment of the incompetence of the London political and bureaucratic establishment) a provincial reaction to a perceived national failure. But, as the activities and rhetoric of the societies demonstrate, they diffused their energy into a variety of different tasks. They sought to educate and inform their local communities through harnessing a new, accessible and practical discipline. They set about enthusing public opinion with specific imperial and commercial objectives. They embarked upon social conciliation, across the fault lines of political parties, between God and Mammon, patricians and professionals, regional aristocracies and a mature bourgoisie. And they set about arousing public excitement by bringing celebrated travellers and propagandists to their localities, luring them with the offer of honorary degrees, substantial fees, help with further expeditions (in the Scottish case) or civic honours. In all these manifestations they became closely bound up with late nineteenth-century concepts of civic pride.

But both municipal imperialism and public enthusiasm had their limits. They started later than the French examples and they declined earlier. They were an expresion of a particular moment, a moment primarily featuring a controversy over a single continent, Africa. Wider public interest was maintained, ironically enough, by a shift from the hot to the cold, from the tropical Dark Continent to the frozen white wastes of the Arctic and Antarctica. The societies which never succeeded in moving beyond these three As, but which stayed locked into a concept of geography as a combination of entertainment and commercial utility, were doomed to failure. The Royal Scottish society had a firm underpinning of scholarship from the start. The Manchester society acquired it. In both cases, acknowledged excellence in regional geography and serious publications helped. In some respects the symbiotic relationship between geography and imperialism survived – in school texts for example – but a reversal occurred. Whereas imperial interests had sought to utilise geography, geography came to seek acceptability through using the imperial framework as a form of mediation to the public and school children. Such a reversal enabled the scientific study to commence the process of moving from one dogma to the next generation of academic paradigms. It would of course be wrong to contrast political compromise with scientific purity. Municipal imperial reaction merely made way for modern intellectual relativities.

Notes

1 William H. Schneider, 'Geographical reform and municipal imperialism in France, 1870–80' in John M. MacKenzie (ed.), *Imperialism and the Natural World* (Manchester 1990), pp. 90–117; J. F. Laffey, 'The roots of French imperialism in the nineteenth century: the case of Lyon', *French Historical Studies*, 6,1 (1969), pp. 78–82; J. F. Laffey, 'Municipal imperialism in nineteenth-century France', *Historical Reflections*, 1,1 (1974), pp. 81–114; Laffey, 'Municipal imperialism in decline: the Lyon Chamber of Commerce 1925–38', *French Historical Studies*, 9,2 (1975), pp. 329–53.

2 T. W. Freeman, 'The Manchester Geographical Society, 1884–1984', *The Manchester Geographer*, new series, 5 (1984), pp 2–19; Elspeth N. Lochhead, 'The Royal Scottish Geographical Society: the setting and sources of its success', *Scottish Geographical Magazine*, 100, (1984), pp. 69–80. See also T. W. Freeman, 'The Manchester and Royal Scottish Geographical Societies', *The Geographical Journal*, 150, (1984), pp. 55–62;

Elspeth N. Lochhead, 'Scotland as the cradle of modern academic geography', *Scott. Geogr. Mag.*, 97 (1981), pp. 98–109; H. R. Mill, 'Recollections of the society's early years', *Scott. Geogr. Mag.*, 50 (1934), pp. 269–80 and the editorials, *ibid.*, in vols 50 and 100 (1934 and 1984); T. N. L. Brown, *The History of the Manchester Geographical Society* (Manchester, 1971).

 3 John M. MacKenzie, 'Geography and imperialism: British provincial geographical societies', in Felix Driver and Gillian Rose (eds), *Nature and Science: essays in the history of geographical knowledge*, Historical Geography Research Series no. 28 (1992), pp. 49–62. See also D. R. Stoddart, 'The RGS and the "New Geography": changing aims and changing roles in nineteenth-century science', *Geogr. J.*, 146 (1980), pp. 190–202 and Horacio Capel, 'The institutionalisation of geography', in D. R. Stoddart (ed.), *Geography, Ideology and Social Concern*, (Oxford, 1981), pp. 37–69.

 4 MacKenzie, 'Geography and imperialism', p. 59, RGS Archives, correspondence block 1881–1910.

 5 *Journal of the Manchester Geographical Society*, 1 (1885), pp. 3–4.

 6 Council Minute, 31 Oct. 1885, RSGS Records, quoted in Roy C. Bridges, 'The foundation and early years of the Aberdeen branch of the Royal Scottish Geographical Society', *Scott. Geogr. Mag.*, 101 (1985), pp. 77–84.

 7 *Liverpool Geographical Society, Report* (1893), p. 24.

 8 Annual Report, *Report, Prospectus and Rules of the Tyneside Geographical Society*, 1899, pp. 24–5.

 9 T. N. L. Brown, 'The Manchester Society of Commercial Geography', *J. Manchr. Geogr. Soc.*, (1952–4), pp. 40–5; 'inaugural address to the society by Mr. J. F. Hutton' and 'a short note on the origin of the society', *J. Manchr. Geogr. Soc.*, 1 (1885), pp. 1, 64–6.

10 *Manchester Guardian*, 13 March 1879, quoted in William G. Hynes, *The Economics of Empire: Britain, Africa and the new imperialism 1870–95* (London, 1979), p. 33.

11 *J. Manchr. Geogr. Soc.*, 1 (1885), p. 65.

12 A full account of the activities of the Manchester pressure group can be found in Hynes, *Economics*, pp. 22–82. Early meetings of the society, harking back to the 1879 project, were addressed by Consul Holmwood and Harry Johnston on the significance of Zanzibar and East Africa.

13 *J. Manchr. Geogr. Soc.*, 1, (1885), p. 7. The addresses of both Hutton and Stanley were reprinted in *The Manchester Geographer* (1984), pp. 20–32.

14 *J. Manchr. Geogr. Soc.*, 1, (1885), pp. 44–56.

15 DNB. and Hynes, *Economics*.

16 *J. Manchr. Geogr. Soc.*, 2 (1886), p. 139.

17 *J. Manchr. Geogr. Soc.*, 5 (1890), p. 7; the obituary occupied pp. 1–8 of the journal.

18 Felix Driver, 'Geography's empire: histories of geographical knowledge', *Environment and Planning D: Society and Space*, 10 (1992), pp. 23–40.

19 The references to Germany and the debates about Africa can all be found in the 'Reports of Meetings' in the first five volumes of the *J. Manchr. Geogr. Soc.*, (1885–90).

20 *J. Manchr. Geogr. Soc.*, 1 (1885), p. 59. He was chairing a meeting addressed by the Revd Chauncy Maples on the Nyasa region (Malawi).

21 Lochhead, 'RSGS, setting and sources'.

22 See, for example, the address of Consul Henry O'Neill before the society on 9 July 1885, *Scott. Geogr. Mag.*, 1 (1885), pp. 337–52 and the Report of Council, 1887–8, *Scott. Geogr. Mag.*, 4 (1888), p. 680. The standard work is John McCracken, *Politics and Christianity in Malawi, 1875–1940* (Cambridge, 1977).

23 Bridges 'The foundation and early years', p. 77.

24 List of Council, *Scott. Geogr. Mag.*, 1 (1885).

25 Quoted in Bridges 'Foundation and early years', p. 78.

26 'Tenth anniversary of the society', *Journal of the Tyneside Geographical Society*, (1897), p. 7.

27 Annual Report, 1889–90, *J. Tyneside Geogr. Soc.*, (1890) p. 22. This was quoted approvingly in the tenth anniversary account of the society's history, p. 8.

28 Annual report, 1889–90, p. 23.

29 Report of the Executive Council for year ending 31st October, 1894, *J. Tyneside Geogr. Soc.*, (1894–7), p. 38.
30 Report of the Executive Council for the year ending October 31st. 1896, *J. Tyneside Geogr. Soc.*, (1894–7), pp. 350–5.
31 Hynes, *Economics*, pp. 124–5. For an account of Liverpool shipping interests in relation to West Africa, see P. N. Davies, *The Trade Makers, Elder Dempster in West Africa 1852–72* (London, 1973). For Jones, see DNB.
32 Liverpool Geogr. Soc. Report, 1893, p. 1.
33 Liverpool Geogr. Soc. Report, 1894, p. 1.
34 *Ibid.*, p. 3.
35 Liverpool Geogr. Soc., Report of Council 1902, p. 6.
36 Liverpool Geogr. Soc., Report of Council 1911, pp. 6–7.
37 Hynes, *Economics*, p. 125.
38 Jones's role in the LGS was first noted by Andrew Porter in 'The Hausa Association: Sir George Goldie, the Bishop of Dover, and the Niger in the 1890s', *Journal of Imperial and Commonwealth History*, 7 (1979), p. 150.
39 The Tyneside society was also notable for its outings to the estates of its senior figures, including those of Lord Armstrong at Cragside, Earl Percy at Alnwick, Albert Grey at Howick, G. O. Trevelyan at Wallington.
40 Liverpool Geogr. Soc., Report of Council 1906, pp. 6–7.
41 Annual Report, 1889–90, *J. Tyneside Geogr. Soc.* (1890) p. 31.
42 Liverpool Geogr. Soc., Report of Council 1913, p. 5.
43 Dea Birkett, *Mary Kingsley, Imperial Adventuress* (London, 1992), p. 67. Mary Kingsley also visited the Scottish society, as did the colonial journalist Flora Shaw and the traveller Mrs Bishop.
44 The society created the Manchester Geographical Society Building Company for the purpose. T. N. L. Brown, *History of Manchester GS*, pp. 35, 47, 49–51.
45 Freeman, 'Manchester and Royal Scottish Societies'.
46 John M. MacKenzie, *Propaganda and Empire* (Manchester, 1984). Chapter 6 for a survey of imperial societies.

Acknowledgements

I am indebted to Terry Barringer, Roy Bridges, Jim Davis, Felix Driver, Paul Hindle, Barbara Johnson, Christine Kelly, Norman McGilvray, Andrew Porter and the editors of this volume for valuable help. The research was partly funded by grants from Lancaster University Research Committee and the British Academy.

CHAPTER SIX

'The Mother of all the Peoples': geographical knowledge and the empowering of Mary Slessor

Cheryl McEwan

Introduction

Many British women found it easier to enter the public sphere and exercise *direct* influence within the British empire than they did at home.[1] During the nineteenth century, the missionary field provided this opportunity. Mary Slessor was disadvantaged in Britain by both gender and class, but she came to exercise influence both in Britain and overseas through her work as a missionary in West Africa. She played a part in the creation of the imagery of West Africa in Scottish missionary literature, and herself became part of the myth of the area. One commentator wrote, 'One cannot quite imagine Calabar without Miss Slessor.'[2] The Church of Scotland made Slessor a heroine, but her biographers have tended to downplay or neglect the part that she played in the British imperial project in West Africa. One biographer wrote:

> We imagine the scene. The dark jungle, overhead the glint of a star through the leaves, the campfire burning to scare the beasts of prey, the hundreds of dark bodies glistening in its light, and the frail white woman sleeping peacefully.[3]

In such portrayals Slessor becomes merely part of the landscape and her impact upon the nature of British imperialism in West Africa, and upon the peoples with whom she lived, is lost. This chapter analyses the significance of Slessor's achievements, both in Africa and in Scotland, in relation to her role in British imperialism.

The geographical setting of West Africa during the nineteenth century is fundamental to an understanding of Slessor's career and her acquisition of influence in both Africa and Scotland. As several authors have discussed,[4] geographical knowledge of overseas lands was often used to further British imperialism by constructing images of the exotic and the 'other'. What becomes apparent in these studies is the complex interplay between Britain and its empire in the creation of images of foreign lands, and also in the creation of Britain's self-image. Imperial and racist theories in Britain inspired a constant feedback of images from Europeans travelling abroad, and this stream of information, in turn, either reinforced existing attitudes about British imperialism and foreign lands, or inspired new ones. In the case of Africa, this complex relationship between Britain and its empire produced the enduring and powerful image of the 'Dark Continent'. During the nineteenth century this image focused on West Africa.[5]

As a participant in this two-way production of imagery, Slessor's life and work assumes significance in the debate surrounding geography, gender and empire. The following illustrates the traditions that influenced her ideas, the imperial frontier in Africa, Scottish missionary literature, and Victorian romanticism. The ways in which Slessor used geographical knowledge to exercise influence are explored, both through her interventions in West Africa, and through her reports and lectures in Britain. Slessor's responses to the tropical environment and African people are discussed, and particular attention is paid to her work with West African women, perhaps the most radical achievement in her forty years in the tropics.

The Missionary Record, the imperial frontier, and Victorian romanticism – early influences on Mary Slessor

Calabar was the centre of the activities of the Presbyterian mission in West Africa from 1848 onwards, and the missionaries played a significant rôle in the production of Calabar's strikingly exotic and romantic reputation. Slessor spent almost forty years of her life in Calabar between 1876 and 1915, and was a significant contributor to the imagery of the area. Of fundamental importance to her own understanding of West Africa were the influences upon her in her formative years.

Slessor was born in Aberdeen on 2 December 1848, the second of seven children and the eldest girl. Central Scotland at this time was making the transition from an agrarian to an industrial society, and life in the burgeoning towns was characterised by all the privations and hardships associated with this transition. Alcoholism cost her father, Robert Slessor, his job as a shoemaker, and he and his family were compelled to move to Dundee in search of employment. Their home was a single-roomed house in the slums of the city, without water, lighting, toilet facilities, or privacy. Slessor's mother found work in the linen factories, and she was left to tend to the children. Her father's growing inability to work forced Slessor to join her mother in the mills at the age of eleven. From the age of fourteen she worked full-time from six in the morning until six at night, and then took night classes. She was still reponsible for household chores and helping with the younger children. Home life was overshadowed by the cruelty and violence of Robert Slessor, who often beat or neglected his children. Dundee at this time had problems with unemployment, alcoholism and street violence, and during the 'hungry forties' more and more people sought solace in the Church. Slessor was twenty-seven years of age when she left Dundee, but in the meantime the Church was to have a significant impact on her life.

Slessor's mother was a devout Protestant, and the local church became a refuge from the brutality of home life. It was also the source of Slessor's education. The Church provided her with a window on a wider world, a world infinitely more romantic than the dreary slums of Dundee. Tales of missionary endeavours in China, India and East Africa fired her imagination, but it was the brave new mission at Calabar that particularly enthralled and inspired her. The lectures from visiting Calabar missionaries, and the monthly reports from West Africa in the *Missionary Record*, ensured that Calabar existed in Slessor's imagination long before it became a reality to her. Missionary literature was thus a vital influence on her early life.

The story of the mission at Calabar was told in sentimentalist style by the *Missionary Record*, and subsequent commentary upon it revolved around nostalgia and heroism. The impetus to evangelise in West Africa emanated from the abolition of slavery in 1833. West Africa was seen to have a teeming population ripe for conversion to Christianity. Calabar, once the centre of the slave trade, would now become the centre of Christian West Africa. The impulse in founding the mission came not from Scotland, but from Jamaica, the destination for thousands of slaves brought from West Africa in the days of the European slave trade. The intention was to raise an African agency to carry out the work of the Church of Scotland in West Africa. The heroic image was of the 'enlightened' descendants of slaves, recognising the benefits of Christianity and remembering their homeland, wishing to spread the benefits of the gospel in the 'Dark Continent'.[6]

The survival of the missions on the coast depended upon the flow of funds from Britain, and for this they relied on powerful images of West Africa and its peoples. The Church of Scotland mission at Calabar was no exception. Conditions in Calabar were undeniably harsh; it had been at the heart of both European and indigenous slave trading and its effects had permeated the whole society. One source estimates that 24,000,000 people were removed from West Africa as a result of slavery.[7] The Efiks had split into houses that quarrelled with each other over greater shares of the slave trade, and had created a feared secret society, Ekpe. They had forbidden European missionaries to travel up the Cross River, and the forest was closed to them by the Okoyong and other warring peoples of the interior.[8] The mission was trapped on the coast. During the first forty years of the Calabar mission thirty Scottish missionaries died of fever, and many more had their health ruined by the climate.[9] Such conditions provided ample scope for revelations in the *Missionary Record*, the mouthpiece and propaganda tool of the Church of Scotland. It was such revelations that helped form the picture of Calabar in the imagination of Mary Slessor.

Initial reports in the *Missionary Record* often painted Calabar and its peoples in a poor light; graphic reports of atrocities were in themselves a justification for the presence of a mission:

> They may emphatically be said to be in a state of darkness, superstition and spiritual death . . . They are enslaved by various sorts of supernatural charms . . . Human sacrifices on a vast scale are exceedingly common. Their fetish rites . . . are of the most sanguinary character. Their priests are frightful monsters, whose weapons of rule are terror and the knife. When a king or a great man dies, when a war is to be averted, or when any important public matter is transacted, hundreds are seized and doomed to cruel and protracted deaths . . . The mangled limbs of the victims are hung on the branches of the horrid fetish tree . . . and the skulls of those that have been sacrificed are seen scattered in all directions . . . Truly do we feel that theirs is a land of death.[10]

Such lurid images were a major feature of the *Missionary Record* at this time. Polygamy was described as 'an evil which affects, debases and pollutes all the relations of life'.[11] In 1850, Ekpe had passed a law banning human sacrifice, but the missionaries argued that murder was still rife in Calabar.[12] Poison-bean and burning oil ordeals were frequent.[13] Twin-murders were widespread, and a favourite subject of missionary reports. Furthermore, the forests hid the 'objects of terror, including snakes [and] leopards';[14] the climate was 'unsuited to permanent residence' and 'injurious to the health of the great majority of those who are exposed to its influence during several consecutive years'.[15] Calabar truly was a dark place within the pages of the Church of Scotland's publication.

Such images had a considerable impact on readers, and were all the more important as the missionaries were amongst the few residents in West Africa. Their knowledge was gained *in situ*, rather than by fleeting visits in the course of travel or trade. This added greater weight to their claims, and validity to their observations. The images they created made a lasting impression upon Slessor. The pioneering work of the missionaries, surrounded by peril and adventure, stirred her imagination. She was imbued with a profound sense of duty, and a speech by a

Calabar missionary to the Dundee congregation so evoked the needs of West Africa that Slessor determined to train as a missionary to Calabar.

The imperial frontier was a prevalent image in missionary literature and a major influence on Slessor. West Africa was seen to be beyond the frontier of civilisation, and it was the responsibility of the mission to spread civilisation into the African wilderness. The tropical environment was the antithesis of civilised landscapes and hazardous to Europeans. The image of Mungo Park was evoked in the *Missionary Record* to stir patriotic pride, and an awareness of the sacrifices that were being made by Scots on the frontier in West Africa.[16] The epic accounts of missionary endeavours had, by the late nineteenth century, become an established European genre, alongside popular travel and exploration narratives. 'This was a literature of the imperial frontier, a colonising discourse that titillated the Western imagination with glimpses of radical otherness which it simultaneously brought under intellectual control.'[17] Beidelman suggests that it was the over-land progression from the coast to the interior that was the *rite de passage* into African reality, and thus a journey constructed in epic terms.[18] The missionary in West Africa added to these heroic narratives. Reports told that, 'The print of the Missionary's foot is as yet only in the sand of Africa's shore. Oh there is not in the entire world a field more needful than this. Its numerous millions, lying in the blood, appeal to us for sympathy and help.'[19] Later, David Livingstone came to embody the spirit of the frontier in West Africa, combining both religion and commerce in his fight against the 'darkness' of Africa. For Slessor the frontier sense of personal freedom and independence in the quest for civilisation was of fundamental importance in her relations with both the Church of Scotland and the colonial government.

The notion of the imperial frontier was particularly powerful in literature on Britain's involvement in West Africa. Even by the early twentieth century, little was known about West Africa beyond the coast. European contact had existed since the fifteenth century, yet it remained a land of mystery. Its extreme climate, high death rate, and dense, unexplored forests ensured that it was perceived by Europeans as the most nefarious region of the 'Dark Continent'. Apart from coastal settlements and tradings posts, formal imperialism in West Africa only occurred after the 1880s; it was seen to have limited commercial value for Britain and was consisdered a burden to the Treasury until its strategic value was recognised later in the century. Europeans were not believed to thrive there, and even after the mapping of the great rivers of the Niger and the Congo, European trade interests remained firmly anchored on the coasts, relying on African middlemen for contact with the interior. Even missionary efforts were confined to the task of establishing a strong base at the coast.[20]

Therefore, little was known of the Calabar area beyond the coast when Slessor arrived there in 1876, much of the interior remaining unexplored by Europeans (Figure 6.1). Even in 1900 the area between the Niger and Cross rivers was still a blank space on the map. In 1894 Roger Casement had attempted to explore the Aro country and open up trade, but was unable to penetrate more than fifteen miles west of the Cross River before being driven off; he was lucky to escape with his life.[21] It was not until the expedition to destroy the Long Ju-ju at Arochuka in 1901 that access to the interior was gained by the British. By 1904 the British had opened up the Cross River to all the peoples of the interior, and had begun to open up the forests on the northern bank, where previously no white person had been allowed to travel. However, the image of the frontier in West Africa persisted into the twentieth century.

The notion of the imperial frontier and the endeavours of missionaries were inherently linked to Victorian romanticism. As Scotland rapidly industrialised, the 'wilderness' became idealised as spiritually pure reflecting eighteenth- and early nineteenth-century romanticism. However, to the Victorian imagination the subjectivity of the likes of Wordsworth, Shelley and Coleridge was to be avoided; Nature should be objectified.[22] Nevertheless, despite the scientists' rigorous definition of Nature, for most Victorians ' "Nature" remained above all a repository of feeling, a sanctuary they were all eager to retain.'[23] The notion of the countryside as a sanctuary, and of the 'wilderness' was an important part of Slessor's thinking. Having lived in a city, she yearned for the pace of the countryside, and initially found it in Calabar. In later years, as George Eliot had bemoaned the increase in the pace of life facilitated by the steam-engine, so Slessor, in 1903, bemoaned the passing of 'the romantic Old Calabar of my youth'. She wrote:

What a rush of memories . . . The changed conditions from the days when we lived among the natives, and they and we were free as the air to pursue calling under more than Bohemian licence, the advent of the steamers, with their old and kindly if rather rough and exaggerated criticism from skipper and man. The great hospitality on hulk, and in Mission House, the general bond of camaraderie and bond of friendship, which a stay on the Coast invariably conveyed. How changed it is now . . . The advent of the British Government has brought all the freedom of the native and the European to an end, and just as if a cannonball had exploded and shattered us and scattered us as so many different particles, and sent us off into different locations, the old community of feelings is at an end, and the rush and competition and conventionality which makes life in Europe such a strain has come here to stay.[24]

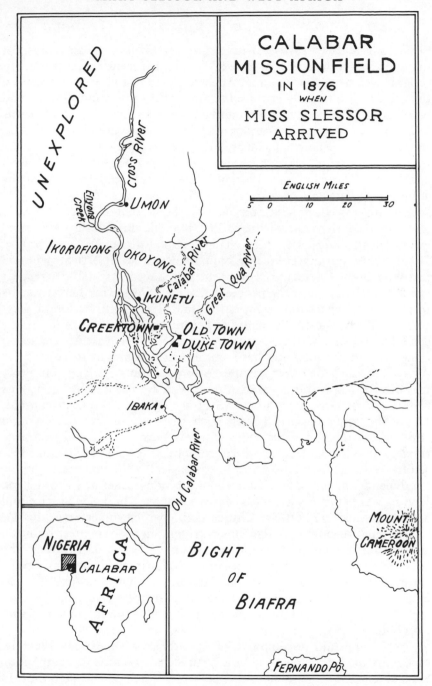

Figure 6.1 Calabar mission field in 1876 when Miss Slessor arrived

Intervention in Calabar, inspiration in Scotland

Slessor's influence in West Africa can best be explored by an analysis of her role within the imperial project. The basis for her interventions in Calabar was her adoption of an African lifestyle, and her proximity to the people. She quickly acquired status within the communities, and was very much a 'benevolent maternal imperialist'.[25] She was known as 'Ma', and her adoption of abandoned children was symbolic of her wider adoption of the people of Calabar. She 'mediated her ethnocentrism . . . through the use of a maternal idiom', making West Africa her 'home' and the West Africans her 'children'.[26] She wrote, 'I spoke to them not as a white woman, but as a mother'.[27] Thus she was assimilated into African life and commanded respect, but the mother–child relationship also involved elements of inequality. She was British, her 'children' were African, and Slessor believed in the superiority of British morality. Despite this, her maternal role brought her close to African women, particularly the elders, with whom she may have felt she shared respect and authority amongst the people of Calabar. Slessor felt her furloughs to be painful separations from her 'home'. The identification of West Africa as home was fundamental to her influence, both within Calabar societies and within British cultural imperialism in West Africa.

Slessor did not question her own, nor Britain's right to be in West Africa, but she believed that Britain's purpose should be one of improving education (albeit defined in ethnocentric terms) and the conditions of life, particularly for women, without damaging West African cultures. She often disagreed with both the Church and the colonial government over their attitudes towards West Africa, and exercised considerable influence on both groups. She attempted to distance herself from the Church by trading, becoming self-sufficient and reducing her dependency on Church funds. Her mission houses at Okoyong and Enyong were built mostly with her own money.[28] In this way she could act independently of official Church policy which she believed did not take enough account of indigenous cultures. She was a pioneer living on the frontier, and her sense of individualism was particularly strong. She believed she understood the needs of the Calabar people better than the Church officials back in Scotland. Her proximity to the Africans, she believed, gave her a greater understanding of their lives; the less her dependency on Church funds, the greater her freedom to implement her own policies.

The Church did not approve of her adopting an African lifestyle, especially her living in a mud-hut,[29] but Slessor ignored its complaints. She exercised considerable influence over Church education policy. In the early days of the mission the least expensive methods of training were sought, avoiding technical and agricultural training which was costly in terms of personnel, money, and equipment. Commercial education was replaced by a quasi-religious system.[30] Slessor believed this abandonment of technical and agricultural training was inappropriate, and worked to rectify it. Her influence can be observed in a letter to her about her plans, which pleads with her to reconsider her threat to sever completely her ties with the mission.[31] The old imperial missionary did not appeal to her since there was too great a distance between themselves and the Africans.

Slessor believed that British control was necessary to establish peace and justice, but she informed the Consul-General of the Oil Rivers Protectorate, Sir Claude MacDonald, that the people at Okoyong were not ready to tolerate a government official in their midst, nor to accept without bloodshed the sudden introduction of new laws.[32] She believed destruction would result from imposing British institutions upon the people without regard for, or understanding of, their customs. Her insight into African law prompted the colonial government to appoint her as the first female Vice-Consul in the British empire in 1892, giving her the powers of a magistrate. However, her sympathy with West African customs brought her into conflict with the colonial administration. She constantly resisted policies that conflicted with West African laws, and was often at odds with colonial officials. For example, in 1910, the people of Akpap complained to her that they were being compelled to build a road through a sacred grove of yams. Slessor was furious, and wrote a letter of complaint to the Assistant District Commissioner, accusing him of violating the sacred grove and ending with the valediction: 'I am 'Not' your obedient servant.'[33] By this time she had a considerable reputation for belligerence amongst the colonial officials.[34]

Despite her resistance to many of the decisions of the colonial government, Slessor's knowledge of the Calabar region was of undoubted importance to the authorities. She was often a mediator between the government and the West Africans. On many occasions she brought officials and chiefs together to negotiate treaties; for example, in 1894 the villagers of Okoyong agreed to hand twin babies over to the Mission instead of killing them. However, her major concern was to spread Christianity without completely destroying indigenous cultures. As a result, she was often critical of British policies. She resented Britain's destruction in 1901 of the sacred grove at Arochuku, the seat of the most powerful oracle in southern Nigeria, and home of the infamous Long Ju-ju. She felt the violence that took place there could have been prevented using less force. She wrote: 'I can't bear those dreadful expeditions, the very sight of force raises their [the Africans'] apprehension and goes to make trouble.'[35]

Slessor also resented the illegal arrests that took place in her districts without the necessary warrants, which were so numerous that the people were afraid to venture out, and which brought previously prosperous markets to a standstill.[36] She referred to this as a 'flagrant breach of law'.[37] Her animosity towards the colonial government erupted in a letter to a friend:

> This land belongs to the native and is worked by the native, tho' our officers do not believe it . . . I am not only writing rank treason, but I am doing so unrepentantly as we live in the bush under bush conditions, and I owe nothing to the government.[38]

At other times she described the government officials as 'iniquitous sometimes in their methods, or want of method, and heart'.[39] She found the lack of consideration for traditional Calabar customs most irksome. She was instructed by the administration to take a census to the chiefs for completion, but scorned the idea. The people were too suspicious to complete it, and were too busy working for the authorities and labouring on the land during the day, and marketing in the evening, to comply. She wrote in her diary, 'So [I] was rather angry of white man's arrogant importunity. It is anything but dignified to natives.'[40]

Slessor acquired positions of authority that would have been unthinkable for a working-class woman in Britain. As well as becoming Vice-Consul and magistrate for Okoyong, she was presented with a fellowship to the Order of St John of Jerusalem in 1913. Perhaps the greatest accolade came from the West Africans, who named her *Eka Kpukpro Owo* – the Mother of all the Peoples.[41] Slessor also attained influence in Scotland through her reports in the *Missionary Record* and the *Women's Missionary Magazine*, and through her many lectures during her furloughs. The magazines were important outlets for Slessor's opinions. Her contributions to the *Missionary Record* consisted of annual reports interspersed with the occasional letter, and she was a prolific contributor to the *Women's Missionary Magazine* from its founding in 1901. As with Violet Markham, Slessor used text to influence others. She constantly emphasised the hardships she endured through the lack of help available to her in Calabar. Her aim was to influence the Church to send out more women missionaries to take over at the stations that she had established, thus freeing her to travel further into the interior. In her personal correspondence her anger is apparent, and she influenced some of her contacts in Scotland (such as Charlotte Crawford, the Assistant Secretary of the Women's Foreign Mission Committee) to help her agitate for more women to be sent out to Calabar.[42] Her strategy in her reports to the missionary magazines relied more on invoking sympathy than expressing her frustration. For example, she wrote in 1897, 'It is the want of sleep, I daresay, which makes me susceptable [to fever], for I get very little sleep with babies to nurse; but . . . I am as well as I would be in Britain, so if no help comes I shall just go on till I feel I *must* give up.'[43] She also evoked the image of West Africa as the frontier in her appeals, in need and desiring of civilisation. She wrote:

> Certainly I cannot come [home] till someone comes out. We are still left, with a wild country gaping in thirsty surprise at our powerlessness . . . [I]f I am able to hold out, it does a little to fill the gap; but, oh, for a score of young women, who would work even for a year and let a weary body or two get a rest![44]

Slessor implied that funds raised at home were not being used for the purposes for which they were given because of the shortfall of workers; she wrote, 'do you at home realise that while you are giving for extension, we are closing stations for want of workers? It is very depressing'.[45]

Overcoming her chronic shyness, Slessor could also be a brilliant public speaker, and people would travel many miles to hear her lectures.[46] One witness wrote, 'her audience was held both by her story and her gift of expression'.[47] The Church used this to full effect by planning extensive lecture tours around Scotland during her furloughs. Unfortunately, transcripts of these lectures were never made or have not survived, but one can assume from the demands placed upon Slessor each time she returned to Scotland that they were popular. Through these lectures and her reports, Slessor may have influenced the policy of the Foreign Missions Board, for during her lifetime the increase in the number of women working in Calabar was considerable. Many of these women followed directly in her footsteps, fulfilling her wish for women to take over her stations and allowing her to travel inland.[48] Furthermore, on one occasion she was interviewed by a representative of Reuters, receiving nationwide exposure for her work and her achievements.[49] She thus enjoyed a degree of celebrity status that extended beyond the Church of Scotland.

'Awful bonnie!' Slessor's responses to the physical environment

Slessor's work mirrors the gradual 'opening up' of the Calabar region to Europeans (Figure 6.2), and very often she was the first white person to arrive in the villages of the interior. Her movement away from the coastal towns into unexplored forests certainly gave her the opportunity to construct epic visions of her local landscapes, and of her own achieve-

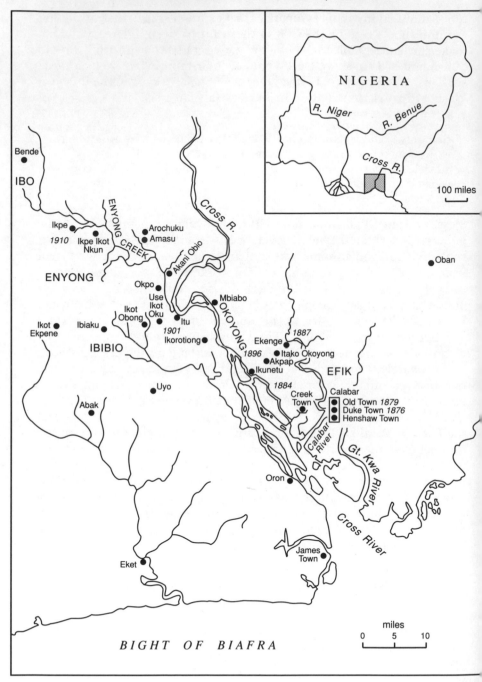

Figure 6.2 Mary Slessor's advance into the Lower Cross River region

ments within these landscapes. However, Slessor's accounts were rarely heroic and the physical environment was not always one of darkness and foreboding. She saw West Africa as a land of magnificent beauty, but one of some hardship for Europeans.

Slessor arrived in Calabar on 11 September 1876 during the rainy season, but the brilliant sunshine and the beauty of her surroundings misled her into describing the climate as 'delightsome'.[50] 'The warm luxuriant beauty of Calabar contrasted strangely with the grey streets of . . . Dundee.'[51] Though she would soon change her opinion of the climate, Slessor remained captivated by the effulgence of the natural environment. During her first months in Calabar she took long walks into the bush, and delighted at the tropical plants, brilliantly plumaged birds, flaming sunsets and the noises from the forest at night. As with Mary Kingsley, the landscapes of West Africa appealed to the romantic side of Slessor's nature, and she was even moved to write poetry about her surroundings.[52] Slessor was lover of nature, and its appeal was deeply rooted in her upbringing. For her the countryside represented an escape, both from the demands placed upon the urban poor, and from the confines of an unhappy home. As a Sunday school teacher in Dundee, she often took her charges into the countryside,[53] which suggests that her training may have exposed her to Victorian ideas about the education of the poor and their relationship with the environment.[54]

Once in West Africa, Slessor revelled in the beauty around her. She compared the sunsets with those in Scotland, familiarising the landscape rather than exoticising it.[55] However, she wrote to friends, 'I feel drawn on and on by the magnetism of this land of dense darkness and mysterious, weird forests.'[56] Clearly, she had contradictory views of the environment. Despite her obvious delight in her surroundings, her descriptions of 'terrors' in the bush, and the 'dark' and 'mysterious' forests are reminiscent of depictions of Africa as the 'Dark Continent', a land of intense beauty concealing hidden dangers to those attempting to penetrate its secrets. Furthermore, when describing her progress into Aro country, Slessor wrote, 'First Itu, then the Creek, then back from Aro, where I had set my heart, to the solitary wilderness of the most forbidding description, where the silence of the bush has never been broken.'[57] Here, Slessor constructed an epic vision of the landscape and framed herself within it. She envisaged herself on the frontier, carrying the torch of enlightenment into the 'wilderness'. However, if West Africa was a 'wilderness' it was also a sanctuary. Scotland had only ever been 'home' while her mother and sisters were alive; by 1895 all were dead, and Slessor's home now, more than ever, became Calabar. On one of her furloughs when ill health caused the Mission Board to doubt her fitness to continue her work, she retorted, 'If ye dinna send me back, I'll swim back.'[58]

Slessor's final destination was Ikpe, in the heart of the interior close to the source of Enyong Creek, a tributary of the Cross River. She found this location particularly beautiful. Ikpe was surrounded on three side by stately palms, which added to its beauty. Slessor wrote, 'These palms are my first joy in the morning when the dawn comes up, pearly grey in the mist and fine rain, fresh and cool and beautiful.'[59] Ikpe, in fact, was located on a mosquito-laden creek and Slessor's house was too close to it. Conditions were often difficult, and Slessor was constantly fever ridden. Her discomfort was exacerbated by periodic affliction with erysipelas (a febrile, inflammatory disease of the skin); on one occasion she lost all her red hair as a result. On another, when she was covered in boils from head to foot, she wrote, 'Only sleeping draughts keep me from going off my head.'[60] Her spirits were often low during the wet season, and at these times her descriptions of the physical environment lost their enthusiasm. She wrote:

> The rain is drizzling down again, and I have a heavy dose of cold on my chest, and must go inside. One blink of sunshine comes, then a drizzle of rain and then a blink, and again a downpour, and all the time we are in a mist for it is never dry under the bush . . . I wish this month were past, to let us get a dry day now and then.[61]

However, the dry season had greater hardships, with the dust-laden harmattans (desert winds) blowing down from the Sahara and draining her energy. She wrote of her longing for a 'wee blink of home', and then declared, 'But though the tears are coming at the thought, you are not to think for one moment that I would take the offer if it were given me! A thousand times no!'[62] Her homesickness never lasted long, and at other times she compared Africa favourably to Britain. She wrote, 'How would you feel if you never had a breath of wind? never a leaf stirring, and everything reeking with heat? It is very trying, but infinitely preferable to your cold.'[63]

Slessor delighted in the Enyong Creek area, perhaps the most luxurious part of Calabar. The banks of the creek were covered in lilies, orchids and ferns, and blossom sometimes cascaded down from the trees above. Slessor particularly enjoyed travelling up and down these waterways. Describing one such occasion she wrote, 'We turned the boat's head and glided into such a lake of equatic plants and flowers as I believe could not be surpassed anywhere . . . On we ran as in an extravagant dream.'[64] In less poetic moments, she described the scene as 'awful bonnie!'[65]

Slessor was greatly disturbed by the outbreak of the Great War, and the notion of 'escape' was even more apparent in her description. Her romantic view of Calabar assumed a new dimension. She contrasted the gathering storm clouds over Europe to the dawning of a new summer in West Africa, far away from the turmoil:

> If you saw the lovely pearl sky in the dawn! The earth all refreshed and cooled, and all the hope and mystery of a new day opening out, you would enjoy it as I am doing. With you and with our poor army and navy it will be darkness and cold, and probably fog, besides all the nerve-racking strain of war and long watching. With us it is all brightness and beauty with long summer months opening out before us.[66]

For Slessor, the 'heart of darkness' was no longer Africa, but Europe and its internecine war.

'True and intelligent friends': West Africans and their customs

The reputation of the people of Calabar was the poorest in Nigeria, they were considered by the British to be intractable, their societies unorganised, and their customs barbarous.[67] This reflected the image of the 'Dark Continent'. Cannibalism was the most sensational of all atrocities associated with West Africa, and a particular favourite of exploration and missionary literature.[68] The missionaries at Calabar were not averse to telling a cannibal tale or two in their reports.[69] Human sacrifice was believed to be widespread, associated with indigenous slavery and the burial rites of kings and chiefs. Widows were subjected to trial-by-ordeal on the death of their husband, as witchcraft was considered to be the primary cause of death. A particularly shocking custom to Europeans was the murder of twin babies and the persecution of twin mothers. Many West Africans believed that babies had a guardian spirit. The occurrence of twins meant that this guardian spirit had assumed human form, an abomination that could only result from the mother having committed adultery with an evil spirit. Since there was no way of telling which was the spirit baby, both were stuffed into calabashes and thrown into the bush to starve to death or be attacked by wild animals.[70] These were the customs, therefore, that Slessor read of before travelling to West Africa.

Slessor's initial impressions of the people of Calabar were not always favourable. She was at first afraid of them, writing, 'Their gesticulations would have frightened me had I not been told they were only trying to make friends.'[71] She wrote to her sister:

> It is such a privilege to get a letter from a christian. This land is dark, and to us, 'dry and parched'. The surrounding heathenism has such a depressing

influence, and the slow progress (which is almost a necessary consequence of their utter debasement) makes one heartsick.[72]

The darkness and debasement for Slessor lay not in the supposed barbarity of West Africa, but in the lack of Christian morality. Her attitude, however, witnessed a subtle change during her first months in Calabar. She became aware of the fundamental importance of the spirit word to the people of West Africa, and this was instrumental in shaping her attitude towards West African cultures. She believed that Europe's superiority lay only in the knowledge of God; if Africans grasped the spiritual 'truths' she offered, this inferiority would no longer exist.

During her first months in Calabar, Slessor visited the compounds of the chiefs, witnessing at first hand the custom of human sacrifice and the 'fattening house' where young girls were prepared for marriage. She was at first appalled and wrote, 'I never thought that my sense of delicacy would be so far blunted. There are scenes we cannot speak or write of.'[73] To people who exalted the life of the 'noble savage' in her hearing, she would retort, 'Let them go and spend a month in a West African harem';[74] her romanticism did not extend to invoking such images. However, Slessor's responses to West Africans and their customs were more complex than a simple condemnation of their lack of Christian morality.

As with her landscape descriptions, Slessor responded on a personal level to the people of Calabar, they were referred to as individuals rather than representatives of race. Sara Mills describes this more personalised form of writing by women as 'going native', arguing that it represents a challenge to male Orientalism and an alternative approach to representing other peoples and countries.[75] Certainly, at times, Slessor attempted to challenge some of the assumptions made about the peoples of West Africa, basing her own descriptions upon a personal knowledge acquired from living in close proximity to them. For example, the Okoyong people had a particularly poor reputation amongst the missionaries, and Slessor's initial impression of them, perhaps coloured by tales she had heard on the coast, was not favourable. She wrote:

> Okoyong is a very dark tribe. They are the princes of drunkards, and smash and hash at each other, and at all and sundry as none of the other tribes do. Calabar people are so frightened of them, that to ask anyone to come to us is to bring a volley of *isungi* (abuse) or *imam* (laughter) down on your head. They would as soon think of going to the moon as going to Okoyong.[76]

However, Slessor made a point of challenging some of the ethnocentric assumptions about the peoples of the interior. On first going inland she wrote: 'I am going to a new tribe up country, a fierce and cruel people, and everyone tells me that they will kill me. But I don't fear any hurt.'[77]

Of the Enyong people she wrote, 'He feels deeply, but cannot express it in ordinary language, or on ordinary times . . . I have not often found deceit.'[78] She wrote of their hospitality and affection, and felt that it was 'a great joy, as well as a privilege to work among them', despite their reputation.[79] Africans became her 'true and intelligent friends', at a time when European perceptions of Africans were generally extreme and crudely racist.[80] Her insight into the customs of Calabar was facilitated by her adoption of an African lifestyle. It was also aided by her knowledge of the Efik language. She appreciated its every shade of inflection and the accompanying gestures, giving her understanding of the many phases of life of the people of Calabar.[81] This provides an important context for her depictions of West African customs, and was also vital to Slessor when she was appointed magistrate. She very often accepted African practices in her rulings at court where European magistrates would have rejected them.[82]

There is no doubt that Slessor, with her fundamental belief in Christian morality, found the customs of ritual sacrifice and twin murder brutal and appalling, and she tried to prevent occurrences of these customs as far as she could. She may have learned from the failed attempts of the early missionaries to eradicate such customs through proselytising, and instead attempted to prevent individual atrocities. For example, in 1888 she intervened at the burial of an Okoyong chief's son to prevent trials-by-ordeal. He was buried after two weeks of pleading and arguing by Slessor without a single attendant death; this was unheard of in Okoyong.[83] Such interventions by Slessor were commonplace. Her understanding of the superstition of twins allowed her to attempt to eradicate the fear of them. She walked miles to rescue twin babies from the bush, adopted them as her own, and raised them to adulthood. The communities in which she lived could see that these babies grew into normal adults, and in this way Slessor made some inroads into the superstition. By the time of her death in 1915 twin murder was considerably reduced in the Calabar area.[84]

It is clear that Slessor's determination and occasional bloody-mindedness won her substantial influence in Calabar, and her fiery nature was often a surprise to many who met her. She was a shy, nervous, and compassionate woman, but at the same time fiercely independent. Maxwell, the chief magistrate of Okoyong highlighted the complexities of her nature on first meeting her in the early 1890s. He wrote:

What sort of woman I expected to see I hardly know; certainly not what I did. A frail old lady with a . . . shawl over her head and shoulders, swaying herself in a rocking-chair and crooning to a black baby in her arms. I remember being struck . . . by the very strong Scottish accent. Her welcome was everything kind and cordial.

He goes on to write:

> Suddenly she jumped up with an angry growl . . . and with a few trenchant words she made for the door where a . . . native stood. In a moment she seized him by the scruff of the neck, boxed his ears, and hustled him out into the yard, telling him quite explicitly what he might expect if he came back again without her consent. I watched him and his followers slink away very crestfallen. Then, as suddenly as it had arisen the tornado subsided, and . . . she was again gently swaying in her chair. The man was a local monarch of sorts, who had been impudent to her, and she had forbidden him to come near her house again until he had not only apologised but done some prescribed penance. Under the pretext of calling on me he had defied her orders – and that was the result.[85]

Slessor's responses to, and portrayals of, the peoples of West Africa were complex. Shirley Foster argues that women travelling in the British empire were conscious of their own marginal status both in Britain and in the empire, and were thus more willing to embrace 'difference', portraying foreign peoples in a more sympathetic light.[86] However, Slessor did not consider herself marginal in the work of the Church of Scotland in Calabar, and her belief in adherence to a strict code of conduct based upon Christian morality led her to condemn many of the customs of the Calabar people. At the same time, because her descriptions were based on her intimate knowledge of both West Africans and their customs, these complaints tended to avoid sweeping condemnations of all West Africans. Slessor believed that West Africans were inherently 'good' and could be 'saved' through their conversion to Christianity. To other Europeans, she seemed eccentric, travelling through the forests barefoot, with cropped hair and the minimum of clothing, drinking unboiled and unfiltered water, and eating African food. However, by adopting an African lifestyle, she was able to pursue her efforts to combat the more bloody of West African customs. It also gave her the opportunity to fight for her favourite cause, that of West African women.

The liberation of West African women: the women's settlement at Use

A feature of nineteenth-century travel writing about Africa is the anonymity of the Africans. African women remained even more anonymous. Very few narratives mention African women, and those that do tend to reinforce twin stereotypes; the image purported by missionaries of the downtrodden wife in a polygamous household, or the racist image reinforced by travellers of the lascivious female, associated with the prostitute in Britain.[87] The Victorian reverence for the virtuous wife led travellers to condemn African societies for selling young girls into marriage, turning child-brides into child-mothers, treating them as slaves when they were adults, and killing them when they were old. There was little effort to portray African women within the context of their cultures and societies.

Slessor's concern for African women had its roots in her upbringing. She had been accustomed to violence and her experiences may have been a significant factor in her attempts to improve the lives of African women. She had been very close to her mother, and had a strong maternal instinct. As with other women who travelled in West Africa,[88] she had a profound insight into the lives of West African women and wrote extensively about them. As a woman she had easy and frequent access to African women and was able to view their lives within the context of their own societies. The wives of missionaries had sometimes attempted pioneering work with West African women, but family life, although intended to set a Christian example, was inclined to keep the missionaries aloof and apart from local populations.[89] However, the presence of a single woman operating independently and living in close proximity with the people had a much more immediate impact. As her familiarity with the women of Calabar grew, Slessor devoted increasing energy to fighting what she thought was their cause and preventing abuses against them. She believed that their lives could only be improved by raising the women's own low view of their status within society.

In her correspondence Slessor was sympathetic to Mary Kingsley's defence of polygamy.[90] However, she wrote, 'God help these poor downtrodden women! The constant cause of palaver and bloodshed here is marriage. It is a dreadful state of society.'[91] She wanted to provide an alternative to polygamy for the women of Calabar:

Our work . . . [must] make the native woman something more than a mere cipher in the community; something more than a mere creature to be exploited and degraded by man . . . Not only must we provide some way of protecting and sheltering women . . . we must create some industry by which these women may earn their living, and thus become independent of the polygamous marriage and the open insult.[92]

She used the strategy of the mission to undermine polygamy in Calabar, allowing only men with one wife to become members of her congregation. She also supported wives who ran away from their polygamous marriages to become Christians. Slessor adhered to ethnocentric attitudes towards polygamy, but attempted to provide practical alternatives. She was appalled at the treatment of women in Calabar, and wrote of one of her court sessions, 'What a crowd of people I have had here today, and how debased. They are just like brutes in regard to women.'[93] There was no place for women outside the compounds, but Slessor was convinced that they could govern their own lives given the opportunity. They did most of the work on the farms and most of the selling in the markets. She believed that if they had their own land they could be self-supporting. They had no rights as citizens and were the property of their husbands. If they were rejected as wives they often starved. In one of her Bibles, against the passage where St Paul lays down the rules for the subjection of wives to husbands, she scribbled in the margin: 'Na! Na! Paul, laddie! This will no do!'[94]

Slessor's proposed solution to the problem of the subjection of West African women was to establish an independent women's settlement in the interior, offering refuge to women who were subjects of the compound master, and, if unmarried, could be legally raped, tortured, assaulted or injured.[95] Such a settlement, she believed, would free Calabar women from the need to marry. Over time her ideas began to assume greater clarity. The women of the settlement would not only be independent, but would also contribute to the local economy, producing food and goods with which to trade with neighbouring villages. Slessor also advocated training schools for women and young girls to acquire trading and farming skills, an independent income and a new self-confidence.

Slessor believed that such pioneering work was best done by women, since they would prove less of a threat to African men and have greater access to the women's yards.[96] However, before she could achieve her aims she had to gain the confidence of the local people. She cemented friendships with women close to the chiefs and elders in the hope that they would use their influence and enable her to pursue her ideas. Her greatest ally in Calabar was Ma Eme, the sister of King Edem at Ekenge, who on the surface appeared to adhere to tribal laws, but often surreptitiously helped her to prevent such atrocities as the flogging of slaves and ritual sacrifices. Slessor perhaps repaid the help of West African women, for when she became Vice-Consul and magistrate it was widely rumoured that no woman ever lost a case 'with "Ma" on the bench'.[97]

In 1908, her dream of creating a women's settlement was realised when a site near Use on the Cross River was cleared for a refuge and training centre.[98] It was an independent unit, self-supporting and unreliant upon outside funding. It had good road links and water supplies, and, most importantly, a sufficient market for the products of the settlement. As well as farming, Slessor encouraged industries that the local people had not yet developed such as basket-making, the weaving of coconut fibre, and cane and bamboo work.[99] The women exchanged pineapples, rubber, bananas and cocoa for books, writing materials and medical supplies. The settlement was known as the Mary Slessor Mission Hospital until it was destroyed in the wars for independence in 1960.[100] Slessor had successfully built a refuge for orphans, twins and their mothers, and refugees from the harems, who would otherwise have been driven out of their villages to starve.

Conclusions

It is difficult to assess the impact of Slessor's work upon public opinion in Scotland. The *Missionary Record* was widely read in the industrial towns, and the *Women's Missionary Magazine* also had a wide circulation,[101] giving her access to a sizeable female audience. It is significant that following her complaints in the magazines over the lack of help she received in Calabar, many more single women travelled to West Africa during the final decade of the nineteenth century. Buchanan informed her that her reports were 'the means of awakening a new interest in Calabar in the minds of many of our people'.[102] The influence that Slessor acquired as a consequence of her experience of Calabar is more easily discerned.

Slessor's knowledge enabled her to attain influence both in Scotland and in the British empire. This discussion has focused upon the power to create images of foreign lands that resulted from such knowledge. It has illustrated that Slessor was very much a product of her time. Her perceptions of Calabar had been shaped by the mythology surrounding West Africa, and by the images in the *Missionary Record*. She was imbued with the Eurocentrism of her time, but her relationship with Calabar and its peoples suggests a complex attitude. She respected African laws and those customs which did not involve bloodshed and cruelty. She possessed a grasp of the Efik language which helped her quest to Christianise Calabar, but which also fostered her respect for African cultures. This, in turn, influenced her descriptions of Calabar.

Slessor's knowledge of Calabar was greater than the earlier pioneer missionaries; West Africa was not new to her in the way that it had been to the founders of the mission. She had grown up with a knowledge of Calabar, albeit a knowledge based on nostalgic and graphic reporting. She spent almost forty years in Calabar at a time when the mortality rate amongst missionaries was exceptionally high. Over this period she assimilated vast amounts of information that had not been available to her predecessors, and which gave her a different insight into West

Africa. Furthermore, the mission was already firmly established by the time Slessor arrived there. She made sure she was less dependent on Church funds and on donations from home than her predecessors, by building her own homes, growing her own food, and setting up her own trading networks. Her reports relied little upon evocative images of barbarity and degradation. An increasing tendency towards prudery at the end of the nineteenth century may have had an impact upon her reports, and may explain the absence of graphic reporting by Slessor. For these reasons, her portrayal of Calabar was often very different from that of earlier missionaries.

The imagery that Slessor created was complex. Her descriptions of the physical environment avoided references to the 'Dark Continent', but were at different times both positive and negative. The landscape was not nefarious, but had a beauty that was striking in comparison with the grey slums of Dundee; her progress within it was not depicted as an epic journey. True, she saw herself as a pioneer on a frontier, but she identified this as 'wilderness' and sanctuary, a refuge both from her upbringing in the Dundee slums, and from the claustrophobia of the Calabar coastal towns where the Church had control. On the frontier, Slessor had control. Her innate romanticism on occasion led to almost poetic reflections upon her surroundings, but the difficulties and hardships of life in the Calabar forests were never far from her mind. The vagaries of the climate had a considerable impact upon Slessor's physical and psychological well-being, and, occasionally, upon her landscape descriptions. Slessor portrayed some West African customs, such as twin murder, human sacrifices and the treatment of women, as barbaric. She often depicted West Africans as infantile, assuming the role of mother, asserting her own authority and punishing even the most respected of chiefs. However, she insisted that the 'inferiority' of West Africans to Europeans lay only in the lack of Christian virtues. Moreover, Slessor insisted that West African women were neither degraded nor lascivious, they had the potential to be independent and influential within their own societies if they were freed from the need to marry. Slessor used her own influence to empower African women and to encourage their independence.

As well as securing her own influence, Slessor inspired others to follow in her footsteps, and may have helped to empower women in Scotland. The *Women's Missionary Magazine* reported her influence on Scottish women:

> In Calabar, more than in any other mission field ... the traditional romance of missionary adventure is not yet a thing of the past. Pioneering is the very forefront of civilisation ... and to an unusual extent it is undertaken by women missionaries. Following an example set by Miss

Slessor, they venture into new districts, and live amongst unfamiliar tribes . . . The country is hardly opened up even now and travelling is therefore a real problem, but . . . the missionaries undauntedly press forward.[103]

Slessor's influence on Church policy in Calabar was substantial, particularly with regard to education and the adoption of ideas on technical and agricultural training. She also had an impact on Britain's economic and political power in this area of Nigeria. She was singularly responsible for negotiating an agreement between the Efik and the Okoyong that allowed trade with, and access to, the interior, but she was at times a belligerent obstacle to British administrators. The so-called 'Native Courts', which as a magistrate she helped establish and run, became the basis upon which a British legal system was imposed upon Nigeria in the late nineteenth and early twentieth century.[104] Slessor, therefore, wielded considerable influence in Calabar which she would not have acquired in Britain. She was undoubtedly part of the imperial project in West Africa, but imperial strategies rarely featured explicitly in her texts. Instead landscapes were described and people referred to as individuals rather than representatives of race. Slessor's texts relied upon personal observations and through these she claimed subjective authority. Her descriptions of West Africa and its peoples thus highlight not only the complexities of her own personality, but also the ambiguous and ambivalent position of white women in the British empire.

Despite having acquired influence and authority through her work in Calabar, Slessor was not a 'new women' in the sense that she did not openly advocate the rights of women in Britain. In a letter to a friend, she wrote, 'I have enjoyed the old world gentlewomen, who after all are more to my taste than the new woman. I'm too old for the new clever independent hand I fear.'[105] However, in defence of women missionaries she wrote, 'women are as eager to share in all the work and sacrifice of the world as men, and it is their privilege to share in it'.[106] Slessor may not have championed the rights of British women, but she encouraged them to follow in her footsteps. She adopted the cause of the women of Calabar, and her work was novel in shedding light upon their lives at a time when this was a much neglected and misunderstood subject.

Notes

1 D. Birkett, *Spinsters Abroad: Victorian Lady Explorers*, Oxford, 1989; S. L. Blake, 'A women's trek: what difference does gender make?', *Women's Studies International Forum*, Vol. 13, (1989), pp. 347–55; H. Callaway, *Gender, Culture and Empire: European Women in Colonial Nigeria*, London, 1987; M. Domosh, 'Towards a feminist historiography of geography', *Transactions of the Institute of British*

Geographers, Vol. 16, (1991), pp. 95–105; C. Oliver, *Western Women in Colonial Africa*, Connecticut, 1982; J. Sharistanian (ed.), *Gender, Ideology and Action: Historical Perspectives on Women's Public Lives*, Connecticut, 1986.

2 J. N. Rattray, 'How the Enyong Creek was entered', *Free Church Record*, (1915), p. 172.

3 Revd McIntyre, *Free Church Record*, (1915), p. 174.

4 For example, P. Brantlinger, 'Victorians and Africans: the genealogy of the myth of the Dark Continent', *Critical Inquiry*, Vol. 1, (1985), pp. 166–203; P. D. Curtin, *The Image of Africa: British Ideas and Action 1780–1850*, London, 1965; D. Hammond and A. Jablow, *The Myth of Africa*, Library of Social Science, 1977; E. Said, *Orientalism*, London, 1978.

5 Brantlinger, 'Victorians and Africans'; Curtin, *The Image of Africa*; Hammond and Jablow, *The Myth of Africa*, London, 1977.

6 W. P. Livingstone, *Mary Slessor of Calabar. Pioneer Missionary*, London, 1916, pp. 15–16.

7 Jubilee Booklet, *The Life of Mary Slessor. Pioneer Missionary, 1840–1915*, n.p., n.d.

8 J. Buchan, *Peacemaker of Calabar: The Story of Mary Slessor*, Exeter, 1984, pp. 5–6.

9 Buchan, *Peacemaker of Calabar*, p. 5.

10 *Missionary Record*, (1846), p. 4.

11 *Missionary Record*, (1851), p. 88.

12 Letter form Hope Waddell, *Missionary Record*, (1857), p. 89.

13 Thomson, *Missionary Record*, (1866), p. 21.

14 Timson, *Missionary Record*, (1866), p. 214.

15 Report of the Medical Committee, *Missionary Record*, (1854), p. 81.

16 *Missionary Record*, (1846), pp. 3–4.

17 J. Comaroff, and J. Comaroff, 'Through the looking-glass: colonial encounters of the first kind', *Journal of Historical Sociology*, Vol. 1, (1988), p. 9.

18 T. O. Beidelman, *Colonial Evangelism: A Socio-Historical Study of an East African Mission at the Grassroots*, Bloomington, 1982.

19 *Missionary Record*, (1846), pp. 3–4.

20 Livingstone, *Mary Slessor*, p. 23.

21 C. Christian, and G. Plummer, *God and One Redhead. Mary Slessor of Calabar*, London, 1970, pp. 124ff.

22 V. C. Knoepflmacher and G. B. Tennyson (eds), *Nature and the Victorian Imagination*, Berkeley, 1977, p. xix.

23 Knoepflmacher and Tennyson, *Nature and the Victorian Imagination*, p. xxi.

24 Slessor to Irvine, Okoyong 12/12/03, St Andrew's Hall Missionary College Library, Selly Oak.

25 B. N. Ramusack, 'Cultural missionaries, maternal imperialists, feminist allies: British women activists in India, 1865–1945', *Women's Studies International Forum*, Vol. 13, (1990), p. 319.

26 N. R. Hunt, ' "Single ladies on the Congo": Protestant missionary tensions and voices', *Women's Studies International Forum*, Vol. 13, 1990, pp. 395–7.

27 Mary Slessor, 'Triumphing over superstition', *Women's Missionary Magazine of the United Free Church of Scotland* (hereafter referred to as *Women's Missionary Magazine*), 1901, p. 109.

28 Livingstone, *Mary Slessor*, pp. 83–6, 210–11, 251–3.

29 Buchanan to Slessor 13/1/05, United Presbyterian Church Papers, Foreign 3, National Library of Scotland.

30 K. K. Nair, *Politics and Society in South-Eastern Nigeria 1841–1906. A Study of Power. Diplomacy and Commerce in Old Calabar*, London, 1972, pp. 66–8.

31 Stevenson to Slessor 4/6/12, United Free Church Papers, Women's Foreign Missions, National Library of Scotland.

32 Christian and Plummer, *God and One Redhead*, p. 90.

33 Slessor to Falk, Assistant District Commissioner at Ikot Ekpene, September 1910, E. M. Falk Correspondence, Rhodes House, Oxford.

34 Falk to District Commissioner at Ikot Ekpene, September 1910; District Commis-

sioner to Provincial Commissioner of the Eastern Province 12/9/10.
35 Slessor to Charles Partridge, District Commissioner for Eastern Nigeria 20/1/08, Slessor Papers, Dundee City Public Library.
36 Slessor to Partridge, 30/5/08.
37 Slessor to Partridge 14/8/08.
38 Slessor to Mrs Findlay 24/8/12, Cairns Papers, National Library of Scotland.
39 Slessor diaries, entry for 5/5/12, Macmanus Galleries, Dundee Museum.
40 Slessor diaries, entry for 3/4/12.
41 West Africans called Slessor *Eka Kpukpro Owo*, the Mother of all the Peoples, see J. Buchan, *The Expendable Mary Slessor*, Edinburgh, 1980, p. xii.
42 Slessor to Partridge 14/7/08; correspondence between Slessor and Charlotte W. Crawford, United Free Church Papers, Women's foreign Missions, mss. 7934, 7941, 7981, National Library of Scotland.
43 *Missionary Record*, May, 1897.
44 Letter from Slessor, *Women's Missionary Magazine*, (1914), p. 79.
45 Mary Slessor, 'Present opportunity in Ibibio', *Women's Missionary Magazine*, (1908), p. 286.
46 Livingstone, *Mary Slessor*, p. 170.
47 Jessie F. Hogg, 'Mary M. Slessor', *Women's Missionary Magazine*, (1915), p. 54.
48 Among these women were her four protégés, Beatrice Welsh, Martha Peacock, Mina Amess and Agnes Arnot.
49 Livingstone, *Mary Slessor*, p. 168.
50 Buchan, 1984, p. 27.
51 D. M. McFarlan, *Calabar. The Church of Scotland Mission, 1846–1946*, London, 1946, p. 93.
52 B. Miller, *Mary Slessor. Heroine of Calabar*, Michigan, 1946, p. 20.
53 Livingstone, *Mary Slessor*, p. 10.
54 See Matless's discussion of Vaughan Cornish in 'Nature, the modern and the mystic: tales from early twentieth century geography', *Transactions of the Institute of British Geographers*, Vol. 16, (1991), pp. 272–86.
55 *Missionary Record*, (1877), p. 377.
56 Miller, *Mary Slessor*, p. 97.
57 Letter from Slessor at Ikot Obon, 28/2/06, *Women's Missionary Magazine*, (1906), p. 142.
58 Christian and Plummer, *God and One Redhead*, p. 109.
59 Miller, *Mary Slessor*, p. 123.
60 Livingstone, *Mary Slessor*, p. 253.
61 Slessor to Mrs Findlay 19/8/13, Cairns Papers, National Library of Scotland.
62 Livingstone, *Mary Slessor*, p. 62; Mary Slessor, ' "A missionary's testimony", extract from a letter to friends', *Women's Missionary Magazine*, (1910), p. 67.
63 Letter from Slessor at Ikot Obon, 25/1/14, *Women's Missionary Magazine*, (1915), p. 78.
64 Christian and Plummer, *God and One Redhead*, pp. 128–9.
65 Buchan, 1980, p. 175.
66 W. P. Livingstone, 'Old Calabar: in memory and in vision', *Free Church Record*, (1915), pp. 101–08.
67 A. I. Nwabughuogu, 'The role of propaganda in the development of Indirect Rule in Nigeria, 1880–1929', *International Journal of African Historical Studies*, Vol. 14, (1981), p. 75.
68 W. Arens, *The Man-eating Myth*, Oxford, 1979, p. 85.
69 Hope Waddell, *Missionary Record*, (1847), p. 7.
70 For a contemporary description of the custom of twin-murder see Mary Kingsley, *Travels in West Africa*, London, 1982, pp. 472–8.
71 Livingstone, 1915, p. 101.
72 Slessor to her sister, 17/4/77, Slessor Papers, Dundee Museum.
73 J. Trollope, *Britannia's Daughters*, London, 1980, p. 195.
74 Christian and Plummer, *God and One Redhead*, p. 67.

75 S. Mills, *Discourses of Difference: An Analysis of Women's Travel Writing and Colonialism*, London, 1991, p. 99.
76 Extract from a letter from Slessor to the *Missionary Record*, October 1890. Slessor soon changed this situation, establishing trade links, procuring the laying down of arms, reducing drunkenness and encouraging work. In doing so she helped reduce the death rate in the Okoyong area.
77 Slessor to Hart from Use Ikot Utu, 6/7/12, Slessor Papers, Dundee Museum.
78 Slessor to Church of Scotland from Use, 20/5/08, Slessor Papers, Dundee Museum.
79 Buchan, 1980, p. 24.
80 Kathleen Goldie, unpublished scripts of broadcasts on Slessor, 1949, p. 8, in Goldie Papers, ms. 8078/401, National Library of Scotland.
81 J. Chappell, *Three Brave Women*, London, 1927, p. 16.
82 M. D. W. Jeffreys, 'Mary Slessor – Magistrate. Parts I and II', *West African Review*, (1950) p. 805.
83 For accounts of this incident see Christian and Plummer, *God and One Redhead*, p. 2; Livingstone, *Mary Slessor*, pp. 91–99; Oliver, *Western Women*, pp. 113–15.
84 Chappell, *Three Brave Women*, p. 41, argues that by 1898, seventeen years before her death, Slessor had rescued fifty-one twins; the final figure must have been much higher.
85 Livingstone, *Mary Slessor*, pp. 129–30.
86 S. Foster, *Across New Worlds: Nineteenth Century Women Travellers and their Writings*, London, 1990, pp. 174–5.
87 S. L. Gilman, 'Black bodies, white bodies: toward an iconography of female sexuality in late nineteenth-century art, medicine and literature', *Critical Inquiry*, Vol. 12, (1985), pp. 204–42.
88 For example, Mary Kingsley and Constance Larymore, see C. McEwan, 'Encounters with West African women', in A. Blunt and G. Rose, *Writing Women and Space: Textual Representations of Difference by White Women Abroad*, New York, 1994, pp. 73–100.
89 *Sunday Mail Story of Scotland*, Vol. 3, (1988), pp. 1091–2, in Slessor Papers, Dundee Museum.
90 Slessor to Irvine, 12/12/03.
91 Letter from Slessor, 7/10/05, *Women's Missionary Magazine*, (1906), p. 20.
92 Mary Slessor, 'Concerning advance work in West Africa', *Women's Missionary Magazine*, (1904), p. 4.
93 Jeffreys, *Mary Slessor*, p. 629.
94 Buchan, 1980, p. 195.
95 Miller, *Mary Slessor*, p. 113.
96 Church of Scotland Overseas Council, *Mary Slessor of Calabar*, 1978.
97 Jeffreys, *Mary Slessor*, p. 629.
98 Jubilee Booklet, p. 13.
99 Livingstone, *Mary Slessor*, p. 252.
100 Kathleen Goldie Papers, 1949, p. 13.
101 The average monthly issue for 1902 was 27,800 see *Women's Missionary Magazine*, (1903), p. 154.
102 Buchan to Slessor 10/11/02, United Presbyterian Church Papers, Foreign 1, National Library of Scotland.
103 Report on Missions, *Women's Missionary Magazine*, (1912), p. 204–5.
104 See Omoniyi Adewoye, *The Judicial System in Southern Nigeria, 1854–1954*, London, 1977, pp. 40–1; J. K. MacGregor, 'Mary Mitchell Slessor', *The East and the West*, Vol. 1, (1917), pp. 161–2; *Nigeria Magazine* (no author cited), 'Mary Slessor', (1958), part 58, pp. 210–11.
105 Slessor to Partridge, Use, 1/1/08.
106 Christian and Plummer, *God and One Redhead*, p. 165.

CHAPTER SEVEN

Historical geographies of the British empire, c. 1887–1925

Robin A. Butlin

Introduction

This chapter is concerned with the 'historical geography' of the British empire; that is with various ways in which 'historical geographies' of the empire were written between 1887 and 1925. Its focus will be on a number of significant texts – a series of 'historical geographies' of the British empire – and the various ways in which these works reflect the deployment of both geography and history in the service of British imperial ideals through the advancement and propagandising of a range of strong and persistent historical and geographical themes. These include the links between empire and antiquity, race and racism, and environmentalism. The publications reviewed here were mainly written by historians and colonial administrators, trained at Oxford, whose subsequent experiences engaged with a wide range of academic and governmental institutions, and which in turn were directly and indirectly implicated in the promotion and critique of the complex process collectively identified as British imperialism.

One of the important features of these historical geographies is the way in which they demonstrate a particular view of geography, seen as a background to history and as a scientific and humanistic justification for a range of imperial ideals and policies. The historical geographies do, however, exhibit sound and at times high standards of geographical narrative and analysis, and deserve a place in the historiographies of geographical thought and knowledge. Their relationship with the 'new' geography, then in its early course of promotion at Oxford by H. J. Mackinder, in London by the Royal Geographical Society, and by the Geographical Association, will be reviewed. C. P. Lucas (1853–1931), the general editor of the *Historical Geography of the British Colonies/Dominions/Dependencies* series, and H. B. George (1838–1910), author of *An Historical Geography of the British Empire*[1] are two central figures in this chapter.

The promotion of imperial ideals by these and other related publications and actions was directed at a broad target, and included government, members of universities, schools, and members of the public. This was a particular and in some ways peculiar historical propaganda, moving toward a new, essentially Anglocentric and Eurocentric, imperial history by using geography to encourage the retrospective and prospective consideration of the bases of empire, and owing much in its inception to the influence of Sir John Seeley's influential *Expansion of England*.[2]

The institutional focus of much of this work was the Royal Colonial Institute, founded in 1868 as a non-political party forum for the gathering of information and the promotion of papers and debates on the British empire, some of whose members as individuals and groups did pressurise the government on imperial questions. Although clearly modelled on the Royal Geographical Society, with one of its stated objects at foundation being to 'occupy as regards the colonies the position filled by the Royal Society with reference to science, or the Royal Geographical Society with regard to geography',[3] and basing its rules on those of the Royal Geographical Society and the Society of Arts, there seems not to have been much formal contact between them. There was more contact, through Lucas, with the Geographical Association, of which he became President, which was founded in 1893 to advance the cause of geography in education.[4]

Geography, history and historical geography in Britain in the late nineteenth and early twentieth centuries

In Britain there remained historically deep-rooted connections between geographical and historical knowledge in the nineteenth century, which formed the basis of a range of scholarly discourses variously described as geographical history, the geography behind history, and historical geography. Geography at that time was beginning to emerge as a separate academic discipline in Britain, and although some of its proponents strove for scientific credibility rather than status within the humanities, it retained significant links with history. Some facets of the nature of the relations between geography and history are identified in J. S. Keltie's report on geographical education to the Royal Geographical Society in 1885:

> I have found in not a few instances a recognition of the indispensability of geography to a thorough understanding of history. Professor Freeman and the late Mr Green have had much to do in demonstrating the intimate connection between history and geography. Professor Freeman's *Historical Atlas* is well known, and the school edition of Spruner's *Historical*

Atlas has a considerable circulation in this country. Probably there is no department of knowledge in which geography is calculated to throw more light than on history, in its completest sense.[5]

Historical geography in Britain was largely practised and published by historians, conspicuously those at Oxford, whose interests were in the geographical contexts and arenas within which the courses of civilisations and territorial claims and gains had been played out, that is in the geography behind history. Maps and atlases were important aids in this type of analysis and description.

Geography and history were inevitably caught up in the process of increasing specialisation and professionalisation of knowledge, and in the political events and social and educational reforms of those changing times. Accounts of the attempts by the Royal Geographical Society to advance the scientific and pedagogic role of geography from the mid-1880s indicate the transitional state of the fledgling discipline: 'Geography itself appeared vague and diffuse, part belonging to history, part to commerce, part to geology. How could some coherence be found which would persuade the universities of its value'.[6] In spite of the earlier interests generated among historians, it was clear that many who sought to promote the cause of the new geography felt it necessary to pursue more vigorously the routes of physical, scientific geography, at the cost of the traditional links with history and commerce. Hence: 'The strategy adopted was to abandon historical and commercial geography at least in the short term, and concentrate on physical or "scientific" geography'.[7] If, however, one looks at the teaching of historical geography in Oxford in the last two decades of the nineteenth century, by historians like E. A. Freeman and H. B. George, and by geographers such as Mackinder, it is clear that it was an established subject, taught to students of history in order to provide background information on classical and more modern civilisations. Its form and context were clear, but unfortunately the subsidiary position of geography in relation to history made for difficulties in advancing the right to separate curricular status for geography. As Scargill indicates, the lectures in geography given by H. B. George (whose writings are discussed later) in the school of modern history 'received little attention since the subject had hardly any status in examinations. In the final honour school of modern history the last paper out of about a dozen had been on historical geography but in 1885 even this had disappeared and was replaced by starred questions in certain of the other papers'.[8]

Another discernible influence on the 'historical geography' texts examined here was that of Liberal Anglicanism. This, as manifest in Victorian historical writing, was a complex amalgam of belief systems, but basically involved a religious perspective with moral and practical purpose which incorporated German idealism, concern (after 1815) with developing social and political crises, a dissatisfaction with utilitarian rationalism, with aspects of romanticist interpretations of the past, and the development of a new philosophy of history.[9] This new philosophy, owing much to the revolutions in methodology and theory in German philosophy effected by Hegel and in history by Niebuhr, Ranke and Burckhardt, incorporated a cyclical concept of the biography of nations, which were conceived as passing through stages of youth, maturity and old age, producing distinctive historical patterns of development but still influenced by the individualistic responses by states to crises and by individuals within the states operating with freedom of action and will.[10] Decline, as with the ancient empires, was possible as well as progress, and the role of major heroic figures in advancing the cause of major empires could be accommodated in this kind of history (here much influenced by the writings of Thomas Carlyle), in which the past strongly influenced the present. Additional features were a belief in the general moral and intellectual progress of humankind, which linked with clear views of ethnocentric superiority: the Teutonic peoples were deemed to be the ultimate races capable of advancing civilisation from the points at which it had been passed to them from the civilisations of the past,[11] and therefore the Victorian liberal Anglicans saw, on account of the Saxons having been Teutonic in origin, 'both the British constitution and European civilization as natural products of racial character. The colonial empires of the nineteenth century could thus be seen as an expression of the European races' potential for world domination.'[12]

The connection between the new German history and the history written by the Liberal Anglicans was initially made through Thomas Arnold and subsequently taken up by E. A. Freeman, A. P. Stanley, and J. R. Green at Oxford, Charles Kingsley and later J. R. Seeley at Cambridge. This was the ethos within which the writing of the historical geographies of empire by Lucas and his collaborators and by H. B. George was initiated, and which consequentially linked into the writing of a new form of imperial history, with emphasis on historical continuity and a strong sense of a manifest destiny for Britain, by virtue of the nature and configuration of its islands and of its history, to lead a major empire. The links between the major academic Victorian historians and the ideas of the writers of the historical geographies and histories of the British empire are, of course, very complex indeed, and they changed through time, though the general influence of Liberal Anglicanism was remarkably persistent, extending beyond World War One.

Figure 7.1 Sir Charles Lucas (1853–1931). Portrait by Walter Stoneman, 1917

Sir Charles Prestwood Lucas

Charles Prestwood Lucas (1853–1931) (Figure 7.1), born at Crickhowel in Breconshire, in the Usk Valley between the Black Mountains and the Brecon Beacons in Wales, was educated at Winchester College and

Balliol College, Oxford, where he had obtained first classes in Classical Moderations and Literae Humaniores (1876). He was first on the Civil Service examination list, and was posted to the Colonial Office. There, according to *The Times* obituary notice of 8 May 1931, in spite of 'a certain modest shyness, which never entirely left him he opened out and developed, and before many years had acquired there a high reputation for knowledge and efficiency. The majesty of the British Empire, the marvellous development of the British Colonial system, fired his imagination. He projected and brought out "An Historical Geography of the British Colonies", the introduction to which and several of the subsequent volumes he wrote himself – a work indispensable to the student of Colonial and Imperial progress'. His most active period at the Colonial Office was during the tenure of the Colonial Secretaryship by Joseph Chamberlain (1895–1903). In 1897 he was promoted to Assistant Under-Secretary of State and head of the Dominions Department, embracing with great enthusiasm Chamberlain's plans for development and consolidation of the empire, including preferential trade tariffs and an Imperial Secretariat. With the coming to power of the Liberals in 1905, however, his fortunes changed, and he was twice passed over for promotion to the headship of the Colonial Office, and took the decision to retire prematurely, at the age of fifty eight, in 1911. Although highly regarded within the Colonial Office, having been appointed to serve in 1886 on the managing council of the Emigrants' Information Office, and in 1907 to supervise a new Dominions' Department, it seems that his use of this department 'to promote imperial unity and preference and to educate the people at home and in the colonies to the value of empire and the problems facing it', and his convictions that 'the dominions actually desired the Colonial Office to exercise stronger leadership'[13] led to strong opposition from within the Colonial Office, which preferred a policy of non-interference, and to his early retirement in 1911. Perhaps Pugh's characterisation of it partly explains Lucas's disengagement:

> In short, the Colonial office staff was too intellectual to be imaginative. Moreover, it was imprisoned by its environment. The culture of London seemed so polished that it was hard to treat with perfect seriousness the aspirations of Toronto or Auckland, Lagos or Belize. To the men of Whitehall the civilisation of the colonies, whether newly contrived by expatriated Britons or the child of the primordial jungle, was not merely different from our own, it was inferior to it. Colonial peoples were like children and were to be treated with all the kindness and severity of the Victorian parent.[14]

Lucas was created a KGMG in 1907, and a KCB in 1912. He spent the rest of his life, supported by a fellowship at All Souls, Oxford, from 1920, writing and lecturing on the British empire. He continued the work that he had earlier begun of teaching and assisting at the Working Men's College in Great Ormond Street in London. He became a member of council of that college at the age of thirty-nine, in 1892, and was Vice-Principal from 1896 to 1903, and Principal from 1912 to 1922. The college, founded in 1854, was an important link between privileged and gifted intellectuals, administrators, writers and artists, and working people in London who wished to advance their knowledge and education. Rooted in Chartism and the Christian Socialist movement, it attracted very early in its life a sparkling array of voluntary lecturers, including F. D. Maurice, the founder, Charles Kingsley, George Grove, John Ruskin, Dante Gabriel Rossetti, and J. R. Seeley, followed later by Lucas and many of his distinguished Oxbridge and Civil Service contemporaries.[15] The college provided a crucial target for much of Lucas's wartime and postwar writings on the nature of the British empire, including his six lectures at the college, published as *The British Empire* in 1915.[16]

In his chapter on the history of the Colonial Office in the third volume of the *Cambridge History of the British Empire*, Pugh, better known perhaps as the editor of the *Victoria County Histories*, says of Lucas that 'his attractive personality had won him many friendships in the Dominions on his not infrequent visits. He was the editor and part author of a comprehensive set of handbooks called *The Historical Geography of the British Empire* – the *vade mecum* of all Imperialists of his day'.[17]

C. P. Lucas's Historical Geography of the British Colonies, Dominions, and Dependencies

In the period from 1887 to 1925, C. P. Lucas edited a series of books under the initial title of *A Historical Geography of the British Colonies*, the general titles of volumes in the series changing respectively to *Historical Geography of the British Dominions* and *Historical Geography of the British Dependencies* after the addition to Canada (already a Dominion) of Australia, New Zealand and South Africa to self-governing dominions status at the beginning of the twentieth century. An *Introduction* to the series, written by Lucas himself, was published in 1887, whose purpose and context was very briefly outlined in the Preface:

> This little book is intended as the first instalment of a Historical Geography of the British Colonies. Any succeeding parts will be more purely geographical and will deal with the separate divisions of the Empire. I know no book which gives quite simply and shortly a connected account

of the Colonies, of the geographical and historical reasons of their belonging to England, and of the special place which each colony holds in the Empire. The present is an attempt to supply the want from materials to hand at the Colonial Office and elsewhere.[18]

This introductory text contains eight chapters, dealing respectively with: 'What is a colony?'; 'Motives of colonisation'; 'Climate and race'; 'Modes of colonising and kinds of colonists'; 'Nations which have colonised – 1. Ancient'; 'Nations which have colonised – 2. Modern'; 'English colonisation'; 'Changes in English colonies during the 19th century'. The book contains eight maps, showing: the principal areas of slavery and the principal convict settlements, past and present; Greek, Phoenician, and Carthaginian colonies, and the Roman empire; Spanish and Portuguese colonies past and present; Dutch colonies past and present; French colonies past and present; English colonies at the end of the 17th century; English colonies in 1814; and English colonies at the present day. The *Introduction* makes no direct methodological mention of the role of geography and history in the understanding and promoting of empire. In the last chapter (VIII, 'Changes in the English colonies during the nineteenth century'), much emphasis is placed on the ways in which technological improvements, particularly in global transport and communication networks, have helped to transform global space and give easier access to empire.[19] Such tendencies would inevitably, in his view, reduce the number and roles of the heroic individual promoters and administrators of empire, such as General Gordon and Sir Bartle Frere.[20] The great social changes of the nineteenth century he identified as the abolition of slavery and the anti-transportation movement, and the major political changes the free-trade movement, the granting of responsible government to the major colonies, and the development of the Confederation movement. Lucas's views on the nature of historical geography are thus largely implicit, and seem to relate to the concept of the geographical background to, and influence on, the historical course of British imperialism. One point at which he gives an indication of the purpose of the series is in the preface to Volume IV, *South and East Africa*, where he says that the object of this and other books in the series is 'simply and solely to try to give a connected and accurate account of British colonisation, its methods, agencies, and results, and of the various provinces of the British Empire, recording facts and avoiding, as far as it is possible to do so, controversial topics'.[21]

Other contributory authors' explicit views on the general methodological approach of this series are rare: for the most part, as with Lucas's writings, they are implicit. One of the few direct opinions on the subject is that ventured by H. E. Egerton, in his preface to *Volume V Canada – Part II Historical*: 'Whatever be meant by historical geography – and I should myself describe the series as histories, laying especial stress on geographical considerations – the practice has been in some, at least, of the previous volumes to separate the purely historical from the geographical portions of the work.'[22]

The major part of the series comprised seven regional volumes, concerned respectively with the Mediterranean and eastern colonies, the West Indian Colonies, West Africa, South and East Africa, Canada, Australia and New Zealand, and India.[23] Volume IV part II and volume V have the title *A Historical Geography of the British Dominions*, and the India volume has the series title of *A Historical Geography of the British Dependencies*. They were published by the Clarendon Press, Oxford. The choice of the Clarendon Press as a vehicle for publication of the series is worth brief comment, for the Lucas historical geographies were initiated at a time when the delegates of Oxford University Press were becoming (especially during the depression of 1892) increasingly concerned at the non-profitability of many of the books published, and increasingly aware of their obligations to the British empire. The Clarendon Press was thus increasingly required by the Oxford University Press Delegates to focus more on books of wider commercial interest, including school texts, and the Lucas series was clearly seen as a potentially profitable venture. Hence, 'C. P. Lucas's *Historical Geography of the British Colonies*, Major Green's *Hindustani Grammar*, and Baden-Powell's *Short Account of the Indian Land Revenue* were accepted as likely to be remunerative (Some thirty volumes of the 'Rulers of India Series' appeared in the 1890s, as the Press assumed its imperial responsibilities).[24]

Most of the books in the series were frequently revised and updated, sometimes by different authors, indicating the impossibility of a single author being able to cope with the speed and complexities of change in the British empire. Lucas himself contributed the historical sections to the first five volumes. Others involved in their writing included J. D. Rogers, a barrister and formerly fellow of University College, Oxford, H. E. Egerton, the first Beit professor of Imperial History at Oxford, and P. E. Roberts, for thirty years teacher of Indian history at Oxford. Many of the contributors were from the Colonial Office, such as R. E. Stubbs, who revised the first volume on the Mediterranean and the eastern Colonies for its second edition in 1906. The whole effort might indeed be characterised as the product of Oxford history, law, and the Colonial Office. The books themselves vary in content, according to the particular interests of the authors, but their basic structures involved the presentation of the facts of the history and geography of the countries and continents concerned, hence the division of most of the volumes into historical and geographical sections. Some examples will provide illustrations of their content, purpose, and style.

Geographical descriptions in Lucas's historical geographies

The quality of geographical description and analysis in the Lucas historical geographies varies considerably. The maps in the series, especially those appearing before the turn of the century, are sometimes poor in quality, a fact noted by a number of reviewers. One, writing in the *Geographical Journal* for 1914, while applauding the continuation of a series which 'is finding an increasingly wide public, for there is no other work which tells the story of the empire so succinctly, so accurately, or with so fine a sense of proportion',[25] strongly bemoans the quality of the maps in the third edition of the West Africa volume. The same reviewer notes an improvement in the topographic representation in A. B. Keith's new edition (1913) of Lucas's historical volume on south Africa, for new maps are presented and 'in some of them the reader can tell whether the country sketched be flat or hilly'.[26] The poorest maps, hand drawn at the Oxford University Press, crowd in much detail, together with hachured topography, which makes for very 'noisy' appearance. There are, however, exceptions, including the clearer maps with printed lettering that appear in J. D. Rogers's 1907 volume VI on Australasia, produced by Emery Walker, though even here the hachuring technique used to represent relief in New Zealand clutters the introductory maps. Coloured maps are used in the volume on the West Indies, and these are of quite good quality.

The historical sections in the volumes written by Lucas himself concentrate very much on the history of colonisation and settlement and the constitutional basis of links with Britain, and the geographical sections are rather basic in form and content, competent without being graphic, and placing, inevitably, great emphasis on new commercial and transportation developments, especially railways.

In a significant observation in the preface of Volume IV: *South and East Africa*, Part I *Historical* and Part II *Geographical*, which was written by Lucas, with extensive help from the Colonial Office and various agents in Africa, he makes the statement, already cited,[27] about the object of this and other books in the series being to give a factual and largely uncontroversial account of the methods agencies and results of British colonisation and of the various provinces of the British empire. This indicates the objective purpose of the series as a whole, reflecting perhaps the attempted even-handedness of Lucas's civil service approach to the use of geography and history in the reporting of the historical geographies of the British empire. Historical frameworks and geographical data had, it seems, been used since the early nineteenth century – and probably earlier – by officials at the Colonial and Foreign Offices when seeking to reconstruct the bases of colonial problems and opportunities for background papers on colonial matters prepared for their ministers and for cabinet.[28] Further research in the archives of the Colonial Office may confirm this stylistic and methodological connection, but for the purpose of this essay it is necessary, in order to fathom the deeper views and feelings of the contributors to the historical geographies, especially Lucas, to refer to their other, often more polemic, writings.

The first part of this volume gives a long account of the history of settlement and war in South and East Africa, the second treats of the various parts of British South Africa, Central Africa and East Africa. The description of the physical geography of, for example, the Cape Colony, has no geological basis of explanation, even where gold and diamonds are concerned. The explanation of its history of settlement follows mildly deterministic lines. This is factual, almost compendium, description, and no attempt is made to give an account of the nature of the landscape and its evolution in natural and human terms. On the whole this type of geography is more of an extended gazetteer, with some historical information added, and although the basic features of physical and human geography are adequately covered, they are not woven into an explanatory narrative of the same quality of Rogers's *Canada*. Lucas lists various sources of geographical information: from friends 'with special knowledge', including the Agent-General of Natal, and the Secretary to the Agent-General for the Cape; statistics for Matabeleland and Mashonaland were provided by the offices of the British South Africa Company; the chapters on British Central Africa were mainly written by Mr H. Lambert of the Colonial Office, and revised and supplemented by Lucas. In the additional reading suggested at the end, for example, of the section on British Central Africa, Livingstone's own accounts of exploration, together with the works of Sir H. H. Johnston, are commended, and the recommendations for geographical reading include Johnston's 1894 report on British Central Africa to the British Parliament, and 'The information contained in Stanford's Compendium of Geography, Africa, vol. ii, [A. H. Keane] is very full and valuable'.[29] It is worth noting that A. H. Keane was professor of Hindustani at University College, London, an active member of the Anthropological Institute (vice-president 1886–90), who, according to Lorimer, 'became in the 1890s an active publicist of geographical and anthropological texts which advocated an extreme racist position. Keane's position stemmed in part from his extremely antagonistic attitude toward blacks'.[30] Additional references given in the historical section include 'Mr. Greswell's *Geography of Africa South of the Zambezi*, 1892, Clarendon Press' (a book written under the auspices of the Royal Colonial Institute as part of an initiative to produce historical and geographical material for the promotion of the British empire for use in schools and universities), and 'Mr. Scott Keltie's *Partition of Africa*, 1893, 2nd. ed. 1894.'[31] He suggests that 'Reference should also be made to the excellent handbooks on the Cape Colony for intending emigrants, edited for the Emigrants' Information Office by Mr Walter B. Paton.'[32]

Lucas's volume on *The Mediterranean and the Eastern Colonies*, revised for a second edition by R. E. Stubbs 'of Corpus Christi College, Oxford and the Colonial Office', takes a similar approach, and deals with 'those dependencies of Great Britain which lie on or near to the great trade routes to India and the Far East' . . . 'The history of these dependencies has followed their geography'.[33] Brief reference is made in the preface to E. A. Freeman's *Historical Geography of Europe*, but the balance of the volume is a compendium of geographical and commercial information.

A major contrast in style and approach is offered by the writing of a freer and more imaginative spirit – J. D. Rogers – on Canada and Australasia in the series. A volume much praised by historical geographers working later in the twentieth century is Rogers's Part III of Volume V: '*Canada-Geographical*'. The preface opens with a somewhat curious statement: 'Geography, with which this volume deals, has only to do with what is present, external, and physical; but Canada is composed of historical as well as geological strata, which do not merely belong to the past, but still remain exposed, visible, or even uncomfortable although a book on geography primarily deals with things, men, though something more than things, are after all things, and cannot be quite left out.'[34] The contents of the book, however, far exceed the modest expectations set by the preface, for the author combines a deep knowledge of the geography of Canada with a powerful narrative of the stages of settlement of its various regions. Inevitably much attention is given to the history of exploration and discovery, but also to the exploitation of land and sea-based resources and patterns of settlement. Accounts, for example of the exploration of Hudson's Bay, are referenced to papers in the *Journal of the Royal Geographical Society*, and one hallmark of this book is the most impressive range of sources cited, including recent geological reports and maps, topographic maps and atlases, a wide range of primary and secondary historical material, and data sources, including census data. The author is cautious but realistic in his use of physical geography as an influence on patterns of settlement and discovery. Speaking of New Brunswick:

> The reader may be weary of seeing rivers and coasts referred to as lines of development, and lines of development described by architectural and mechanical metaphors such as passages, props, bands, bonds and the like; but these metaphors recur irressistibly to those who realise that if there is one essential truth which has persisted through the ages, it is that New Brunswick is the province with two corridors to Quebec Province, two bands and bonds between the St. Lawrence and the Atlantic, two props or pillars upon which Quebec Province rests, and must rest during half the year, unless it is to depend on the United States.[35]

In the chapter on Quebec there is an interesting account of the development and layout of townships and seigneuries and the nature of land speculation. The description of the scenery of the regions and provinces of Canada is graphic and of a very high standard. An example is the opening section of Chapter VIII: *The Middle West*:

> Somewhere on the threshold of Manitoba woods vanish, rough places are made smooth, the earth is a level lawn, lakes and rivers are not what they were, and the horizon widens. To the east an infinite series of wooded hills, watery hollows, lakes, swamps, and rocks, cramps while it diversifies the scenery, and perplexes while it enchants the imagination; and as we move westward the maze becomes more intricate and stone-strewn or wet up to a point, beyond which there is utter change; but the point is not definite nor is the change sudden.[36]

This is very close to being an outstanding piece of 'modern' historical geography, and is undoubtedly an excellent exemplar for regional geographical writing. Rogers was also author of the volume on Australasia, first published in 1907, which exhibits similar characteristics. In his *Preface and Introduction* to Australasia he spells out the premises on which the book is based. He says that more could have been included in the book, but in that event 'it may be doubted whether it would be a book on Historical Geography',[37] although characteristically he does not define what he means by historical geography. He goes on to say that 'Military and personal details have been discarded as irrelevant; and I have abstained from expressing any opinion on the tendency towards imperial and colonial union and on other tendencies which might be regarded as within the domain of politics, partly from want of sufficient knowledge, partly from inclination, and partly because politics lie outside the plan on which this series of books is based'.[38] Part I is devoted to *History* (308 pages) and Part II to *Australasian Geography* (132 pages). As with the Canadian volumes, Rogers relies on a very wide range of printed sources, including government reports and censuses, and given that his evidence is secondary (there is no indication that he had visited Australasia) he produces again a very lively account. The make-up of the two parts is shown as follows:

Structure of J. D. Rogers' Australasia, Vol. VI (1907) of C. P. Lucas's Historical Geography of the British Colonies

Part I: History

Chapter I.	The Old Pacific.
Chapter II.	The Natives of Australia.
Chapter III.	The South Sea Islanders.
Chapter IV.	The Plan of a Colony in Botany Bay.
Chapter V.	Australia in the First Epoch.
Chapter VI.	Second Epoch of Australian History – Dispersion.
Chapter VII.	Second Epoch of Australian History – Extension.
Chapter VIII.	Convicts and Emigrants; Land Laws and Constitution.
Chapter IX.	New Zealand in the Second Epoch.
Chapter X.	Transition: The Age of Gold.
Chapter XI.	Transition: The Golden Age.
Chapter XII.	Australia in the Third Epoch.
Chapter XIII.	Australian Extension and its Effects.
Chapter XIV.	Extension in New Zealand and its Effects.
Chapter XV.	The Modern History of the Pacific.
Appendix I.	The New Hebrides.
Appendix II.	The Constitution of the Australian Commonwealth.

9 maps

Part II: Australian Geography

Chapter I.	Pacific Islands.
Chapter II.	New Guinea Geography.
Chapter III.	Geography of New Zealand.
Chapter IV.	Australian Geography.

13 maps

Reactions to the historical geography series within the geographical literature inevitably were varied. A review in the *Scottish Geographical Magazine* for 1887 of Lucas's *Introduction* to the series shows contemporary expectations:

> Mr. Lucas has commenced a work which promises to be one of some importance . . . has made a distinct contribution to geographical science, and the dependence of historical events on geographical position is clearly

shown on his pages. Read side by side with Dr. Freeman's *Historical Geography of Europe*, the student will find Mr. Lucas's book a valuable auxiliary by bringing up to date the colonial movements of European nations.[39]

Another reviewer wrote in the *Scottish Geographical Magazine* in 1894 that in the volume on West Africa,

> Mr. Lucas shows that he has formed an admirable conception of the range and aims of historical geography. His method is thoroughly sound. Physical features are carefully described for the purpose of showing their bearing on exploration and settlement, on the relations between the settlers and the natives, and on the subsequent development of the resources of the countries dealt with.[40]

At the end of the review there is strong commendation of Lucas's clear literary style. A review in the *Geographical Journal* for 1915, of Lucas's Part II of Volume IV, *South Africa. History to the Union of South Africa*, indicates that in his general summary 'Sir Charles Lucas deals with the geographical factors which have influenced the development of South Africa, and draws contrasts and comparisons between it and Canada and Australia, in reference to such matters as area, accessibility, climate, resources, and the lie of the land', and also praises the improved quality of maps in this book from those in earlier books in the series.[41] A review of Volume I, *The Introduction*, in 1889, speaks of the manner in which 'The happy combination of University culture and official information in Mr. Lucas, and his acceptance of assistance from specialists on the various colonies, makes his book a very oasis in the desert of detailed geographical literature.'[42]

Later commentators viewed the series from different perspectives. A. J. Christopher, for example, in *The British Empire at its Zenith*, evaluates Lucas's approach as 'highly historical, seeking to create a measure of continuity from classical times to the nineteenth century'.[43] Harris and Warkentin, in their study of the historical geography of Canada, *Canada before Confederation*, refer to Egerton's preface to the second part of the Canada volume as an indication that 'whatever the term historical geography had meant to Sir Charles, the volumes in the series were histories, albeit histories laying particular stress on the influence of physical factors such as climate and landform in the course of events'.[44]

Lucas, geography and imperialism

There are many indications of Lucas's involvement with and use of geography to help promote knowledge about the empire. Lucas was obviously an advocate of the teaching of geography at advanced level in schools, because of the insights he felt it could give, in conjunction with history, to the understanding and advancement of the imperial cause. His principal agency to this end was the Royal Colonial Institute, to which in 1894 the Geographical Association had sent copies of replies to a questionnaire about the teaching of geography in schools, and at whose headquarters the Geographical Association had held its first annual meeting (on 21 December 1894). Lucas was also active within the Geographical Association, of which he was President in 1920, and subsequently Vice-President. Balchin has shown that the Geographical Association in the period immediately following World War One experienced a dramatic increase in membership, rising from 1,458 in 1918 to 2,379 in 1919, 3,965 in 1920 to 4,159 in 1921 and also that in 1920, while Lucas was President, the association 'went international' and invited membership from, in particular, overseas territories of the British Empire, with the result that members were recruited from Accra, Lagos, Nairobi, South Africa, Egypt, Madras, Burma, Ceylon, Straits Settlements, Hong Kong, Canada, Bahamas, Trinidad, Leeward Islands, Australia, Tasmania and New Zealand. 'One of the objectives of this overseas recruitment was not only membership as such but also the hope of securing authoritative first hand geographical contributions for the journal'.[45]

Lucas contributed three articles to *The Geographical Teacher*, the journal of the Geographical Association: 'Islands, peninsulas and empires', his presidential address;[46] 'The place names of the Empire;'[47] and 'The islands and the Empire'.[48] In the first of these he reviews the roles of peninsulas and islands as points for the diffusion of plants, animals and humans, and of peninsulas and islands as 'mothers of empires'.[49] Referring to E. A. Freeman's observations on the roles of peninsulas and islands in European history (presumably, although not explicitly stated, his section on the effect of geography on history in the introduction to his *Historical Geography of Europe*),[50] he pursues the idea of the coherence of the British empire as an 'island empire', that is, an empire with an island core. Thus:

> We have seen the characteristics of islands as jumping-off places and footholds, and we have also seen that islands are not great fusing places of races and peoples, and given our own island as illustrations. Similarly, a leading feature of the British Empire is that it is a non-fused and non-fusing empire, and there is within it a total absence of uniformity. We commonly regard our policy of granting self-government as solving the problem of how to hold together an empire by other means than force. But should it not be regarded rather in the light of wise recognition that the problem is insoluble, and that the island has gradually abandoned the

empire idea in favour of that of a League of Nations having a common sovereign? Attempts as fusion have been abandoned. The argument is, that if an island makes an empire it must be an overseas empire without geographical continuity, and that it will have the characteristic non-fusion of races and peoples which has been noticed in the case of islands.[51]

He suggests that both islands and peninsulas had completed their historic roles, particularly as air power would soon replace land and sea power. Some of these arguments and analyses are repeated in Lucas's paper 'The islands and the empire', given to the Imperial Education conference in 1923, describing his task as being that of speaking on the teaching of the history and geography of the [British] empire. Hence 'All will agree that history cannot be properly taught without the aid of the map, and if history should be always coupled with geography, especially is this true of the British Empire which, as I shall try to show, is geography's own child.'[52] Suggesting that the history and geography of empire should be taught to the young without omitting reference to mistakes and blunders, with generosity of reference to rivals in imperial ambition and without blatant jingoism, he nevertheless also advocates a very positive attitude to the achievements of the British empire: 'The British empire is not a colossal expansion of original sin, as some seem to think. It is a great and glorious achievement. It has set up a standard for empires and provided a text on recognition of responsibility. Instead of letting Empire give us a bad name we should go on giving it a reformed character.'[53] He concludes:

> In this rare combination of diversity and continuity, I find the key to the riddle of the Empire. The old order changeth. A limit has apparently been set alike to territorial expansion and to predominance of sea power. Distance is fading away. The powers of the air take no count of the inviolate sea which encompasses the island. Yet the breed has been but yesterday supremely tried and found the same as ever. A world-wide commonwealth enriched by many elements has grown up on the island's lines.[54]

One additional historiographic question deserves brief analysis. This is the influence on Lucas and such fellow writers of the historical geographies of J. R. Seeley's *Expansion of England*.[55] There is no doubt that they were strongly influenced by his ideas on the history and nature of British imperial expansion, but Greenlee, for example, has suggested that they did not fully understand Seeley's philosophy, in the sense that they followed a credo of a formal empire as an organic whole which pulled together in a cultural and practical sense the diverse members into an imperial nation or federation, and whose existence was to be understood primarily as a result of an inexorable historical process of

[167]

evolution through three centuries of expansion, at the centre of which was, of course, Great Britain.[56] Seeley, according to Greenlee, was not, however, entirely pro-imperialist, and has to be viewed in a more Euro-centric ideological context. The Oxford imperial historians and historical geographers were, as indicated, supporters of the idea of the British empire as the outcome of natural processes of organic growth of a superior society but, as Symonds suggests,[57] their emotional approach to the racial virtues of the British and the growth of empire contrasts with Seeley's more dispassionate methods of analysis. Greenlee also points out that the world of these 'lesser Seeley's' was a more rapidly changing and more charged and anxious one than that in which Seeley had written his major work. The frictions of imperial territorial and commercial competition, with France, Germany and the United States in particular, combined with the threat felt by the ruling classes in Britain in the face of increasing political democracy, seem to have given energetic and fevered impetus to the likes of Lucas, Egerton and A. P. Newton, friends sharing common ideals of empire, to the production of histories and geographies of empire. Their sensitivities and concerns were hightened by the South African war, which emphasised the military weakness of the empire and questioned its moral roots, and effected much radical, humanitarian and socialist critique.[58] This accounted for a strident tone in the writings of Lucas's group, and a determination to further educate the nation to the importance of the empire, including the continuity from past to present, and to prepare a better educational basis through the promotion of imperial history and geography. H. E. Egerton, a writer in the historical geography series, was the first Beit professor of colonial history at Oxford, holding the chair from 1905 to 1920, combining a functionalist approach with a moral tone and a belief for closer imperial union in his teaching. In the opinion of Symonds,[59] 'This moral tone and concern for the welfare of the colonies was to become the hallmark of Oxford Imperial historians', as also, according to Madden, was the sense of continuity from British imperial experience from the past.[60]

Lucas was not a lone or a minority voice. Large numbers of books, many aimed at schools and universities, were published in Britain in the early twentieth century linking the teaching of geography and history with the understanding of imperialism and colonialism, the majority dealing with commercial and the political aspects. They differ, of course, in balance and tone, and a contrasting example of another kind of historical geography of empire will now be considered.

The Reverend H. B. George

Hereford Brooke George (1838–1910) had read classical and mathematical moderations at Oxford in the early 1860s. A man of private means, having inherited money from his father and a directorship of the West of England and South Wales Bank at Bristol (which 'failed' in 1880), he was called to the bar in 1884 and practised on the western circuit until 1867 when he returned to New College, Oxford, as tutor in the combined school of law and history, becoming a tutor in history when the two schools separated in 1872. His biographer in the *Dictionary of National Biography* describes his work in the following terms:

> His historical writing and teaching were chiefly concerned with military history (in which he was a pioneer at Oxford) and with the correlation of history and geography. His chief publications, 'Battles of English History' (1895), 'Napoleon's Invasion of Russia' (1899), 'Relations of Geography and History' (1901, 4th ed. 1910), and 'Historical Evidence' (1909), all show critical acumen and fertility of illustration, if no recondite research. His 'Genealogical Tables illustrative of Modern History' (1874; 4th ed. 1904) and 'Historical Geography of the British Empire' (1904; 4th ed. 1909) are useful compilations.[61]

George was a pioneer climber and alpinist from 1860 onwards, joining the Alpine Club in 1861 and editing the first three volumes of the *Alpine Journal*. The link between geography and climbing is a significant one, climbing being in the late nineteenth century an activity pursued by many pioneer geographers, including Mackinder and Douglas Freshfield (secretary of the Royal Geographical Society from 1881 to 1894 and vice-president and president in the period 1910–24). The challenges of the Alpine peaks and mountain chains elsewhere offered the opportunity of a form of pioneer exploration, associated with a credo of health and fitness, and linked with geographical interests in military strategy, in which mountains played a crucial part. Blouet has described how George, like Mackinder, was much involved in military training and war games, having been a member, when Mackinder was an undergraduate at Oxford, of the Oxford University Rifle Volunteers, and a leader of the University Kriegspiel (War-Games) Club, and also refers to his pioneering role in the teaching of military history at Oxford.[62] Sir Charles Oman, in his autobiography, refers to George's involvement in the University Kriegspiel Club, his own interest having been 'due to my tutor at New College, H. B. George, who, finding me interested in military history, was good enough to introduce me to this very serious association, of which he was the president. It played war games in the German style on the old set of Prussian official maps for the campaign of Sadowa, and occasionally, for a change, on the ordnance survey maps of

Oxfordshire, or the vicinity of Aldershot'.[63] George was a fellow of the Royal Geographical Society, and he, together with Douglas Freshfield and Reginald Cocks, proposed Mackinder for election to membership of the Society in March 1886.[64] He lectured on geography in the Modern History School at Oxford, but, as Scargill has shown, his lectures received little interest from students because of the low status afforded to geography on the examination syllabus, students preferring, as George wrote, to attend 'the hardest to things which paid them best, and if they were short of time they did not take great pains to obtain a knowledge of geography'.[65]

George's book on the *Relations of Geography and History* (1901) gives the basis of his understanding of this important link. In its preface he states:

> Every reader of history is aware that he must learn some geography, if he would understand what he reads. Comparatively few, however, if one may judge from experience, seem to realize how much light geography throws on history. Geographical influences account for much that has happened. Geographical knowledge affords valuable data for solving historical problems. At the same time human action alters the aspects of those things of which geography takes cognizance: man cuts canals and tunnels mountains, drains marshes and constructs artificial harbours, though it must be admitted that these things are trifles compared to the steady operation of geographical causes all history through.[66]

The first eight chapters deal with general themes – the general nature of geographical influences, frontiers, towns, nomenclature, fallacies of the map, sea power in peace and war, and geography in war, followed by twelve chapters on regional aspects of Europoean history, and two chapters on India and America. His general view may be described as moderately determinist, and is much focused on the strategic aspects of geographical conditions, notably warfare on land: 'War, in the modern sense of the word, is altogether based on geography',[67] he says at the opening of his chapter on the subject, and basic topographic influences are illustrated by reference to British experience in the Khartoum campaign and the Boer War.

H. B. George's Historical Geography of the British Empire

H. B. George's *Historical Geography of the British Empire*, was of different in character to Lucas's historical geographies. It was first published in 1904, and had reached a seventh edition by 1924, revisions of the work after his death in 1910 being undertaken by R. W. Jeffery, Fellow of Brasenose College, Oxford. In his preface he indicates that: 'My object in writing this little book has been to present a general survey of the British Empire as a whole, with the historical conditions, at least so far as they depend on geography, which have contributed to produce the present state of things.'[68] He also says in the preface, that 'I have burdened my pages with few names and fewer figures, my object being less to state facts than to elicit their meaning.'[69] There is only one map – a predictable world map, based on Mercator's projection, showing the British empire coloured in red. No photographs are used – surprising in the sense that George had 'hunted' mountains and glaciers with a camera.

In the introductory Part I he details his perspective: 'The character of every state, still more of every empire, must be greatly affected by geographical considerations, and this is pre-eminently true of the British empire, which is unlike any other that the world has yet seen. Her position on the globe, close to Europe yet separate from it, gave England the best opportunity for maritime and commercial greatness.'[70] he follows the familiar metaphor of the empire as a family, with the parents sending out the children into the world full grown, through with a shift to the perspective of 'each member of a family group caring for the welfare of the whole'.[71] He sees the British empire as something which was not the outcome of deliberate policy but which grew, partly through rivalry, especially with the French in Canada, though the main motive of growth was commercial. As far as the administration of the empire is concerned, he stresses the variety and flexibility of donors and recipients: 'an extraordinary variety of systems, from highly developed codes down to the rude customs of African semi-savages, are administered under the English flag . . . The British empire exhibits the dominant race in almost every possible relation to other races'.[72]

The style of the introduction immediately shows us that we are dealing with a very different approach from that revealed by the texts of Lucas's historical geographies. Here we have a confident personal assertion about the rightness and justifiability of the British empire, and a robust defence of criticism by Britain's opponents:

> The enemies of England [sic] abroad are in the habit of sneering at her greed of territory, fulminating against her endless intrigues, threatening her with the vengeance of outraged humanity, and so forth. Englishmen in general know that their country has a clearer national conscience than most others, and that the last thing their government would dream of doing is intrigue. It is perhaps expedient for those who would understand the complicated inter-workings of historical and geographical influences upon the formation of empire, and its maintenance when formed, to enquire what is the basis of such ill-will, and whether our conduct justifies or excuses.[73]

The envy of other nations derived, according to George, from the acquisition by Britain of 'many of the fairest portions of the earth', notwithstanding the fact that Britain derived very little material benefit from their control, especially on account of the costs of defending them.

The way in which the book is organised is on the basis of the status of each section of empire, starting with the British Isles, then moving to the 'stepping-stones' (including Gibraltar, Malta and Cyprus, St. Helena, Mauritius, Ceylon, Singapore, and the Falkland islands), the 'daughter nations' (Canada, Australia and New Zealand), the 'dependencies' (India and the West Indies), the protectorates, and the British dominions in Africa. In each case there is a sensible and well informed description of the strategic and physical geography of the places and of their economic basis and historical evolution in the context of empire. Thus the section on Australia starts with a description of the geography of Australia, followed by a history of its colonisation from Captain Cook onwards, followed by individual descriptions of the geography of the individual states of the commonwealth. Basic schemes of regional division are used: Canada is divided into four parts (the maritime region; the great lakes and the valley of the St. Lawrence; the central plain or series of plateaux from west of Lake Superior to the Rocky Mountains; and the region west of the Rockies) each of which is expanded upon in sections dealing with their colonisation and their prospects for economic development.

What is distinctive about this particular book? There are several significant features. The first is the soundness for its time of the geographical and historical information and the persuasive nature of the narrative. The second is the caution shown at making too ready connections between geographical character and the sequence of historical events in a particular territory, even though the influence of climate on settlement and economic development is a factor consistently touched on. The third is the ethnocentric perspective adopted, illustrated, for example, by his section on the indigenous inhabitants of Australia, which takes a very superior white European view of the potential of the aboriginal population. There is also a strong penchant for analysis of the relationships between wars and geographical terrain.

There is far less factual detail in George's *Historical Geography* than in the Lucas works, and no recommendations for further reading. This may have to do with the market for which the book was produced. Though no specific information is given by the author in this regard, the publishers (Methuen) append at the end of the book a 'selection from the list of schools that have adopted *A Historical Geography of the British Empire*'. Presumably undergraduates also used the text, but it is quite different from the colonial handbooks provided by Lucas.

Oxford imperial history: empires ancient and modern

A characteristic of Oxford imperial history, reflected in both historical geographies described above and consonant with its embrace of the theory of organic growth of the British empire, was the frequency of appeal and reference to the Roman empire, this especially strong in the writings of C. P. Lucas, who had also published a book on the subject – *Greater Rome and Greater Britain* – in 1912, in which he attempted to compare the differences and similarities between the Roman and the British empires. The circumstances of the contemporary appeal of such allusions to Roman republican precedents and contrasts in the late nineteenth and early twentieth centuries has been discussed by Betts, who points to the conjunction of the Royal Titles Act of 1876, which gave Queen Victoria the title 'Regina et Imperatrix', the parallels sought in Seeley's influential *The Expansion of England*, published in 1883, and the rise of classical studies in Britain's older schools and universities. In the latter case the influence of German scholarship on Roman history was particularly influential, and the message of the Roman experience was clearly instilled in the minds, for example, of Cecil Rhodes, Lord Curzon, and generations of imperial administrators. Contrasts were made, however;

> Throughout the salad years of imperialism, the common rhetorical defence of the British Empire was not based on analogies to the sword but rather on appeals to custom, race, to responsibility. Even with the advent of Social Darwinism ideas, force was admitted as necessary rather than as desirable in expansion. When comparisons were made between the ideologies of Rome and British Imperialism, it was frequently asserted that Rome had been tyrannical and exploitive, whereas Britain was generally humanitarian and commercial.[74]

The geographical contrasts between the territorial contiguity of the Roman empire and the scattered, oceanic, nature of the British empire were also pointed out in a number of these studies.

The use of the Roman allusion and metaphor was not, of course, confined to Britain: it was extensively used by the French, especially in relation to north Africa, by the Germans (usually with reference to the Holy Roman Empire), and the United States.[75] This general theme is also developed in Quinn's (1976) study of the influence of Roman and Renaissance humanist experience on the formulators of British colonial policy and experience in and beyond the Renaissance,[76] and in Ranger's analysis of the transition, in nineteenth- and early twentieth-century colonialism, 'from resort to classical precedent towards the deployment of the new sciences of man'.[77]

There is little evidence in the historical geographies of empire produced by Lucas and his co-authors of a move away from this type of classical precedent and a Seeleyan sense of historic imperial continuity towards a mode of analysis grounded in the views of the new social sciences: these were not part of their training and seem not to have crossed their paths, and they adhered tenaciously to the classical imperial models, which provided for them a firm justification for the historic growth of the British empire.

Ethnology, ethnicity, race and environment in the historical geographies of empire

The historical geographies of empire reflect two significant characteristics of mid-Victorian geography and anthropology: the idea of the natural inequality of humans and the tendency to make broad 'unscientific' (though increasingly thought to be scientific) generalisations about ethnic groups, characteristics sustained in the later nineteenth century by expansion of European colonial possessions and the increasing interest of the middle class in such questions.[78]

The question of the influence of climate on the possibility of white settlement figures in these writings. The stronger evidence comes from George, who constantly reviews the prospects for white settlement in tropical and equatorial regions, though giving clear evidence of the uncertainty of scientific opinion on the subject. In a section on South Africa in his *Historical Geography of the British Empire* he makes a statement which, although balanced in one sense, indicates his broader view on the question of racial superiority on the other:

> The history of colonisation contains instances of native inhabitants being displaced by European settlers, under every variety of moral conditions. Sometimes they were cheated, sometimes dispossessed by force. Sometimes they welcomed the Europeans, and more often than not repented later of their confidence. In few cases could the colonies be justified except on the principle that civilised man is not bound to let the fairest portions of the earth be wasted on savages – a theory which can be defended in this general form, but which is obviously liable to abuse in specific instances.[79]

His most strongly ethnocentric observations concern the aboriginal populations of Australia, the Maori of New Zealand, and the Chinese.

> Australia is, as a matter of fact, a white man's country. The aboriginal blacks have never shown any aptitude for civilisation. They simply die out before the white man, and are comparatively few in number: in Tasmania

indeed they are entirely extinct. There is not, and now never can be, any half-bred race, arising from intermixture between the original natives and the white newcomers.[80]

Proceeding to the Chinese in Australia:

They are industrious and willing to work for small pay, which makes employers welcome their advent wherever labour is scarce. But in most other respects they are undesirable members of a community. The most zealous opponents of socialism [including, presumably, himself] would, however, agree in thinking that in a country inhabited by white European Christians, the fewer Chinese the better.[81]

Of the Maoris of New Zealand:

Though far higher in the scale than any other natives of the southern hemisphere, and having among them many individuals who rise to a high level of education and intelligence, they seem unable, as a race, to assimilate civilisation. Like the red men of North America, though at a slower rate, the Maoris are apparently destined at no very distant date to disappear.[82]

The Lucas historical geographies adopt a more cautious and even-handed tone. In the volume on the West Indies, for example, mention is made of white bond-servants in the seventeenth century being less valuable to the planter than the black slave, partly 'as not being so well fitted to work under tropical suns.'[83] In a later section dealing with slavery and the slave trade, however, while condemning the treatment of the slaves by the slave owners, Lucas makes the curious statement that

The great mass of literature on the subject of slavery naturally and rightly deals in the main with the position and sufferings of the slaves. From the point of view of colonization, however, it is still more important to consider the system as affecting the slave owners. Leaving the moral side of slavery out of sight, it is conceivably arguable that under certain extraordinary circumstances it may be good for a man to be a slave, but it is not conceivable that under any circumstances it can be good for man to be a slave-holder; in reviewing British colonization in the West Indies, it is not an unfair statement of the case to say that here British colonists were, by coming to a hot climate in a half-civilized age, with a special industry [sugar] presenting itself to them, and with another continent, so to speak, offering a plentiful labour supply, doomed to become slave-holders?[84]

Later he says, of the determination of the economy of British plantations in the West Indies, that: 'A tropical climate has much to account for, and geography fashions history. Heat proved too much, but only just too much, for British colonization, Africa was rather too near to America, and sugar was too remunerative and grew too well.'[85]

While Lucas in his other writings consistently identifies and generally criticises the experiences of prejudice based on colour and race, he nonetheless strays on occasions towards more than a hint of supporting theories of racial superiority and of climatic determinism. Thus, in the chapter (VII) on 'Class, colour and race' in his *Greater Rome and Greater Britain*, he speaks of what he calls the 'colour question' in the following terms:

so far as it concerns the relations between England and her dependencies at the present day . . . the feeling on the subject is not merely the result of prejudice, but the result also of practical experience. In other words, colour prejudice is one thing, and what may be called colour discrimination another. The white man may be, and usually is, prejudiced against the coloured man, because he himself is white, while the other is coloured; and the prejudice is probably mutual. But the white man, or at any rate the Englishman, also finds more rational ground for discrimination, in that the qualities, character, and upbringing of most coloured men are not those which are in demand for a ruling race, and are not, except in rare individual cases, eliminated by education on the white man's lines. The same discrimination is made by coloured races themselves, or some of them.[86]

It is interesting to compare this with a later work by Lucas – his book, *The Partition and Colonization of Africa*, published in 1922 and based on lectures given by him at the Royal Colonial Institute in 1921 for a study circle of teachers of the London County Council. It inevitably has a long section of the period before the scramble for Africa in the late nineteenth century, but concludes with an interesting and revealing chapter on the effects of the Great War on the map of Africa. Towards the end of the chapter, Lucas gives a balanced and wholly reasonable prospectus for the future of relations between black and white, especially in east and south Africa, favouring the idea of encouraging among the indigenous inhabitants the growth of products for export, permanent cultivation, land ownership and self-employment. He highlights the dilemma of differing views on the question of colour and labour:

The traditional British view is that in principle colour should be no bar to equality. But there is another British view, which has developed with the advent of labour democracy, that coloured labour shall not compete with white labour . . . All that can be said is that we cannot substantially go back, and most certainly ought not to go back if we could. The right view of Africa and the Africans is not to regret that Europeans came in, but to deplore that, having come in, they were guilty of so many abuses instead of shouldering their rightful job, which is to be trustees of the black men until in some distant future (if ever) the black men have become able to stand by themselves.[87]

The whole question of Victorian and Edwardian views of Britian's role in empire, particularly as related to the governance and, as some saw it, the naturally endowed or limited economic, social and political capabilities of the indigenous inhabitants of the territories controlled, is extremely complex. There were many atempts, through anthro- ⨯ pological and philological studies, for example, to provide a scientific basis for assumptions of racial superiority and inferiority – scientific racism. From the middle of the nineteenth century, beliefs in the natural inequality of human beings and a widespread tendency to generalise about race and ethnicity were the two dominant features of Victorian racism.[88] Racism, initially sustained through a mixture of cultural prejudice and information from travellers' reports, from the 1880s onwards underwent a strengthening of its quasi-scientific basis by the expansion of European empires, and the increase in professional middle-class interest, some born of colonial experience, in the work of such societies as the Anthropological Institute, which also provided better means for the rapid diffusion of opinion. The adoption in a social context of Charles Darwin's theories of natural selection and competi- tion – Social Darwinism – strengthened the 'scientific' hand and also provided a further justification for imperialistic expansionism, con- sonant with 'natural' principles of territorial gain and the means to achieve this cnd.

Bolt sets the scene:

> During the Victorian epoch, with the advance of world trade and the improvements in world communications, as literacy increased, and travellers, geographers, settlers and anthropologists enlarged the sum of knowledge about distant peoples, there was a unique degree of popular interest about race and racial theories. This interest was in turn significant because of the importance of public opinion in Britain.[89]

The publications and debates of the learned societies established in mid-century, such as the Aborigines Protection Society (1837), the English Ethnological Society (1843), the ethnology section of the British Association for the Advancement of Science (1847), and the Anthropological Society of London (1863), and indeed the Royal Geo- graphical Society, are clear indicators of both the intensity of interest and the degrees of confusion. An imperial ethnocentrism prevailed which allowed Victorians 'to think in terms of a cultural hierarchy, in which Western civilizations occupied first place, followed by those of the East, and with the stagnant, technologically backward cultures of Africa and the Pacific at the bottom',[90] a view which was not very seriously challenged until after World War One. Recognition, however, must be made of the variations in the meanings and contexts of the term 'race' in nineteenth-century Europe, and of the existence of non-racist and anti-racist views.[91] Discussions of race by anthropologists were accentuated by T. H. Huxley's paper, on the classification of racial types through physical characteristics, to the Ethnological Society in 1870, and by the explorations in the Pacific from the mid-1870s to the mid-1880s, and of Africa from the mid-1880s, leading to renewed debates on monogenist/polygenist systems of racial evolution, and on the relationships between intelligence and perceived degrees of civilisation and biological and physical environmental endowments, highlighted in the 'nature versus nurture' question posed by Francis Galton (1822–1911), President of the Anthropological Institute from 1885 to 1889, and Honorary Secretary of the Royal Geographical Society in the 1860s.[92]

The Royal Geographical Society's meetings, recorded in its proceedings and subsequently its journal and also widely reported in the popular press, helped to feed the public's appetite for information on the varied natural and social characteristics of a wide range of far-off peoples. The question of human acclimatisation was also a matter of growing concern for Victorian and Edwardian science, medicine and geography, as was that of the commercial implications of the climatic limits of crops. Issues of tropical settlement and labour are touched upon by George and Lucas, and focus on the question of is now called tropical anti-acclimatisation, that is a pessimistic view about the health and demographic consequences of attempted white settlement and work in the tropics. While reviewing the labour problem in Queensland, George, for example, says:

> Recently the working classes of Queensland, being politically preponderant, have sought to exclude all black labour, as injurious to their own prospects, though they have shown no readiness themselves to undertake the tasks involved in sugar growing. Whether white men can really do the work in tropical heat is a question not yet conclusively answered: good judges say that they can but the amount of experience is small.[93]

The possibility of white labour carrying the sugar industry in the region would, he thought, give a

> reason for reconsidering many of the current opinions as to what the white races can do in the tropics. The grave doubt however presents itself, not whether they can but whether they will. Such work, even if not destructive to health, must needs be trying: it remains to be seen whether

white men in sufficient numbers can be induced to undertake it, so long as they can find other and less exhausting occupation elsewhere.[94]

Similar comments about the apparent unsuitability of tropical climates for white settlement are made about, for example, the eastern part of the Tanganyika Territory in British East Africa and of India outside the hill stations, the latter instance involving a climate that in modern times had rendered it 'virtually impossible for the present rulers to colonise the country. The governing classes can hardly bring up their children in India, even by the most lavish use of the hill stations.'[95] Initially related, from c. 1840, to anxiety about the health consequences of settlement in India, and the apparent evidence of white population decline in other areas, Livingstone has shown that 'By the 1890s acclimatization phobia had largely shifted its focus from India to tropical Africa', with many medical experts concluding that 'Climatically, agriculturally, and temperamentally, Africa was unsuited to European colonization',[96] a notion subsequently modified in relation to tropical highland areas, especially in East Africa. Debates on these issues are to be found in the publications of most of the major and provincial geographical societies and medical and scientific journals, and, writers on the geography of empire had absorbed them, rather uncritically, into their publications.

Linked to these issues is that of environmental determinism, which in broad terms sees human action as having been strongly controlled by the environment, but which in recent literature on the history of geographical thought has been linked in the late nineteenth and early twentieth centuries with Social Darwinism and the legitimation of a strongly racist imperialism.[97] There has been extensive critique of this perspective, partly on grounds of its naive abstract simplicity and idealism, in face of the complexity of the constitutions of many different discourses of colonialism, partly because of a focus on the ideas of major figures (men) rather than on the social and cultural constitution of geographical knowledge, and a failure to acknowledge the depth of the historic roots of environmental theory.[98] In the absence from most of the works on historical geography of empire with which we are concerned of the scholarly apparatus of extensive references or footnotes, and of major general statements on environmental influence, one can only surmise on the basis of some of the climatic influences and racial issues pronounced upon, that they are characterised by a weak determinism which seems not to have been extensively informed by contemporary debate, nor which in turn had any profound influence on other works on the historical and commercial geographies of empire. George's statement at the beginning of the chapter (II) on 'The general nature of geographical influences', in his *The Relations of Geography and*

History, confirms this view. There is a bold opening: 'No one will deny, however firmly he insists on believing in free will, that the destinies of men are very largely determined by their environment. Among the many influences covered by this very wide modern phrase, the most obvious, for mankind as a whole as distinguished from individuals, are the geographical'.[98] He goes on, however, to qualify the statement with reference to other factors, which are not specified, but leading to the view that 'in setting forth the geographical influences which have guided or modified history, it is necessary to guard against overstating its force'.[100] Lucas, although relating to such environmental questions as acclimatisation in the historical geographies, develops a more modern view in his later writings.

Further insights into Lucas's views on environmental questions are found in his paper 'Man as a geographical agency', originally delivered as a Presidental address to the Geographical Section of the British Association meeting in Australia in August 1914, and published in the *Geographical Journal* in the same year. His basic premise is clear and straightforward:

> I have no expert geographical knowledge, and am wholly unversed in science, but I am emboldened to try and say a few words because of my profound belief in geographical studies. I believe in their value partly on general grounds and largely because a study of the British Empire leads an Englishman, whether born in England or Australia, to the inevitable conclusion that statecraft in the past would have been better if there had been more accurate knowledge of geography. I am encouraged, too, to speak because the field of geography is more open to the man in the street than are the sciences more strictly so-called.[101]

He contends that with the recent advances in Polar exploration, the exploration and discovery side of geography is fading, but is being replaced by the descriptive role of geography in accounting for the many modifications to the earth's surface made by human beings, and cites and critically assesses some of the themes in G. P. Marsh's *Man and Nature* (1864), revised as *The Earth as Modified by Human Action* (1874). The principal themes which he addresses are: deforestation and afforestation, land drainage, irrigation, land, water and air transport, climatic change and modification, and the elimination of the friction of distance. In his analysis of the clearing of European and African forests and of the draining of the Dutch marshes and lakes and of the Fens of East Anglia he anticipates some of the themes of landscape change taken up later by historical geographers in Britain, H. C. Darby in particular. Concluding, he suggests that one of the greatest difficulties consequent on the overcoming of problems of distance is that of racism: 'The most difficult and dangerous of all Imperial problems at this moment is the colour problem, and this has been entirely created by human agency, scientific agency, bringing the lands of the coloured and the white men closer together.'[102]

What emerges from the above discussion is that Lucas and George represent different perspectives on critical aspects of imperialism and the way they are represented under the heading of historical geography, particularly regarding issues of race. Lucas's voice (and that of his collaborators), while reflecting accepted nostrums of his day, is always cautious and qualifying, George's is more confident and outspoken and clearly supportive of notions of racial superiority and the need for discrimination in matters of colonial administration and immigration policies. His is a militaristic and strategic perspective, Lucas's that of the balanced even-handedness of an experienced senior civil servant and modest individual.

The historical geographers, geography and the promotion of the British empire

There are many indications that, during the late nineteenth and early twentieth centuries, when British imperialism underwent a series of major crises, including diminution of her world trade status, the support of geography and history were actively sought and employed in promoting the ideals and practical advantages of empire. The case has been effectively and convincingly made in the literature, and need not be extensively repeated here. MacKenzie has detailed and analysed the chronology, ideologies and institutions involved in propaganda for empire, emphasising the Victorian 'ideological cluster (of monarchism, militarism and Social Darwinism] which formed out of the intellectual, national, and world-wide conditions of the later Victorian era, and which came to infuse and be propagated by every organ of British life in the period'.[103] He shows how many of the imperialist societies promoted the teaching of empire geography and history, a process accelerated by the Royal Colonial Institute (founded in 1861) in the 1880s and 1890s. The focus on educational propaganda for empire can be seen in the work of a wide range of departments and institutions, including the Colonial Office and the League of the Empire, in whose projects many of the authors of Lucas's *Historical Geographies* were involved. One of the significant projects of this period closely linked to geography was the 'one instance of formal, official, imperial propaganda'.[104] Initiated by the Colonial Office, namely the Colonial Office Visual Instruction Committee, promoted by M. E. Sadler and Sir Charles Lucas, established to produce printed lectures on and illustrative slides of the United Kingdom, first made in 1903. It later expanded to include materials on many parts of the tropical zones of empire, such as Mauritius, Sierra Leone, the Gold Coast, Southern Nigeria, Trinidad, British Guiana, Jamaica, and ten of the Provinces of India. The geographer Halford Mackinder, having provided a specimen lecture for the earlier United Kingdom project was also engaged to prepare the lectures on India, while photographic illustrations were prepared by A. Hugh Fisher of the *Illustrated London News*, whose travels were provided by free passage offered by the Royal Mail Line.[105] The promotion of empire through books, illustrative materials, and educational syllabuses was widespread, part of an education policy geared to cultural imperialism, the imperial curriculum and the materials produced providing what Mangan has called 'images for confident control'[106] and racial stereotypes used in imperial discourse.

The range of textbooks and periodical articles incorporating geographical perspectives into the promotion of imperial ideals was extensive, and involved many of the leading British geographers, especially the early twentieth century.[107] There are numerous parallel texts written by historians, and the engagement of geography and history with the promotion of empire is also evidenced in studies produced in other countries of the empire.

In a paper on 'Geography, determinism, and Empire, an Australian episode 1902–1912', M. M. Roche analyses the interests and contexts of J. Gregory's books *The Geography of Victoria, Historical, Physical and Political* (1903), and *The Imperial Geography for New Zealand Schools* (1905), together with P. Marshall's *Geography of New Zealand – Historical, Physical, Political and Commercial* (1905). These books, Roche suggests, reflect 'a confidence in geography as a socially important body of knowledge', and imperialist underpinnings to these works are also reflected in their perspectives of an environmental determinism whose associations of race and climate elevated the position of whites in colonial societies, in their stress on the importance of resource inventories, of minerals and farming potential for example, and in the consciousness of the relations between the geography and identity of white settler colonies within the empire.[108] Doubtless many other similar examples could be found for other regions of empire.

The purpose of this manipulation of the images of empire and the positions of its people through the presentation of particular racial images was 'to inculcate in the children of the British Empire appropriate attitudes of dominance and deference',[109] a form of cultural stereotyping to accommodate and justify a perception of the world and Britain's purpose in it, which went in fact well beyond the education of children, and included, one way and another, very large sectors of the population, young and old, and in widely different social classes.

Conclusion

The motivations for the production of these and other historical geographies were an amalgam of a desire to combine history and geography for pedagogic and propagandist promotion of empire, involving the incorporation of notions of racial superiority through environmental and biological influences, of the need to use the resources of empire to commercial advantage, and the need to maintain strong imperial links for military and political purposes at a time of increasing territorial and strategic rivalry between the major imperialist world powers. Concern at the growing international competition for territory and commerce and response to what seemed to them, however mistakenly, to have been Seeley's clarion call to historians to advance the cause of empire through the use of history, led the likes of Lucas and his Colonial Office and academic historian friends to emphasise the historical strength of foundation of empire in their writings and activities, and to promote the establishment of imperial geography and history in universities and schools, as for example in the promotion of the Imperial Studies Movement through the agency of the Royal Colonial Institute. Greenlee sums up their approach as that of a striving for a sense of imperial unity through a sense of historic organic growth by a group which 'merely put scholarly flesh on the bones of their own generation's imperial assumptions', giving to Seeley's concepts 'a new twist with the injection of the notions of organic evolution, inbred racial instinct, providential design, and optimistic fatalism'.[110] George, as evidenced by his writings on historical geography, seems to belong to a rather different tradition: on the one hand aware of developments at Oxford and the Royal Geographical Society, yet on the other seemingly adhering tenaciously to a militaristic image of empire which admitted no critique of the less acceptable canons of belief culturally constructed in the late nineteenth and early twentieth centuries.

Geography, in the sense in which they understood it as the background to historical events and as the basis of economic and strategic resources, was undoubtedly an important part of their academic weaponry, but seems only to a very limited extent to be reflective of the 'new' geography at Oxford and increasingly advocated by the Royal Geographical Society. George, as has been shown, knew Halford Mackinder at Oxford, but he died in 1910, before the initiatives for more geographical teaching and research had progressed far. Lucas himself was not a Fellow of the Royal Geographical society, nor it would seem of the Anthropological Institute, both of which, like other societies at the time, were important channels of communication for news of travel and exploration of the remoter parts of empire.[111]

There are two letters from Lucas to Keltie, the secretary of the Royal Geographical Society, in that society's archives, concerned with an invitation to Lucas to lecture to the Society on German colonies (which he did not accept), the detaining of the German geomorphologist Penck in London at the start of the Great War in 1914, and the commendation by Keltie of an individual for a post as interpreter in France,[112] but no indication that, unlike George, he was a member. It seems that the Royal Colonial Institute was his main institutional base, though his promotion of joint initiatives for the teaching of the causes of empire via history and geography, through both the Historical Association and the Geographical Association, show a sensitivity to the potential influence of a certain kind of geography.

What a scrutiny of the texts and contexts of these historical geographies demonstrates is the extent to which historical geography, geography and history were inextricably bound up in the complex belief-systems and culturally constructed contexts and discourses of empire. These cannot be linked to a single image of the nature of geographical or historical knowledge at the time, but offer particular perspectives which, on further research, will undoubtedly reveal contrasts with other works directed at the justification of British imperialism. We would be well advised, therefore, when constructing future research programmes in this field, to heed Driver's advice: 'Any viable attempt to place geographical knowledge within the discourses of colonialism must surely acknowledge that the "age of empire" was constituted in complex ways, culturally, politically, as well as economically.'[113]

Notes

1 H. B. George, *Historical Geography of the British Empire*, Oxford, 1st edition, 1904; citations hereafter are from the 7th edition, 1924.
2 Sir J. R. Seeley, *The Expansion of England*, London, 1883.
3 C. E. Carrington, *An Exposition of Empire*, Cambridge, 1947, p. 85; T. R. Reese, *The History of the Royal Commonwealth Society*, London, 1968, p. 15.
4 W. G. V. Balchin, *The Geographical Association: the first hundred years 1893–1993*, Sheffield, 1993.
5 J. S. Keltie, *Geographical Education. Report to the Council of the Royal Geographical Society*, London, 1885, p. 19.
6 D. R. Stoddart, 'The RGS and the "new" geography: changing aims and changing roles in nineteenth century science', *The Geographical Journal*, 146 (1980), p. 192.
7 *Ibid.*, p. 197.
8 D. I. Scargill, 'The RGS and the foundations of geography at Oxford', *The Geographical Journal*, 142 (1976), p. 442.
9 C. Parker, *The English Historical Tradition since 1850*, Edinburgh, 1990, p. 2.
10 *Ibid.*, p. 2.
11 *Ibid.*, pp. 2–3.
12 P. J. Bowler, *The Victorians and the Past*, Oxford, 1989, p. 51.
13 B. L. Blakely, *The Colonial Office 1862–1892*, Durham, North Carolina, 1971, p. 89.

14 R. B. Pugh, 'The Colonial Office, 1801–1925', in E. A. Benians, J. Butler, and C. E. Carrington (eds), *The Cambridge History of the British Empire, III. The Empire–Commonwealth, 1870–1919,* Cambridge, 1967, p. 768.
15 J. F. C. Harrison, *A History of the Working Men's College 1854–1954,* London, 1954.
16 C. P. Lucas, *The British Empire: six lectures,* London, 1915.
17 Pugh, 'The Colonial Office', p. 766.
18 C. P. Lucas, *Introduction to a Historical Geography of the British Colonies,* Oxford, 1887, p. v.
19 *Ibid.,* pp. 118–19.
20 *Ibid.,* p. 126.
21 C. P. Lucas, *Historical Geography of the British Colonies, vol iv. South and East Africa, Part I, Historical,* Oxford, 1897, preface.
22 H. E. Egerton, *Historical Geography of the British Colonies, vol v. Part II Historical,* Oxford, 1908, p. iii.
23 C. P. Lucas (ed.) *An Historical Geography of the British Colonies,* Oxford: vol. I, *The Mediterranean and Eastern Colonies* (C. P. Lucas, 1888, 2nd edition, revised by R. E. Stubbs, 1906); II, *The West Indian Colonies* (C. P. Lucas, 1890, 2nd edition, revised by C. Atchley, 1905); III, *West Africa* (C. P. Lucas, 1894, 3rd edition, revised by A. B. Keith, 1913); IV, *South and East Africa* (C. P. Lucas, 1897, *Part I – Historical,* by C. P. Lucas, new edition, *Historical,* to 1895, by C. P. Lucas, 1913; revised, as Part II, *Historical to the Union of South Africa,* 1915; 1897, *Part II – Geographical,* by C. P. Lucas and H. Lambert, with input by H. H. Johnston; revised, as Part III, *Geographical,* by A. B. Keith, 1913); V, *Canada* (Part I, 1901, *New France* by C. P. Lucas, 2nd edition, 1916; Part II, *Canada – Historical,* by H. E. Egerton, 1908, 3rd edition, *Historical, 1763–1921,* 1923; Part III, *Geographical,* by J. D. Rogers, 1911; Part IV, *Newfoundland* by J. D. Rogers, 1911); VI, *Australasia* (1907, by J. D. Rogers, Parts I and II, 2nd editions, 1925), and VII, *India,* by P. E. Roberts, 1916, *Part I–Historical, to the end of the East India Company,* 1916, and *II, History under the Government of the Crown,* 1920.
24 P. Sutcliffe, *The Oxford University Press: an informal history,* Oxford, 1978, p. 86.
25 'F.R.C.', *The Geographical Journal,* 44 (1914), pp. 90–1.
26 *Ibid.,* p. 91.
27 C. P. Lucas, *Historical Geography of the British Colonies, vol iv. South and East Africa, Part I. Historical,* Oxford, 1897, preface.
28 I am grateful to Professor J. M. Cameron for this information.
29 Lucas, *South and East Africa,* 1897, p. 105.
30 D. Lorimer, 'Theoretical racism in late-Victorian anthropology 1870–1900', *Victorian Studies,* 31 (1988), p. 415.
31 C. P. Lucas, *Historical Geography of the British Colonies, vol iv. South and East Africa. Part I, Historical,* Oxford, 1897, p. 339.
32 *Ibid.*
33 C. P. Lucas (ed.) *An Historical Geography of the British Colonies,* Oxford: vol. I, *The Mediterranean and Eastern Colonies* (C. P. Lucas, 1888, 2nd edition, revised by R. E. Stubbs, 1906), pp. 1, 3.
34 C. P. Lucas (ed.) *An Historical Geography of the British Colonies,* Oxford: V, *Canada* (Part III, *Geographical,* by J. D. Rogers), 1911, pp. iii, v.
35 *Ibid.,* pp. 87–8.
36 *Ibid.,* p. 197.
37 C. P. Lucas (ed.) *An Historical Geography of the British Colonies,* Oxford: VI, *Australasia* (1907, by J. D. Rogers, *Parts I and II),* p. iii.
38 *Ibid.,* pp. iv–v.
39 Anon., *Scottish Geographical Magazine,* 3 (1887), p. 607.
40 Anon., *Scottish Geographical Magazine,* 10 (1894), p. 272.
41 'R.F.C.', *The Geographical Journal,* 46 (1915), pp. 231–2.
42 Anon., *Scottish Geographical Magazine,* 5 (1889), p. 109.
43 A. J. Christopher, *The British Empire at Its Zenith,* London, 1988, p. 14.
44 R. C. Harris and J. Warkentin (eds), *Canada before Confederation,* Oxford, 1974, p. v.

45 W. G. V. Balchin, *The Geographical Association: the first hundred years 1893–1993*, Sheffield, 1993, p. 21.
46 C. P. Lucas, 'Islands, peninsulas and empires', *The Geographical Teacher*, 10 (1920), pp. 126–30.
47 C. P. Lucas, 'The place names of the empire', *The Geographical Teacher*, 10 (1919–20), pp. 192–200.
48 C. P. Lucas, 'The islands and the empire', *The Geographical Teacher*, 12 (1923–4), p. 164–71.
49 Lucas, 'Islands, peninsulas and empires', p. 129.
50 E. A. Freeman, *The Historical Geography of Europe*, London, 1881.
51 Lucas, 'Islands, peninsulas and empires', p. 130.
52 Lucas, 'The islands and the empire', p. 165.
53 *Ibid.*, pp. 165–6.
54 *Ibid.*, pp. 170–1.
55 Sir J. R. Seeley, *The Expansion of England*, London, 1883.
56 J. G. Greenlee, ' "A succession of Seeleys": the "Old School" re-examined', *Journal of Imperial and Commonwealth History*, 4 (1975–6), p. 268.
57 R. Symonds, *Oxford and Empire. The last lost cause?* London, 1986, p. 52.
58 Greenlee, ' "A succession of Seeleys" ', p. 269.
59 Symonds, *Oxford and Empire*, p. 52.
60 F. Madden, 'The Commonwealth, Commonwealth history, and Oxford, 1905–71', in F. Madden and D. K. Fieldhouse (eds), *Oxford and the Idea of the Commonwealth*, London, 1982, p. 11.
61 *Dictionary of National Biography*, Oxford, 1912, pp. 97–8.
62 B. W. Blouet, *Halford Mackinder: a biography*, College Station Texas, 1987, p. 21–2; Symonds, *Oxford and Empire*, p. 145.
63 C. Oman, *Memories of Victorian Oxford and of Some Early Years*, London, 1941, p. 108.
64 Blouet, *Halford Mackinder*, p. 22.
65 D. I. Scargill, 'The RGS and the foundations of geography at Oxford', *The Geographical Journal*, 142 (1976), p. 442–3.
66 H. B. George, *The Relations of Geography and History*, Oxford, 1901; 4th edition, from which citations in this chapter are made, Oxford, 1910, p. iii.
67 *Ibid.*, p. 95.
68 H. B. George, *A Historical Geography of the British Empire*, London, 1904 (citations from 7th edition, 1924), p. v.
69 George, *A Historical Geography of the British Empire*, p. vi.
70 *Ibid.*, p. 1.
71 *Ibid.*, p. 2.
72 *Ibid.*, pp. 6, 8.
73 *Ibid.*, pp. 3–4.
74 R. F. Betts, 'The allusion to Rome in British imperialist thought of the late nineteenth and early twentieth centuries', *Victorian Studies*, 15 (1971), pp. 153–4.
75 Betts, 'The allusion to Rome', p. 159.
76 D. B. Quinn, 'Renaissance influences in English Colonization', *Transactions of the Royal Historical Society*, 5th Series, 6 (1976), 73–94.
77 T. O. Ranger, 'From humanism to the science of man: colonialism in Africa and the understanding of alien societies', *Transactions of the Royal Historical Society*, 5th Series, 6 (1976), p. 116.
78 D. Lorimer, 'Theoretical racism', p. 428.
79 George, *Historical Geography of the British Empire*, p. 270.
80 *Ibid.*, p. 188.
81 *Ibid.*, pp. 133–4.
82 *Ibid.*, p. 198.
83 C. P. Lucas, *Historical Geography of the British Colonies, II, The West Indian Colonies*, Oxford, 1890; citation is from the 2nd edition, revised by C. Atchley, 1905, p. 47.

84 *Ibid.*, p. 69.
85 *Ibid.*, p. 70.
86 C. P. Lucas, *Greater Rome and Greater Britain*, Oxford, 1912, p. 99.
87 C. P. Lucas, *The Partition and Colonization of Africa*, Oxford, 1922, pp. 206–7.
88 Lorimer, 'Theoretical racism', p. 428.
89 C. Bolt, *Victorian Attitudes to Race*, London, 1971, p. 217.
90 *Ibid.*, p. 27.
91 M. Adas, *Machines as the Measure of Men. Science, technology, and ideologies of Western dominance*, Ithaca and London, 1989, p. 339.
92 Lorimer, 'Theoretical racism'.
93 George, *Historical Geography of the British Empire*, p. 189.
94 *Ibid.*, p. 190.
95 *Ibid.*, pp. 213–14.
96 D. Livingstone, 'Human acclimatization: perspectives on a contested field of inquiry in science, medicine and geography', *History of Science*, 25 (1987), p. 372.
97 R. Peet, 'The social origins of environmental determinism', *Annals, Association of American Geographers*, 75 (1985), pp. 309–33.
98 See, for example, J. Kay, 'Commentary on "The social origins of environmental determinism" '. *Annals, Association of American Geographers*, 76 (1986), pp. 275–83; J. M. Hunter, 'Commentary on "The social origins of environmental determinism" ', *Annals, Association of American Geographers*, 76 (1986), pp. 277–81; R. Peet, 'Reply: or confessions of an unrepentant Marxist', *Annals, Association of American Geographers*, 76 (1986), pp. 281–3.
99 George, *The Relations of Geography and History*, p. 7.
100 *Ibid.*, p. 8.
101 C. P. Lucas, 'Man as a geographical agency', *The Geographical Journal*, 44 (1914), p. 477.
102 *Ibid.*, p. 492.
103 J. M. MacKenzie, *Propaganda and Empire. The manipulation of British public opinion, 1880–1960*, Manchester, 1984, p. 7.
104 *Ibid.*, p. 162.
105 *Ibid.*, pp. 162–5.
106 J. A Mangan (ed.), *The Imperial Curriculum. Racial images and education in the British colonial experience.* London, 1993, p. 8.
107 Examples include: L. W. Lyde, *Commercial Geography of the British Empire* (1922); M. I. Newbigin, *British Empire beyond the Seas: an introduction to world geography (1914)*; J. Hewitt, *Geography of the British Colonies and Dependencies, Physical, Political, Commercial and Historical* (1894); F. D. Herbertson, *The British Empire* (1910); C. B. Thurston, *Economic Geography of the British Empire* (1916); A. J. Herbertson and R. L. Thompson, *Geography of the British Empire*, (1924); A. J. Berry, *Britannia's Growth and Greatness: an historical geography of the British Empire* (1913); W. R. Kermack, *The Expansion of Britain from the Age of Discoveries: a geographical history* (1922).
108 M. M. Roche, 'Geography, determinism, and empire. An Australian episode 1902–1912'. Paper presented to the New Zealand Geography Society/Institute of Australian Geographers conference, Auckland, 1992, p. 12. I am grateful to Dr Roche for providing me with a copy of his manuscript, which is to be published in the conference proceedings.
109 J. A. Mangan (ed.), *The Imperial Curriculum. Racial images and education in the British colonial experience.* London, 1993, p. 6.
110 Greenlee, ' "A succession of Seeleys" ', p. 279.
111 Lorimer, 'Theoretical racism', p. 407.
112 RGS Archives, Correspondence 1911–20, Letters, C. P. Lucas.
113 F. Driver, 'Geography's empire: histories of geographical knowledge', *Environment and Planning D: Society and Space*, 10 (1992), p. 27.

Acknowledgements

Helpful comments on drafts of this chapter have been made by Dr Morag Bell, Dr Mike Heffernan, Professor John MacKenzie, James Ryan, and Professor James Cameron. Particular pieces of information were kindly provided by Professor W. G. V. Balchin, Miss J. Legg of the Geographical Association, Miss J. C. Turner (assistant librarian) and Mrs P. Lucas (archivist) of the Royal Geographical Society.

CHAPTER EIGHT

'Citizenship not charity': Violet Markham on nature, society and the state in Britain and South Africa

Morag Bell

Introduction

The end of the nineteenth century witnessed a marked change in the position of women in British society. The empire played a contradictory role in this process; at once reinforcing traditional notions of female domesticity and assisting many women to cross the apparently unbridgeable divide between their private, domestic world and the public domain. Studies devoted to the newly fashionable genre of women travellers and travel writing illustrate this point.[1] Many travellers regarded themselves as honorary men rather than promoters of the feminist cause. Nevertheless, when added together, their intriguing individual stories offer a formidable challenge to the more conventional masculine images of empire. They demonstrate that the geography of women's overseas journeys extended far beyond those territories deemed in scientific and political circles to be appropriate for women. Survival in the wilds of untamed nature and the publication of often controversial travel narratives won them critical acclaim as enterprising, if eccentric, pioneers. As publicists for Africa they addressed the Geography division of the British Association for the Advancement of Science, became members of the Scottish Geographical Society and campaigned for access to the restricted space of the Royal Geographical Society.

Studies which seek to reinterpret the relations between geography, gender and empire would benefit from a shift away from women travellers as a category, to specific regions in their imperial setting and to the participation of particular women in these regions. This geographical approach offers scope to investigate the links between women's identities and their representations of the colonial realm. Close attention can be given to women's contribution to the recursive linkages between Britain and particular territories, to the methods used to express their views and to their influence upon imperial thought and practice in Britain.[2] Whilst British women were part of imperial history, they were also part of a broader cultural history, the middle-class search for personal liberty, public power and professional status. By combining people and place there is scope to explore in detail how women's geographical knowledge of empire contributed to the achievement of this broader goal.

Until her departure for South Africa in 1899 at the age of twenty-seven years, Violet Rosa Markham (1872–1959) was little known beyond her East Midlands home in Chesterfield. Domestic 'duty', the religion of the Victorian spinster, and social work in the slums of Chesterfield, preoccupied her. On return to Britain four months later, her public career was profoundly influenced by this colonial encounter. Markham represented one of the early twentieth-century progressives who sought to redefine the role of the state in society. In her view, the development of Britain's overseas empire and municipal reform at home were intertwined. Markham became known for her commitment to the conditions of labour and employment both in Britain and overseas, and combined professional public service with writing, public speaking and international travel (Figure 8.1). By 1913 she had written three books and numerous articles on South Africa, corresponded with leading statesmen and women, become active in societies linked both to South Africa and to public work in Britain and had joined the fellowship of the Royal Geographical Society (RGS).

Markham's life and work illustrate the relations between ideas of progress in Britain and the colonial periphery. Her public life spans the first fifty years of this century. It offers a commentary not only upon the national and imperial values of late-Victorian and Edwardian England but also the internationalism of the inter-war years and the twilight of empire after World War Two.[3] Her career evolved at a time when Britain's industrial supremacy was being challenged by Germany and the United States. A demographic and economic shift was taking place within Britain from the northern coalfields to the south east and the metropolis. In response to the problems of urbanisation, initiatives in municipal government sought to promote principles of urban welfare and social citizenship. In the wake of recession and social unrest at home, a Greater Britain was being promoted overseas through emigration. Within this political and social environment, Markham built a public career; first, by her published geographies of South Africa, a territory deemed to be ideal for empire settlement; second, by the wide social and geographical network which she developed.

Attention will focus here on the assumptions about the relations between environment and society on which Markham's activities were

Figure 8.1 Violet Rosa Markham, Deputy Chairman, Assistance Board, in 1938

based. As social moralist, she initially believed in the environmental resolution of social problems. As a committed nationalist, her fears and aspirations for Britain were played out in her interpretations of South Africa's landscapes. In formulating her views, Markham drew upon a range of intellectual and political debates on race and class differences. She appealed to the supposedly universal laws of nature and the progressive skills of science and technology to achieve social harmony. But this enthusiasm was tempered by a concern over the moral degeneration of urban industrial squalor and affluence. As social activist, Markham's public duties were initially divided between municipal reform in Chesterfield and national imperial interests in London. By the end of World War One, experience at home and abroad had profoundly altered her views of society-environment relations. She had become an advocate of active government in the service of radical social change while her work for the League of Nations reflected a commitment to international social justice and a common humanity. Contemporary representations of South Africa in Britain provide the context for Markham's early life and work.

Emigration, settlement and patriotism in South Africa

In the years surrounding 1900 exploration and informal control in Africa were being replaced by formal rule. The presiding metaphors were conquest and development. Travel through the continent could be combined with new opportunities for civilian emigration to a formal settler community; a timely solution to economic problems at home and an opportunity to extend and enhance Britain's 'progressive' influence in a Greater Britain overseas.[4] This more interventionist imperialism also coincided with important changes in the position and aspirations of many women in Britain.[5] Enjoying a measure of freedom and independence in British society not available to the majority, middle-class women with intellectual curiosity, resources and enterprise formed part of a new generation for whom experience in Africa and in Britain became intertwined. Facilitated by improvements in steamship technology and advances in medical knowledge, overseas journeys were becoming less tedious and hazardous than formerly.[6] Thus paradoxically, while the 'civilising' role of British wives in colonial Africa had just begun, a prominent minority of women were entering into public debate over Britain's engagement with her new territories. For these women, experience in the colonies became intergrated with, and sustained, their personal interests and public activities in Britain. They contributed to what Pratt[7] has described as the process of transculturation between metropolis and periphery.

South Africa was quite distinctive in the opportunities and challenges it presented. It was hardly part of the Dark Continent. Close associations with Europe long predated those of the interior to the north. By the early nineteenth century Cape Colony had become implicated in the politics of the European state system. The Cape was of strategic importance to Britain en route to India. With the transfer of power from the mercantilist Duch East India Company in 1795, and formal acquisition of the territory at the Congress of Vienna in 1814, the industrialising and expansive British government encouraged immigration to the eastern frontier of the colony. The Albany Settlement, established at the Cape of Good Hope during the 1820s, represented one of the early planned settlement schemes in Africa.[8] The region drew missionaries, traders, speculators and foreign economic investment.[9] It also attracted women settlers. These emigrant communities prided themselves on being less formal and more egalitarian than the British at home.[10] But in seeking to attract rich and skilled immigrants, it was the similarities with Britain which were emphasised.[11] Cape Colony was deemed to be neither distant nor exotic but, in the words of one woman settler, simply an extension of 'our own country, Britain'.[12]

By the end of the century, through the influence of prominent personalities like Cecil Rhodes, South Africa had become a major centre of primary production in Africa.[13] Its image as an 'incomparable Eldorado' was not, however, matched by emigrant numbers. In 1902 it was reported that:

> South Africa is not, and never has been, a favourite field for British emigration. The whole white population, English and Dutch, of Cape Colony and its dependencies was no more than 376,987 in the year 1891, and that of Natal in 1898 was 53,688 – together less than the population of Birmingham.[14]

Nevertheless, in the period of reconstruction following the South African War (1899–1902), the region was included in an intensive programme of female emigration to Britain's white settler states including Australia, Canada and New Zealand.[15] In the case of South Africa, this sponsored phase of gender-specific emigration formed part of a broad programme of social engineering promoted by Sir Alfred Milner to Anglicise the Boer Colonies of Orange River and Transvaal.[16] In its strongly racial flavour, this emigration became bound up with the eugenics movement and the survival of the British race overseas. Emigration of women to South Africa was 'a question of national importance'. 'Loyal British women as well as British men' were crucial in the peopling of the territory if it was 'to become one of the great self-governing colonies of the British Empire, warm in sympathy and attach-

[193]

ment to the mother country'.[17] Moreover, imperial motherhood necessitated the sound selection of women 'of good character, health and capability'.[18] Numerous articles in popular journals like *The Imperial Colonist* and *The Nineteenth Century* actively promoted the emigration of women 'possessed of common sense and a sound constitution' to this 'sunny, prosperous' land which offered physical and social freedom for the fettered British maid.[19]

South Africa could satisfy the personal interests and professional ambitions of many British women other than simply as mothers of the empire. Going abroad was becoming a feature of the young, clever and literate, and South Africa was frequently included on an international grand tour. Visitors came to the region for the purposes of short-term travel; it was an ideal environment within which to enjoy a vacation, to maintain family ties and, in view of the equable climate, to seek improved health.[20] Indeed here lay the fascination of South Africa, a landscape at once exciting and threatening but, in the case of Cape Colony, a climate and social environment remarkably familiar. For professional women, it offered scope to prove themselves in entrenched male spheres of scientific research and journalism. By providing a laboratory for systematic field investigation, work in this region could advance their academic disciplines and increase their professional status. South Africa attracted fellows of the Royal Geographical Society and, from 1893, the pages of *The Geographical Journal* featured articles on its archaeology, ethnography, physical and economic geography. A formal geographical community quickly established itself locally; the South African Geographical Society, formed in 1917, was the first in anglophone Africa.

Combining environmental similarity with Britain and difference, familiar and alien, this southern point of the continent was highly marketable at home. But, *fin-de-siècle* images of South Africa and the emotions which the region aroused, were particularly strident; it had become a battle ground for European conflict. For many groups in Britain, war with the Boers had a significance beyond the realms of economics and the voting rights of new immigrants.[21] South Africa was central to Britain's political and cultural imperialism. The imagery of popular art, poetry and fiction, as well as war reporting, reinforced this national myth.[22] Conflict in the region fired the public imagination at home, it dominated the affairs of statesmen and masculine heroes. Women too participated in the critical intellectual and popular debates which the region aroused. Whilst excluded from the arena of formal public power, Britain's imperial ambitions in South Africa offered an outlet and a focus for their own political and social agendas.

Until the 1890s Olive Schreiner, active both in Britain and South

Africa, expressed a faith in the apparently unique commitment of Britain to the preservation and diffusion of human freedom, the protection of weaker states and global peace.[23] Her subsequent disillusionment with what she saw as Britain's stark betrayal of these principles in Ireland, India and South Africa, in the interests of racial hegemony, led to an outspoken critique of both the empire and the South African War (1899–1902). Her condemnation was shared by many South African and British liberal, socialist and feminist contemporaries. In their capacity as official correspondents and political commentators for a new and buoyant popular press, British women could also make their views known. Florence Dixie, employed by the *Morning Post* to cover the first Anglo-Boer War, shared Schreiner's concern over Britain's treatment of the African population.[24] Flora Shaw (later Lady Lugard) also used journalism to express her views.[25] Commissioned by *The Times* to visit South Africa in 1892 as part of a tour of the dominions, her letters to the newspaper were published in book form at the special request of 'the most prominent public men in South Africa'.[26] One year later, Shaw became the first Colonial Editor of *The Times*.

From industrial East Midlands to Africa's 'inland wastes'

Violet Markham, like Flora Shaw and a growing number of women with personal experience of Britain's overseas possessions, used the popular press to offer public comment upon the country's imperial role. Markham was brought up in a middle-class family near Chesterfield. Her father, Charles Markham, was a local colliery owner. His younger daughter and fifth child was therefore familiar with the industrial landscapes of Derbyshire and grew up in a household where 'mining problems were daily discussed'.[27] Her maternal grandfather, Sir Joseph Paxton, was head gardener at Chatsworth and designer of the Crystal Palace.[28] Formally educated for only eighteen months at West Heath, a private school near Richmond, Yorkshire, Markham regreted the limitations of her liberal education.

> We were taught no science, no economics, no social history. I cannot remember that the problems of poverty was ever mentioned beyond an occasional perfunctory address on some charitable subject.[29]

In national politics, from an early age she was socialised at home in Liberal ideas. Her second brother, Arthur, was active in the Liberal Party and represented Mansfield in Parliament from 1900 until his death in 1916. Through discussions at home, Violet was acutely aware of contemporary imperial issues and, like many articulate middle-class women in Britain, enjoyed participating in domestic debate. But her family also reflected the Liberal split over empire which was apparent at this time. Whilst sharing her brother's interest in South Africa, its 'natural wealth . . . its political and racial problems', she challenged his anti-imperial position.[30] Nor was this commitment to empire weakened by her initial experience overseas.

Markham's first trip to empire was to Egypt in 1895. Poor health drove her to the 'exhilarating' climate of South Africa in the spring of 1899.[31] Following a convalescence of four months, she returned to Chesterfield seven days after the declaration of the South African War. She revisited Soth Africa in 1912 to restore her mental and emotional state following the death of her mother. Released thereafter from the family obligations which confined many women of her generation, in 1915 she married Lieutenant-Colonel James Carruthers, whom she had met on her 1912 visit. They had no children. Markham returned to South Africa in 1924 and on several subsequent occasions. She also travelled in Asia, North and South America. But it was the first trip to South Africa which 'had far-reaching consequences' for her life.[32] It transformed a genuine enthusiasm for a distant territory into a commitment of time and energy to South African affairs. Her decision to speak and write about the environment and society in the region and to foster a range of social networks between the two countries marked the beginning of a change in self-perception. Markham's view of herself, her sense of identity, subsequently crystallised into that of an active public figure; the politics of the public domain, including international politics, came to occupy a central part of her adult life.

But while for Markham the 'proper' sphere was no longer the domestic and conjugal, she cannot be described as one of the glamorous or heroic female personalities of the time. Neither feminist, intrepid explorer nor ardent campaigner for the rights of oppressed peoples, she had no intention of restructuring popular racist images of Britain's colonised peoples. Nevertheless, from her representations of South Africa we glimpse the complexity of her public identity – its multi-faceted national, race and class forms. Markham's publications offer the first and most immediate evidence. She wrote one of the many promotional books on South Africa published at the time of the Boer War. By 1913 she had produced two more based on material recorded in personal diaries during her visits. *South Africa: past and present*, published in 1900, and *The New Era in South Africa with an Examination of the Chinese Labour Question*, published in 1904, were written following her initial trip in 1899. Sensitive to the remarks of her Cape Town friends about English visitors and their volumes on African experiences, she had not intended to publish on the region. However, on return to Britain, she was aware of the controversies created by the

South African War and resolved to inform and enliven the debate by publishing material which would appeal to both popular and intellectual audiences. Sections of her books also appeared in leading national and provincial newspapers including *The Westminster Gazette* and *The Sheffield Daily Telegraph*. Markham's third book *The South African Scene*, was published in 1913.

Her writings on South Africa reflect a fascination with both the physical and cultural landscapes of the region. Markham's awe at the sheer scale of the continent shares much in common with Victorian travel accounts. She wrote that the 'desolation' of Africa held a 'strange charm' to the British traveller. The country was 'silent and barren'; the population seemed 'nil'. While at first sight the interior was 'dreary and monotonous', these 'inland wastes of Africa are grand and overpowering, magnificent in their desolation, all-compelling in their vastness'.[33] Markham recognised their poetic appeal compared to the intimate landscapes of Europe:

> All travellers know that strange and irksome feeling which is inspired after a time by the small perspectives of Europe (the English lane, the sunny French valley, the romantic Italian hillside). A desire awakens for more mental elbow-room, so to speak. And then Arctic fever, or veldt fever, or the 'East's a-calling', as Kipling has it, seizes one, and like a magnet the deserts of Africa or the snows of Greenland draw the wanderer back.[34]

The magnetic power of these vast natural environments 'where Nature rules alone', reflected Victorian romantic fascination with the sublime.[35]

But her texts were not intended as travel books. The relationship between author, text and audience was, in Markham's mind, clearly defined. She sought to promote the region among the public at large, the scientific and the political community, by outlining Britain's colonial responsibilities to transform its 'desolation'. Markham's history and geography of South Africa offered a defence of British interests; the region provided Britain with the physical and moral terrain within which to ameliorate its social problems at home and to fulfil its imperial obligations. At a time when emigration to South Africa was being actively promoted, Markham described her book of 1900 as one which offered geographical descriptions of the scenery and the principal towns for the benefit of 'the large number of people at present interested in different South African localities'.[36] But it also provided a 'carefully researched history' of conflict in the region since the seventeenth century and a diagnosis of the region's 'problems'. Thus, according to its author, the text combined an advocacy of Britain's right to colonise with a record of the context on which this authority was based.

Three issues dominating Markham's writings on South Africa are discussed here: European settlement; the South African war; white supremacy and 'race' relations. Of particular interest are the geographies of South Africa which her discussion constructs and on which her interpretation of these issues is based. A single theme underpins the landscapes she creates, the relationship between science, social progress and environmental transformation. Drawing on a range of contemporary intellectual and political debates, Markham's position on this crucial imperial theme shaped her attitude to South African society. It found expression in a particular and distinctive evaluation of the country's rural and urban landscapes.

Order and disorder: South Africa's landscapes of progress

Markham's ideology of empire was not a coarse or aggressive jingoism. Britain's domination overseas was justified when it conformed to an environmental and cultural model of progress grounded in apparently universal laws of nature. Mindful of links between race and place, her imperial vision depended first upon a physiological model of human adaptation. British intervention and settlement in South Africa was logical in view of the region's physical geography and climate in particular:

> Owing to the height above sea-level of the great central plateau of South Africa, owing also to the extreme dryness of the atmosphere, territories situated in tropical and sub-tropical latitudes are healthy and salubrious contries in which the children of Europeans can thrive and grow up.

Markham supported late nineteenth-century racial biology that tropical areas were inappropriate for permanent European settlement since children in particular, were unable to adapt:

> Climate is important not so much for its effects on adults as for its effect on infants. A country where children cannot live is necessarily a country without home ties and the question of home ties indirectly involves weighty considerations of state. The geographical conditions [of this region], however, place it among the category of what are known as 'a white man's country'.[37]

Under the influence of the contemporary science of human adaptation and survival the equable climate south of Capricorn fully justified Britain's expansion in the region.[38]

Beyond this, eurocentric preferences shaped her interpretation of the physical and social conditions. The 'monotony' of Africa's physical environment had little appeal. By contrast, approaching Table Bay for the first time from the sea, this 'land's end of the African continent' was surely 'one of the beauty spots of the world' (Figure 8.2). The 30-mile coastal strip to the west of Table Mountain was delightfully different from the 'unending stretches of veldt, karoo or bush' which typify the South African scenery.[39] It offered 'an enchanting combination of sea and mountain, wood and garden, rock and forest, sometimes impressive, sometimes alluring'.[40] Visible expressions of progress were the familiar, intimate and varied; a tamed physical environment on the European model. Within the urban setting of Cape Town, while noting the cultural mix of African, coloured and Malay peoples, Markham was neither surprised nor dismayed to find that the overwhelming impression was one of an Anglo-Saxon community. It was this group which drew her attention including their carefully planned and managed agricultural landscape of the hinterland surrounding Cape Town comprising temperate and sub-tropical fruit farms. Order in the landscape came not from uniformity but from the harmony and unity underlying its diversity. Describing the journey from Cape Town to Kimberley she noted that:

> This corner of the Dark Continent is a smiling land with vineyards, wheatfields and snug farmhouses surrounded by lofty trees. It is as though Nature had prepared a last peep of pastoral Europe for the homesick traveller before he plunges into the deserts of the north.[41]

Figure 8.2 'Cape Town and Table Mountain'

To Markham, British colonisers were honourable. They came not to exploit but to enlighten; to bring order out of chaos. South Africa marked the entry point from which Britain's 'good government', superior science and technology, as instruments of social justice, spatial organisation and economic policy, would transform the continent. The work of Sir David Gill, HM Astronomer, on the transcontinental triangulation was to be admired; so too Cecil Rhodes's enthusiasm for a transcontinental railway.[42] It was 'engineering skill' which allowed 'railway enterprise' to overcome the 'obstacle to progress provided by the barrier ridge of Hex River Mountains'.[43] Equally, it was irrigation which worked 'miracles in the desolate landscape of the Karroo bush . . . If water is turned on this waste of stones the desert blossoms like a garden'.[44] Some twenty years later, Markham noted how the steamship, railway and motor road also facilitated British tourism to a climate far 'better behaved' than the 'vaunted Cote D'Azur'[45] (Figure 8.3). But around 1900 there were 'threats' to this rational order and economic prosperity from conflict between British and Boers. An imperial

Figure 8.3 'The mountain drive'

nationalist spirit found expression in Markham's defence of Britain's responsibility in South Africa including the South African War.

Although personal motives first drew her to South Africa, Markham arrived during the critical months of negotiation prior to the outbreak of war (1899). Unlike the men of empire, freedom from the political and economic responsibilities of overseas residence and employment required no formal imperial duty. Observer and commentator, rather than participant in this British colonial culture, Markham enjoyed opportunities to analyse different opinions and to study dispassionately the 'problems' of the country.[46] But her position as both outsider and insider was ambiguous. An articulate visitor with means and political connections, she remained close to the imperial establishment throughout her stay. Letters of introduction to Sir Alfred Milner and the family of Colonel John Hanbury-Williams, military secretary to the Governor, and access to Government House and the Cape Legislative Assembly, enabled her to observe at close quarters debates at the centre of British political power in South Africa. It also confirmed an admiration for the 'men of destiny', with whom she came into contact including sharing the conviction of Joseph Chamberlain and Lord Milner in the symbolic and strategic importance of the region to the survival of Britain's African Empire.[47] Writing in later years Markham admits to having been 'a wholehearted, and it may be uncritical supporter of the British side in the South African War and of the policy followed by Sir Alfred Milner'.

But Markham's interpretation of South Africa's 'problems' was complex. Her imperial pride did not support a uniform condemnation of the entire Boer population which was widespread among the British at home and in South Africa at the time.[48] She respected the courteous and cultured South African Dutch, in spite of 'their anti-British prejudices', and admired the built environment of Boer settlements in the Cape based upon 'the architectural legacy of the beautiful gabled houses built by the Dutch in the eighteenth and early nineteenth centuries'.[49] This sophisticated, urban-based, social group, with their 'civilised' European material culture and arable economy were entirely acceptable; not so Boer pastoral farmers of the interior 'whose mentality . . . had changed little since the days of Van Riebeek'.[50] Their 'base' instincts derived, in part at least, from the physical environment to which they were confined:

> The more one sees of the South African secenery . . . the more compre-
> hensible become certain characteristics of the Dutch farmers who have
> inhabited this country . . . Natural features react powerfully on the
> disposition of a people . . . The vastness, the solemnity of Africa would
> press heavily upon the most enlightened of men. The tendency is to grow
> sad, introspective; to lose the sense of human worth and dignity, because

all dignity and greatness seem swallowed up by Nature herself. It is well for man in his pride of intellect that from time to time he should go out into the wilderness and humble himself before its great mysteries. We read that it was the habit even of our Great Exemplar. But a daily life in the wilderness, cut adrift from every noble and refining influence, makes man in time brutish.[51]

This desolate karoo in the interior, this wilderness beyond the control of Europe's social order, could offer only a temporary escape from the constraints of civilisation; otherwise degeneration would be the tragic outcome. For 'the first Boer settlers' with their 'ignorant and brutish nature', it was hardly surprising that:

> in the course of generations such an existence should have fostered instincts of savagery rather than civilisation among these emigrant peasants fom Northern Europe . . . No one denies that their characters are hardy and robust, but unfortunately it is a robustness purchased entirely at the cost of the nobler, the finer, the higher side of human feeling and sensibility.[52]

Markham's disdain for Boer pastoral farmers reflected growing differences within Europe between peasant society and an increasingly powerful arable economy combined with urban-based science and technology. Contempt was also tinged with resentment of their genuine determination to challenge 'British supremacy in South Africa and to "drive the English into the sea" '.[53] Writing in 1900 she claimed that 'The Boer is an anomaly in the progressive history of South Africa: a block in the way of liberty, righteousness and good government'.[54] National pride overruled Markham's non-violent morality; war was 'necessary and inevitable'.

But imperial power brought with it duties and responsibilities to subject peoples. Here Markham's faith in the order of science was not matched by unqualified faith in the inevitability of moral progress. On her departure form the Cape three days prior to the outbreak of war, Markham anticipated the racism based on colour which was later to underlie political domination. Writing in her diary, she achnowledged that the 'the real and abiding issue in South Africa' was 'the relationship of a minority of white people and a preponderating and ever-growing Bantu population'.[55] For sure, her views of this relationship reflected prevailing ethnocentric biases. As an advocate of evolutionary human development, it was right and inevitable that Britain should intervene in African affairs to assist their 'progress'. But Markham was also concerned with the manner of this incorporation. A supporter of pro-imperial humanitarianism, she rejected the cruder forms of racism which rationalised injustices against 'lesser' groups as natural and acceptable. This conviction found expression in her attitude to the industrial environment.

Industrialism, decadence and degeneration

Markham's familiar mining landscape of Derbyshire shared many features in common with the emerging industrial centres of the Witwatersrand. They sustained Britain's economic and productive power. But it was not merely the imperatives of economic growth which linked the two regions. Observations from her home environment that urban industrialism corrupted man's 'natural' morality, were confirmed by evidence of unrestrained industrial development in South Africa. The physical appearance of the 'unlovely cities of gold and diamonds . . . to which men hasten without thought or regret' reflected ugly labour relations.[56] Industrial paternalism, including humane intervention in the living and working environment of labour, was essential to ensure harmonious labour relations. But Markham's interpretation of South Africa's industrial landscape reflected a social morality in which national, class and colour difference interacted.

The Afrikaners' apparent lack of interest in the welfare of their workforce reinforced her dislike of them. Like her brother Arthur, she was particularly critical of Boer mine managers; notably, what she described as their selfish, narrow-sighted concern for commercial gain which took precedence over the humane treatment of their employees and brought 'discredit on the mine-owners at home'. In effect, race and class difference fused across the imperial divide. But the British racial record in South Africa was 'by no means a spotless one' either. Owners of capital were liable to a 'self-seeking ethic' which spawned human exploitation and corrupt industrial cities.

> Our treatment of the Free State as regards the Diamond Fields is one of the gravest blots on our colonial history . . . natives have been ill-treated by Englishmen and the Dutch despoiled by British traders. Some of the worst qualities we condemn in the Boers are to be found in Johannesburg. Avarice, greed, corruption, lack of honour and lack of principle are evils which have followed in the wake of the gold industry.[57]

Industrial and moral progress did not necessarily coincide. Markham was referring here to the dispute over the boundaries at what was to become Kimberley in the late 1860s and the ways in which the British arrogated the area to themselves.

Noting that 'the reflex action of Johannesburg makes itself felt in London', Markham was not uncritical of the new business wealth and the rise of an international metropolitan financial network at this time.

[203]

Like many groups in British society, she had little sympathy for the aristocratic families forced, through a fall in land values, to invest in 'new' money.[58] There was a pathetic irony about the noblemen with 'no business experience' who were compelled to, 'so to speak, capitalise their forbears and turn the Crusader in the family chapel into £ s. d.'. Markham was equally scornful of the speculators who sought to purchase the aristocratic lifestyle which they could not inherit.[59] She noted with disdain the alliances between speculator and needy aristocrat through which the former could gain access to the exclusive social networks of the latter.[60] By 'paying for a large house' the speculator might enjoy the 'social advantages' of entertaining 'other people's friends who are barely civil to him'.[61] Here Markham shared common ground with Beatrice Webb. In her schemes to promote technical education, Beatrice sought the support of 'public-spirited and scientific-minded' millionaires like Sir Julius Wernher, South African diamond magnate and joint owner of Wernher, Beit and Co. She neverthelsss despised the pretentious social climbing of these self-made Randlords and their ostentatious displays of wealth.[62]

Markham's views of the 'labouring classes' combined industrial paternalism with national pride. In assessing white 'working class' conditions in Johannesburg, her bias against Boer 'bad government and the denial of all political rights' fuelled a condemnation of the deplorable housing and lack of education which resulted in 'hundreds of children' growing up in the 'blackest' ignorance and crime.[63] Writing on indigenous labour in the mid-1870s, Anthony Trollope had seen Africans toiling in the diamond mines as 'growing Christians'.[64] In the 1880s, the more radical Florence Dixie had, by contrast, observed incipient revolutionaries 'fashioning out of the pit [diamonds] . . . to bring gain to the white man'. The money from their toil could be 'quickly converted into arms, ammunition . . . to wrest back . . . their own fair land'.[65] To Markham, like Trollope, indigenous wage labour could have a civilising influence; but moral order also depended upon a landscape of social control. Urban racial segregation was an essential feature of her industrial paternalism.[66] She noted with due approbation the model village built at Kenilworth, a suburb of Kimberley, for De Beers employees where no effort had been 'spared to bing the amenities of life within reach of the European staff'.[67] Equally, Markham's support for the 'benignant despotism' of De Beers under Cecil Rhodes, lay in the apparently 'kindly' rule which the organisation exercised over its 'own dependants'. In defence of the 'world-renowned compounds in which the natives live . . . unique and peculiar to the diamond fields', Markham's interpretation of racial difference rationalised this segregation as entirely natural. She argued that they 'are well cared for, and are the most good-tempered, merry-looking set of men . . . for when the native is ruled with justice and kindness he is as amenable as a big child'.[68]

In applying the concept, 'benign rule', to owner–working relations and the imagery of the 'native' as a 'child', Markham's moral order depended upon the extension of imperial 'good government' to the monopoly interests which controlled urban industrialism.[69] In her memoirs, some fifty years later, she wrote:

> Let me here protest that there was nothing base or bullying in that much-abused word 'Imperialism' as many of us conceived it at the time. To us Imperialism was based upon a fine enthusiasm for what is known as the English way of life and had in it a great deal of missionary zeal for the welfare of races who had no experience in managing their own affairs and were living under the rule of incompetent tyrants. We earnestly desired to promote British standards of justice and integrity and good government in such places.[70]

For Markham, and many of her liberal-minded contemporaries, an urban landscape characterised by humanely managed social and racial segregation symbolised the exercise of good government in the workplace. To ensure that such high standards were rigorously applied, she also agreed with a dissenting minority in Britain and South Africa, that the state had a duty to enshrine workers' rights in the law.[71]

Markham's assessment of South Africa's industrial cities further illuminates her rational 'progressive' views. Gold as the medium of international exchange, was a commercial 'necessity, not a luxury, in life' and therefore justified the 'mines, the machinery, the vast and complicated financial organisation' of Johannesburg.[72] Kimberley was very different. In this Diamond City the darker side of industrial progress found expression not only in poor labour relations but also in the degeneration of decadence. Of the city she wrote that it:

> owes it existence to no necessity more potent than that of a woman's vanity . . . a feminine love of display. For her and for her alone, exist the mighty ramifications of the De Beers Company; for her pleasure countless Kaffirs toil in the bowels of the earth.

Mindful of concerns in Britain over the corrupting influence of new wealth and luxury, Markham expected modest and prudent ethics from both British men and women.[73] Her condemnation of the feminine vanity and extravagance underpinning this particular form of urban industrial development was as strident as her dislike of the feminist cause at this time:

> Men have suffered, and endured, and perished, to place that tiara on her dainty head. A large town and a large population are visible emblems of her sovereignty. For it is the old, old story illustrated here at Kimberley –

what women ardently desire, the men who love them will search the ends of the earth to discover . . . It takes two or three days at Kimberley to see all the different stages through with the diamonds pass in their course of extrication.

She proceeded to warn Britain's feminists:

The spectacle of so many thousand human beings wholly engaged in a work which results from man's desire to gratify a woman's fancy is a very striking object-lesson. It is one apparently ignored by those worthy ladies who talk so loudly of the brutal treatment we endure at the hands of our lords and masters. I offer it respectfully to their consideration.[74]

In anticipating future changes in South Africa's cultural environment, Markham's belief in the principles of natural and moral law fused. The superficial values and unethical practices underpinning urban industrial development served to highlight the enduring qualities to be found in the country's future rural economy and society. Although Cape Town, Durban, Kimberley and Johannesburg were 'all more or less cosmopolitan . . . the future of a country or the greatness of a race depends upon other qualities than those peculiar to' large towns and mining centres. Industry and cities were transitory and corrupted natural human morality:

The mining interest is not the permanent interest in South Africa. In course of time both gold and diamonds will be exhausted. The land will remain . . . The commercial wealth of South Africa may come from Kimberley and Johannesburg: its moral stamina must be derived from other sources. In every age the soil has invariably turned out a finer race of men than the city.

But it was based not on the primitive pastoralism of Boer farmers; rather, on the British farming model of the Eastern Province:

The best blood of European emigration has been poured into the Eastern Province, and it is from these men and their descendants that the real leaven of the Anglo-Africander race must come.[75]

This faith in the progressive qualities of the arable economy also echoed the widespread British anti-industrialism when Johannesburg and Kimberley were being built.

Natural law dictated that South Africa's progress would continue to be European inspired. Markham acknowledged that in this region 'south of the Zambezi [where] the climatic and geographical conditions are favourable to black and white alike', it was demography linked to politics which lay at the base of future racial problems.[76] Differential rates of black and white population growth challenged European paternalism. The desire among many Africans for a western education might produce 'a good citizen' but it also threatened the franchise and white political control.[77] Nevertheless, by appealing to supposedly natural differences in the quality of races, future change to the white political landscape was unimaginable:

> The ethics of colonisation bluntly resolve themselves into the fact that a strong white race will always take possession of any land where its surplus population can flourish, quite irrespective of the wishes and feelings of the original inhabitants. History has proved [that] . . . the dogma of the original proprietorship of the soil must be dropped in consequence.

Social morality was ultimately subordinated to the laws of nature:

> We are sometimes able to perceive more or less dimly how the working out of a great natural law creates confusion among our more or less artificial systems of politics and morality. The doctrine of proprietorship is a very firmly rooted one in every society, but now and again it comes into collision with that natural law which decrees that the fittest alone shall survive and enjoy the fruits of the earth. Nature makes no *contract social*; she cares nothing for ethics of primogeniture; she cares for one thing only, the strong; and she ordains that her gifts are not for the first comer, but for the one who can make best use of them.[78]

Whilst accepting that in the hierarchy of races, the strong would take over the weak, Markham's moral order in social relations demanded that the strong should also support and assist lesser races to 'progess'. But her interventionist humanitarianism, her evolutionary development model, did not produce universal political equality. Notwithstanding the entitlement of lesser races to civil and workers' rights, the weak could never match the strong. In a country, 'the climatic conditions of which are equally favourable to black and white . . . the brutal law of the stronger forces the black to bow to the white man's will and adapt himself to the governing systems of the latter, or else perish off the face of the earth'.[79]

Markham's geographical knowledge of South Africa drew upon a range of contemporary debates on science, social progress and environmental transformation. The peoples of the world and the landscapes with which they were associated, conformed to a racial and geographic hierarchy derived from her commitment to Britain's science, technology and 'superior' humanity. Like many of her contemporaries, the problems of South Africa were explained by the relations within two social categories, nationhood (British/Dutch) and colour (white/black). Neither pacifist not egalitarian, her views on the right to rule in South Africa drew upon the biological deductions of Social Darwinism. She specifically acknowledged the work of Thomas Henry Huxley on the hierarchical ordering of nations and races.[80] But she avoided the use of monolithic social categories; the Boer population was not uniformly uncivilised. Nevertheless, in line with nationalist sentiment, Britain's political and cultural imperialism remained superior. Appropriate British interventions would inscribe a civilised social order on the country's savage wilderness.

But Markham's fears as well as aspirations for British society were played out in South Africa. Evidence could be found of the brutality and decadence of urban industrialism which she abhorred at home. The enduring qualities of a civilised society lay not in these corrupt industrial cities. Rather, Markham's social and environmental ideal conformed to the tamed, cultivated model of rural England. In the spirit of Edwardian nostalgia and the new ruralism which was sweeping Britain, these comfortable, intimate environments, both at home and abroad, embodied the finest qualities of British/English national culture.[81]

Text and audience

Markham reinforced Britain's hegemonic world vision and defined knowledge according to European intellectual traditions. But her writings not only responded to the intellectual and social context at home and abroad with which she was familiar. Like a growing number of her female contemporaries, the colonial encounter profoundly affected Markham's views of, and place within, British society. Following the initial journey to South Africa she records that it 'meant a great widening of experience to me. Boundaries were pushed back and new horizons glimpsed'.[82] In a letter to Sir David Gill dated 15 February 1901, she wrote, 'I can honestly say that it is South Africa alone which has ever made me long for the influence which comes from place and power'.[83] Although outside the sphere of formal political power and without the professional qualifications increasingly required in a new age of the expert, Markham's authority to represent the region and its peoples was based, in the first instance, upon personal experience and literary skills.

The content of her texts illustrates two sets of social and geographic relations; between author and audience in Britain and between writer with the power to represent societies beyond Britain and those she claimed to represent. In both instances she addressed a white, pro-imperial public. Markham notes that on her return to Britain, she plunged 'wholeheartedly into war work and did a good deal of speaking and writing in support of the British cause'.[84] In defending Britain's interest in South Africa, she confesses to having been closer to the Conservatives than many Liberals in these early days. Her first book was favourably reviewed in the national and provincial press. *The Spectator* commented that:

> Miss Markham's admirable contribution to our knowledge of South Africa deserves to be classed very high among the flood of publication to which the war and its allied problems have given birth.

As an informed and widely reviewed imperial commentator, Markham contributed to women's increasing public prominence. In contributing to a buoyant popular literature, her ability to gain recognition was facilitated by the already huge appeal of imperial themes across British society.[85] She also aimed to influence public opinion.

Through the inclusion of detail on race relations in particular, her writings sought to modify prevailing views on the propriety of Britain's racial record; an objective amply fulfilled, according to several press reviews. *The Spectator* drew particular attention to the informative 'chapters on the native question'. So too did *The Manchester Guardian* which noted that 'She contributes a good deal of really useful information and many judgements which are material evidence to be pondered.' The journal, *Truth*, echoed this assessment, a reflection of the manner in which Markham's social imperialism had become acceptable to radical publishers in this period. *Outlook* described Markham's book as:

> One of the clearest and most pleasantly written works that have come before us . . . Specially suggestive throughout are her observations on the question of the political relations of the black and white races.

Markham herself stressed that the chapters on the 'native question' were designed to inform 'the 19 out of 20 people [who] do not know that a colour problem exists in South Africa' and the twentieth who 'probably has but an academic knowledge of the fact'.[86]

Markham's third book, *The South African Scene*, was politely reviewed in *The Geographical Journal* in 1914,[87] a mark of her acceptance into this scientific community. In the review it was pointed out that:

> if her travel sketches do not teach geography formally, they give the reader a distinct and accurate picture of the general characteristics of the land and the people who dwell therein.

On the section, 'Some polices and problems, again the reviewer commended the author 'for her plain speaking on subjects often slurred over and misunderstood'. A year before this review Markham had been awarded an RGS fellowship. One of 140 women fellows elected in 1913, Markham was among a new generation of professional women, travellers and wives of colonial servants who participated in the work of the society. Sponsored by Gertrude Bell, a personal friend, and David

George Hogarth, archaeologist and arabist,[88] her nomination papers cited the following qualifications; 'author of *South Africa, Past and Present*, and interested in the geographical problems of South Africa'.

But publishing was not an isolated activity; authority to comment publicly on other societies depended upon more than personal skills. Markham's imperial nationalism and the activities which followed from it, including publication, fellowship of the RGS and public service, depended upon an extensive and diverse social and geographical network. This network, both formal and informal, offered the outlets though which her views and energies could be channelled and her position in the public domain secured. Two dimensions of this network are mentioned here, namely, Markham's South African connections close to Lord Milner and her Liberal Imperial associates for whom social reform at home and commitment to the empire were intertwined.[89] Through these personal and political relations she participated in imperial debate and in a range of social actions designed to promote social progress. They found a dual focus in Chesterfield, the parental home, and in London where she secured a second home following her return from South Africa.

Transnational networks: from race to class and gender

Under pressure from South African acquaintances, Markham participated in a range of pro-imperial women's associations and served on the London committee of the Daughters of the Empire Guild. Its South African affiliate, the Guild of Loyal Women, was established in 1901 to combat 'the teaching of not a few women' who had declared 'their undying hatred of Great Britain and their wicked resolve to bring their children to swear vengeance against her'. Re-creating imperial harmony appeared to be a natural role for loyal 'daughters of the British Empire'. At a meeting of the Guild in Bloemfontein on 2 February 1901, it was noted:

> What a glorious task for sweet womanhood the Guild sets its members! The pacification of South Africa, the closing of old wounds, the healing of all sores, and the bringing together of all nationalities in one common brotherhood.[90]

Markham declined the request of a prominent South African member of the guild, Mrs K. H. R. Stuart, to establish a branch in Chesterfield.[91] In her work for the London branch, she encouraged the formulation of 'a definite programme . . . some more practical aim than . . . mere loyalty to the Crown and a union of the colonies'.[92] This included promoting schemes for women's emigration as 'a practical exhibition of sympathy between the women of the Empire . . . One wants to make the individual women work; not set up an organisation with a lot of paid clerks and secretaries'.[93] Markham also supported the Victoria League. Founded in London in 1901 in memory of Queen Victoria, its object was to foster friendship throughout the British Empire, to alleviate conditions in the concentration camps of South Africa and among British war refugees.[94] The wives of successive colonial secretaries were also closely involved.[95] These included Mrs Alfred Lyttelton with whom Markham became acquainted in South Africa.

Experience in South Africa at this high stage of empire also sharpened Markham's focus on British society. Her subsequent career as public servant was profoundly influenced by May Tennant, fellow Liberal Imperialist and national campaigner for improvements in the conditions of labour.[96] They shared the conviction that Britain's imperial commitment to 'good government, welfare, education and health', should be properly respected at home. Unlike many in Britain who sympathised with the poor at home but not the exploited abroad, for Markham, moral, humanitarian and welfare objectives cut across national boundaries.[97] In this respect she anticipated the internationalism of many inter-war women and men.[98] In was in Markham's attempts to reshape the urban environment in particular, that her interests in race and class, national and civic pride came together. Committed to municipal improvement in health, housing and education, she sought to ensure that a moral urban landscape should exist not only in South Africa but also in her home town. Industrial Chesterfield and the slum conditions of mining communities awaited 'social betterment'.[99] To this end, Markham became active in local government, sitting on the Chesterfield Education Authority from 1899 to 1934. She was a town councillor in 1924 and mayor in 1927.

But the breadth of Markham's social actions illustrates the subtle and changing relations between government and society at this time. One of the early twentieth-century progressives who abhored the 'odious spirit of patronage' attached to charity, she pressured 'the state into doing its obvious and often neglected duty' in social service.[100] Markham was deeply resentful of those who retained a vested interest in perpetuating 'my poor' and condemned the manner in which it subsidised and perpetuated social evils like bad health, bad housing and low wages. On these matters she shared comon ground and conducted a correspondence with Beatrice Webb.[101] Seeking further practical expression for her views, on the advice of May Tennant, Markham invested time, financial resources and energy into building a Settlement. Funded through her personal legacy and constructed in 1902 in a part of Chesterfield 'where civilising influences are few and far between', it followed the teaching of

John Ruskin and Arnold Toynbee in supporting 'citizenship not charity'.[102] This education and recreation centre reflected her belief in a partnership between voluntary work and the state; that the former should be a supplement to, not a substitute for, the latter.[103] Among her assistants in this endeavour was Hilda Cashmore, suffragist and supporter of the rising Socialist movement. Personal friendship and a shared commitment to active citizenship cut across their political differences. During the 1930s, Cashmore took the Settlement principle to India.[104]

Imperial connections also enabled Markham to expand her welfare activities in scale and location. Through acquaintance in South Africa with Mrs Alfred Lyttelton, Markham secured her first post in London as Honorary Secretary of the Personal Service Association, a voluntary body established in 1908 to relieve unemployment in the capital city. Markham also followed the Edwardian practice of establishing a Liberal Dinner Club to exchange ideas about empire and to express and promote liberal imperialist views. In this endeavour, she acknowledged the model of Beatrice and Sydney Webb who cultivated friendship with the 'Limps', not through shared political ideas, but because of the support which they provided for the Webb's social initiatives.[105] Markham was also guided by Edward Tyas Cook, a prominent London writer and journalist with whom Markham shared many similar views.[106] As a liberal imperialist, he was a close associate of Sir Alfred Milner, supported the Victoria League and admired the writings of John Ruskin. Fellow members of Markham's club included Jack and May Tennant.

But while Markham's interest in the civilisation of colonial societies effectively complemented a commitment to urban improvement and the conditions of labour at home, it was not until World War One that her concern for race and class difference extended to gender. A believer in the natural inferiority of women and initially opposed to women's suffrage, it was through experience in World War One that her attitude to the feminist cause in Britain began to change.[107] With May Tennant and Mary Macarthur, an activist for the rights of women workers, Violet served on several committees including the Central Committee on Women's Employment, the executive committee of the National Relief Fund and, in an effort to promote the Dominions for war widows, the Relief Fund emigration sub-committee.[108] In 1917 she became a Companion of Honour. In the same year, at the age of forty-four years, she finally acknowledged that 'Lincoln's principle – "a country cannot be part serf and part free" – applied to the relations of men and women as well as to those of black and white'.[109] Committed to equality in the franchise, Markham made her conversion public at the Albert Hall.[110] One year later, in the first general election at which women were entitled to stand, she unsuccessfully contested her brother's former Mansfield constituency for the Liberal Party.

During the inter-war years Markham's attention turned to local government in Chesterfield and outward to the Dominions including one of the newly formed international bodies, the International Labour Organisation. Her views on South Africa also hardened. A second visit in 1912 had already demonstrated how selectively the liberal package could be implemented overseas. She was conscious that 'the outlook on native affairs' had deteriorated, particularly in Cape Province where their pre-Union liberality 'had been swallowed up by a harsher policy dictated from the north', notably the Transvaal. Subsequent visits confirmed her opposition to racism[111] and to the apartheid system, formally established following the Nationalist Party victory in 1948. In her South African diary for 1953 she recorded that:

> Under apartheid the natives are skilfully encouraged to think they can be lawyers, administrators, judges and complete freedom to expand their own civilisation. But of course it is absurd to think that natives capable of those functions and who have attained their degree of civilisation will be content with the perpetual stamp of inferiority.

While remaining a firm believer in evolutionary human development, her youthful scepticism over the inevitability of moral progress was confirmed. Writing in 1953 she claimed that:

> There has been a moral recession all over the world, but in no place has that recession been more painful or fraught with such potential disaster as in South Africa.[112]

Conclusion

Women participated in empire in many different ways and by the turn of this century prominent public figures had emerged for whom an interest in social progress in Britain and the empire were intertwined. Personal networks acquired overseas provided the outlets through which their views and energies could be channelled. As diarists, reporters and public speakers, they gained a new public authority in Britain and were acknowledged by scientific societies like the RGS. With literary skills and overseas experience they became acceptable additions to the fellowship. For many of these women, South Africa offered an attractive focus. Violet Markham provides one perspective on these linkages between metropolis and colonies. Marking the transition from Victorian adventure to twentieth-century colonial rule, her interests in the region extended beyond the drama of the physical environment or the 'strangeness' and 'curiosity' of the indigenous population. Here the supposedly universal principles underpinning natural racial hierarchies fused with the climatically selective process of racial adaptation. Thus Britain's moral, scientific and technical right to transform the territory was unchallenged. As a personal statement, Markham's geographical interpretation of South Africa also displays the complexity of British imperial ideas. In interpreting the country's physical, social and economic conditions, she drew upon a series of intellectual debates and a range of political doctrines about race, class and environment. It was upon these that her vision of the country's pastoral, arable and urban landscapes was based.

Industrial Derbyshire, Markham's home region, shared some common ground with the Witwatersrand. Both had been colonised to sustain and enhance Britain's economic supremacy. In visual appearance, institutional arrangements and social organisation, indicators could be found of the physical, social and moral degeneration of an avaricious, urban, industrial society. These problem landscapes also lacked the enduring qualities of a tamed wilderness. Like many social critics of her generation, Markham's image of a cultivated rural environment symbolised the harmony of humans and nature and represented the source of a morally just society. But this faith in the progressive power of natural law did not diminish the importance of institutional and individual responsibilities. Markham participated in, and helped forge, a range of social networks and organisations at municipal, national and international levels. At their core was Britain's social order and place within the world in which the interplay between rights, duties and responsibilities of citizens was central.

In the course of her female odyssey, Markham was drawn to marginal groups – to Britain's working classes, South Africa's indigenous population and finally to women. Institutional responsibilities to these groups lay in promoting 'British standards of justice, integrity and good government'. They applied not only at the lofty scale of metropolitan and colonial state but also at municipal level through the commitment both of employers to the welfare of their labour force and of local government to social services. That these institutional responsibilities should be matched by active citizenship is apparent in the dynamics of Markham's social consciousness and in her social actions over the course of her life. At municipal level, the Chesterfield settlement marked the clearest initial expression of this commitment. By the end of World War One her paternalism and liberal imperialism, relatively indistinguishable from her male counterparts, had become more radical and feminist. This was reflected in her support for the extension of the franchise to South Africa's black population and to women in Britain. Similarly, in defence of the rights of labour, her practical help for the unemployed in London before 1914, had, by the inter-war years, extended to support for national organisations and international institutions like the International Labour Organisation. But respecting the rights of working people also required the reciprocal commitment of labour to national interests. Women's emigration would not only fulfil their employment needs but also their duty as active citizens in the service of the empire.

Here was a progressive vision which looked outward beyond Britain and inward to the workings of British society. It enshrined views of the ideal state, city and community. Shaped by influential transcultural networks, it aimed to modify and sustain particular imperial and later international interests; it also sought to rework and renegotiate imperial nationalism within British society and to implement its principles at home. While Markham inherited the Victorian middle-class duty to serve, she formed one of a new generation of women for whom social problems could be confronted by means other than individual social work within the family and community. For sure, science and technology were required to fashion the ideal environment. But so was social action based upon a partnership between voluntary sector and the state. With access to a range of outlets for their views, from the early years of this century prominent women like Violet Markham not only participated in debates about progress, they also helped to shape the practices which followed them in Britain and overseas.

Notes

1 See, for example, D. Birkett, *Spinsters Abroad: Victorian lady explorers*, London, 1989; B. Melman, *Women's Orients: English women in the Middle East, 1718–1918*, London, 1992; S. Mills, *Discourses of Difference: an analysis of women's travel writing and colonialism*, London, 1991. This literature builds upon the work of Dorothy Middleton, *Victorian Lady Travellers*, London, 1965.

2 By imposing the category, traveller, on women who journeyed overseas, there is a tendency to assume that they did nothing else. For many who wrote and spoke about other societies, traveller is an unworthy label since it describes only a small part of what were often rich, varied and difficult lives in Britain.

3 While much attention has been given to the heroines of travel in the Victorian era, few studies look forward to the twentieth century.

4 S. Constantine (ed.), *Emigrants and Empire: British settlement in the Dominions between the wars*, Manchester, 1990; J. Mangan (ed.), *Making Imperial Mentalities: socialisation and British imperialism*, Manchester, 1990.

5 E. J. Hobsbawm, *The Age of Empire, 1875–1914*, London, 1987.

6 D. Denoon, 'Temperate medicine and settler capitalism: on the reception of western medical ideas', in R. Macleod, and M. Lewis (eds), *Disease Medicine and Empire: perspectives on western medicine and the experience of European expansion*, London, 1988, pp. 121–38; A. Porter, *Victorian Shipping, Business and Imperial Policy: Donald Currie, the Castle Line and Southern Africa*, Woodbridge, 1986.

7 M. L. Pratt, *Imperial Eyes: travel writing and transculturation*, London, 1992.

8 The Cape was part of the Batavian Republic from 1802 to 1806 but was British thereafter. On British immigration to, and settlement in, the region see C. Crais,

White Supremacy and Black Resistance in Pre-industrial South Africa: the making of the colonial order in the Eastern Cape, 1770–1865, Cambridge, 1992; K. Williams, 'A way out of troubles': the politics of Empire settlement, 1900–1922', in S. Constantine (ed.), *Emigrants and Empire: British settlement in the Dominions between the wars*, Manchester, 1990, pp. 22–44; M. Winer, and J. Deetz, 'The transformation of British culture in the Eastern Cape, 1820–1860', *Social Dynamics*, Vol. 16 (1990), pp. 55–75.

9 The region was also the focus of leading environmental research. R. H. Grove, 'Colonial conservation, ecological hegemony and popular resistance: towards a global synthesis', in J. M. MacKenzie (ed.), *Imperialism and the Natural World*, Manchester, 1990, pp. 15–50.

10 Sir Roderick Murchison, a founder and President of the Royal Geographical Society, fostered mineral exploration in the region for scientific and imperial commercial purposes. R. Stafford, *Scientist of Empire: Sir Roderick Murchison, scientific exploration and Victorian imperialism*, Cambridge, 1989.

11 E. Bradlow, 'The culture of a colonial elite: the Cape of Good Hope in the 1850s', *Victorian Studies*, Vol. 29 (1986), p. 391.

12 W. Irons, *The Settler's Guide to the Cape of Good Hope and the Colony of Natal*, London, 1858.

13 E. Bradlow, 'The culture of a colonial elite', p. 391.

14 L. Griffin, 'South Africa and India', *The Nineteenth Century*, May (1902), p. 709.

15 J. Van Helten and K. Williams, 'The crying need of South Africa', *Journal of Southern African Studies*, Vol. 10 (1983), pp. 17–38.

16 As British High Commissioner for South Africa and Governor of Cape Colony and Transvaal from 1897 to 1905, Milner played a significant role in shaping the South African state to meet the demands of twentieth-century British imperialism in the region. S. Marks and S. Trapido, 'Lord Milner and the South African State', *History Workshop Journal*, Vol. 8 (1979), pp. 50–80.

17 A. Cecil, 'The needs of South Africa', *The Nineteenth Century*, Vol. 50 (1902), p. 683.

18 His Grace the Duke of Argyll, 'Emigration', in L. Creswicke (ed.), *South Africa and Its Future*, London, 1903, p. 14.

19 Editor of *Rhodesia* 'Women's life in South Africa', *The Imperial Colonist*, Vol. 1 (1902), pp. 70–2.

20 J. A. Ross, *Consumption and Its Treatment by Climate with Reference especially to the Health Resorts of the South African Colonies*, London, 1876.

21 A. Porter, 'The South African War (1899–1902): context and motive reconsidered', *Journal of African History*, Vol. 31 (1990), pp. 43–57.

22 J. Hichberger, *Images of the Army: the military in British art, 1815–1914*, Manchester, 1988; W. Reader, *As Duty's Call: a study in obsolete patriotism*, Manchester, 1988; J. O. Springhall, ' "Up Guards and At Them!": British imperialism and popular art, 1880–1914', in J. MacKenzie (ed.), *Imperialism and Popular Culture*, Manchester, 1986, pp. 49–72; M. van Wyk Smith, *Drummer Hodge: the poetry of the Anglo–Boer War (1899–1902)*, Oxford, 1978.

23 J. A. Berkman, *The Healing Imagination of Olive Schreiner: beyond South African Colonialism*, Amherst, 1989.

24 B. Roberts, *Ladies in the Veld*, London, 1965; C. B. Stevenson, *Victorian Women Travel Writers in Africa*, Boston, 1982.

25 E. Moberly Bell, *Flora Shaw*, London, 1947.

26 *The Times*, Special Correspondent, *Letters from South Africa* (frontispiece), London, 1893.

27 *Dictionary of National Biography*, 1951–60, p. 692.

28 Joseph Paxton was instrumental in developing glass-house technology for gardening and arranged for the import from the colonies of a range of exotic plants. As part of a government committee, established in 1838 to investigate the management of Kew Gardens, he played an important part in retaining the Gardens for the nation. V. R. Markham, *The New Era in South Africa with an Examination of the Chinese labour question*, London, 1904.

29 V. R. Markham, *Return Passage. The autobiography of Violet R. Markham, C.H.*, Oxford, 1953, p. 43.
30 V. R. Markham, *Return Passage*, p. 51.
31 Like Markham's Victorian predecessors and many women of her generation, liberty to travel came from the resources of a middle-class family. D. Middleton, 1965; M. Domosh, 'Towards a feminist historiography of geography', *Trans. Inst. Br. Geogr.*, Vol. 16 (1991), pp. 95–104; P. Moorey, 'Kathleen Kenyon in retrospect', *Palestine Exploration Quarterly*, July–December (1992), pp. 91–100,
32 V. R. Markham, *Return Passage*, p. 51.
33 V. R. Markham, *South Africa: past and present*, London, 1900, pp. 341–3.
34 V. R. Markham, *South Africa*, p. 343.
35 Britain's highland landscapes shared a similar magnetic appeal to Africa's 'inland wastes'. See J. M. MacKenzie, (ed.), *Imperialism and the Natural World*, Manchester, 1990; T. C. Smout, 'The Highlands and the roots of green consciousness, 1750–1900', *Proceedings of the Royal Academy*, Vol. 76 (1991), pp. 237–63.
36 V. R. Markham, *South Africa*, preface.
37 V. R. Markham, *South Africa*, p. 240.
38 M. Bell, ' "The pestilence that walketh in darkness": imperial health, gender and images of South Africa', *Trans. Inst. Br. Geogr.*, Vol. 18 (1993), pp. 327–41; R. W. Felkin, *On the Geographical Distribution of Some Tropical Diseases and Their Relation to Physical Phenomena*, Edinburgh, 1889; S. White, 'On the comparative value of African lands', *The Scottish Geographical Magazine*, Vol. 7 (1891), pp. 191–5.
39 V. R. Markham, *Return Passage*, p. 52.
40 V. R. Markham, *Return Passage*, p. 52.
41 V. R. Markham, *South Africa*, p. 327.
42 In an appreciation of Cecil Rhodes, Violet described him as the last conquistador. LSE Markham Papers Pt. 4 25/85.
43 V. R. Markham, *South Africa*, p. 340.
44 V. R. Markham, *South Africa*, p. 340.
45 V. R. Markham, 'The South African riveria', *Southern African Railways and Harbours Magazine*, Vol. 19 (1925), p. 201.
46 V. R. Markham, *South Africa*.
47 Markham established a close friendship with Sir Alfred Milner and conducted a correspondence with him over many years. LSE Markham Papers Pt. 4 25/65.
48 P. Brantlinger, 'Victorians and Africans: the genealogy of the myth of the Dark Continent', *Critical Inquiry*, Vol. 12 (1985), p. 194.
49 V. R. Markham, *Return Passage*, p. 53.
50 V. R. Markham, *Return Passage*, p. 52.
51 V. R. Markham, *Return Passage*, p. 52.
52 V. R. Markham, *South Africa*, pp. 346–7.
53 V. R. Markham, *Return Passage*, p. 56.
54 V. R. Markham, *South Africa*, p. 217.
55 V. R. Markham, *Return Passage*, p. 59.
56 V. R. Markham, *South Africa*, p. 339.
57 V. R. Markham, *South Africa*, p. 218.
58 A. Howkins, *Reshaping Rural England. A social history 1850–1925*, London, 1991.
59 W. Rubenstein, *Men of Property: the very wealthy in Britain since the industrial revolution*, London, 1981.
60 L. Davidoff, *Best Circles*, London, 1973; F. Thompson, *English Landed Society*, London, 1963.
61 V. R. Markham, *South Africa*, p. 372,
62 C. Aslet, *The Last Country Houses*, New Haven, 1982; B. Drake, and M. I. Cole, *Our Partnership. The diary of Beatrice Webb*, London, 1948, p. 311.
63 V. R. Markham, *South Africa*, pp. 374–6.
64 A. Trollope, *South Africa*, Cape Town, 1878, (reprint 1973), p. 369.
65 Quoted in Stevenson, *Victorian Women*, p. 56.

66 Unlike Trollope, Markham was writing at a time of increasing order in the built form of Kimberley. R. V. Turrell, *Capital and Labour on the Kimberley Diamond Fields 1871–1890*. Cambridge, 1987; W. H. Worger, *South Africa's City of Diamonds. Mine workers and monopoly capitalism in Kimberley, 1867–1895*. New Haven, 1987.
67 V. R. Markham, *South Africa*, p. 354.
68 V. R. Markham, *South Africa*, p. 360.
69 Markham's belief in the responsibilities of mining capital to labour may have been influenced by her mine-owning family experience in Derbyshire and her knowledge of company towns in Britain.
70 V. R. Markham, *Return Passage*, p. 48.
71 The barrack-like compounds for black employees were frequently over-crowded, squalid and cruelly disciplined. See C. van Onselen, *Studies in the Social and Economic History of the Witwaterstrand 1886–1914, Volume 1, New Babylon, Volume 2 New Ninevah*, London, 1982.
72 V. R. Markham, *South Africa*, p. 349.
73 P. Bade, 'Art and degeneration: Visual icons of corruption', in J. Chamberlin and S. Gilman (eds), *Degeneration. The dark side of progress*, New York, 1985, pp. 220–40; S. Siegel, 'Literature and degeneration: the representation of "decadence" ', in J. Chamberlin and S. Gilman (eds), *Degeneration. The dark side of progress*, New York, 1985, pp. 199–219.
74 V. R. Markham, *South Africa*, p. 350.
75 V. R. Markham, *South Africa*, p. 404.
76 V. R. Markham, *South Africa*, p. 240.
77 V. R. Markham, *South Africa*, p. 279. Markham distinguished between the 'tribal' and 'non-tribal kaffirs' (*South Africa*, p. 258). The former lived in communities under tribal control; the later had adopted European customs, were subject to European law and, in the British territories, were entitled to exercise the franchise.
78 V. R. Markham, *South Africa*, pp. 252–3.
79 V. R. Markham, *South Africa*, p. 253.
80 Markham's analysis built upon the naturalistic or scientific approaches to the relations between society and nature which were debated at this time. Notwithstanding the cultural dominance of science and the secularisation of society, her discussion also lent support to the view that the 'brutish' character of natural law must be tempered by social morality. Markham, *South Africa*, p. 249; R. Barton, 'Evolution: the Whitworth gun in Huxley's war for the liberation of science from theology', in D. Oldroyd and I. Langham (eds), *The Wider Domain of Evolutionary Thought*, Dordrecht, 1983, pp. 261–88.
81 R. Colls and P. Dodd, *Englishness. Politics and culture 1880–1920*, London, 1986; P. Taylor 'The English and their Englishness: "a curiously mysterious, elusive and little understood people" ', *Scottish Geographical Magazine*, Vol. 107 (1991), pp. 146–61.
82 V. R. Markham, *Return Passage*, p. 60.
83 RGS Markham papers.
84 V. R. Markham, *Return Passage*, p. 60.
85 J. M. MacKenzie, *Propaganda and Empire: the manipulation of British public opinion, 1880–1960*, Manchester, 1984.
86 V. R. Markham, *South Africa*, p. 231.
87 Review of V. R. Markham, *The South African scene*, London, 1913, in *The Geographical Journal*, Vol. 43 (1914), p. 30.
88 Janet Hogarth, his sister, was a close friend of Gertrude Bell.
89 B. Semmel, *Imperialism and Social Reform. English social-imperial thought 1895–1914*, London, 1960. Social imperialism emphasised the relationship between the maintenance of colonial possessions abroad, security and social reform at home. It was variously interpreted and, in extreme form, stressed national self-interest over obligation to colonised peoples. Through its links with ideas of national efficiency and fitness, it became associated both with the control of territory and the production of an imperial race.
90 The Guild of Loyal Women, 1901.

91 Two letters from South African born, Mrs Stuart, to the Editor of *The Times* were published on 2 July and 7 September 1901. In these she challenged critics in Britain of conditions in the refugee camps and the concentration camps established for the Boer population.

92 Letter to David Gill, 15 February 1901, RGS Markham Papers.

93 Letter to David Gill, 3 March 1901, RGS Markham Papers.

94 D. J. Potgieter, editor-in-chief, *Standard Encyclopedia of Southern Africa*, Cape, 1970.

95 MacKenzie, *Propaganda and Empire*.

96 May Tennant was active on a range of national bodies including the Central Consultative Council of Voluntary Organisations; Assistant Commissioner of the Royal Labour Commission; HM Superintending Inspector of Factories; Member of the Royal Commission on Divorce; Director of the Women's Department of National Service; Chief Adviser on Women's Welfare, Ministry of Munitions; Member of the Health of Munition Workers and other related War Committees. She became a Companion of Honour in 1917, the same year as Violet Markham.

97 Under the auspices of the Victoria League, Markham produced a handbook on industrial laws in the British Dominions. With a preface by May Tennant, it was designed to inform and to assist in alleviating common industrial problems both in Britain and the Empire. V. R. Markham, *The Factory and Shop Acts of the British Dominions. A handbook*, London, undated.

98 B. Harrison, *Prudent Revolutionaries. Portraits of British feminists between the wars*, Oxford, 1987.

99 In this Markham built upon her youthful experience of individual social work among the town's poor, a form of personal service which was the 'duty' of middle-class women. J. Lewis, *Women and Social Action in Victorian and Edwardian England*, London, 1991.

100 V. R. Markham, *Return Passage*, p. 67. As the South African War retreated, the General Election of 1906 brought in a strong Liberal government committed to social reforms through legislation.

101 Issues discussed were social reform in Britain including diet and local government. LSE Markham Papers Pt. 4 25/87.

102 V. R. Markham, *Return Passage*, p. 65.

103 J. Alberti, 'Elizabeth Haldane as a women's suffrage survivor in the 1920s and 1930s', *Women's Studies International Forum*, Vol. 13 (1990), pp. 117–25.

104 V. R. Markham, *Friendship's Harvest*, London, 1956. Albertini, 'Elizabeth Haldane as a women's suffrage survivor in the 1920s and 1930s, *Women's Studies International Forum*, Vol. 13 (1990), pp. 117–25. Settlements were designed to be independent meeting places for people of different classes to exchange views and participate in social activities of common interest including adult education, child welfare, art and music. They gradually merged into community centres which were given formal status under the Butler Education Act (1944).

105 In 1902 Beatrice established the Coefficients Club to discuss the aims, policies and methods of national and imperial efficiency. Among its members was Sir Halford Mackinder. B. Blouet, *Sir Halford Mackinder: a biography*, College Station, 1987.

106 E. T. Cook edited the *Pall Mall Gazette* (1890–92). He was a founding member and editor of the liberal-oriented *Westminster Gazette* (1893–95) and, from 1895 to 1901, edited the *Daily News*. He resigned as a result of divisions of opinion among the proprietors over the South African War.

107 For many British women, the demands placed upon them by World War One presented unexpected opportunities in the public domain and profoundly affected their attitude to citizenship rights. See Lewis, *Women and Social Action*.

108 The National Relief Fund was established in August 1914, among other things, to assist officers' widows left in poor circumstances owing to the war. The Emigration Sub-Committee was appointed in February 1916 to examine and report on a proposal by the Salvation Army, in connection with their *Darkest England Scheme*, for the emigration of widows of soldiers to the Overseas Dominions. The sub-committee

recommended that a grant be given and that the scheme be modified to include unmarried women as well as widows. LSE Markham Papers Pt. 1 1/11.

109 V. R. Markham, *Return Passage*, p. 99.
110 J. Alberti, 'Elizabeth Haldane'.
111 V. R. Markham, Racial and political issues in South Africa, *Edinburgh Review*, October (1924), pp. 243–60.
112 V. R. Markham, *Return Passage*, p. 60.

Acknowledgements

My thanks to Denis Cosgrove, Jonathan Crush, Robin Butlin, Mike Heffernan and John MacKenzie for their constructive comments and advice on this chapter.

CHAPTER NINE

The spoils of war:
the Société de Géographie de Paris
and the French empire, 1914–1919
Michael Heffernan

Introduction

This essay considers one short episode in the history of French geography. It is an episode which exemplifies the intimate relationship between the production of geographical knowledge and the exercise of imperial power. The context is the final phase of French imperial expansion during and after the 1914–18 war. Though the outbreak of war brought the major period of imperial expansion to a close, its uneasy resolution led to a redistribution of colonial territory between the victorious Allied powers. Under the terms of wartime accords and post-war settlements, the Ottoman empire in the Middle East and the million square miles of Africa and Asia that constituted the German *Kolonialreich* were dismantled. Nominally, the new League of Nations assumed administrative responsibility for these territories but in practice 'trustee' colonial authority passed to one or other of the Allies under the mandates system established at the Paris Peace Conference. The lion's share went to Britain, the Dominions and France.

Preparations for post-war negotiations had been gathering pace in all belligerent nations throughout the latter part of the war and the services of leading academics – including geographers – were much in demand to help formulate war aims and territorial demands. This chapter examines how French geographers, particularly those in the Société de Géographie de Paris (SGP), became involved in developing their country's imperial war aims between 1914 and 1919.[1]

The SGP and French imperialism

French geography and French imperialism were related intellectual and ideological projects. During the Napoleonic First Empire, geography (particularly cartography and topographic survey) occupied a significant

[221]

place within a reformed educational system primarily because of its strategic and military importance. After the fall of the empire, well travelled Napoleonic veterans formed a central constituency in the clientele of the SGP, the world's first geographical society which held its opening session in July 1821.[2] Initially, the society's rise was less than meteoric. Membership increased to 300 in 1827 but slumped to 279 in 1830, the year in which the Royal Geographical Society (RGS) was launched in London with a foundation fellowship of 460. The SGP struggled through the next twenty years and could boast just 100 members in 1850 compared to over 500 in the RGS.[3] This was partly due to a lack of scientific and ideological coherence. While the RGS concentrated on exploration (a topic of endless fascination for the British middle class) the SGP's objectives were vague and its activities electic.[4] Although committed to imperial expansion,[5] the SGP's smaller size was illustrative of French uncertainty about the benefits of overseas expansion. The memory of Napoleonic imperial overstretch was still fresh in many minds.[6] Mid-nineteenth-century French imperial expansion in north and west Africa, beginning with the seizure of Algiers in 1830, was dictated not by a popular commitment to empire but rather by short-term attempts to divert attention from domestic problems through military 'heroics' overseas. At this stage, the only really committed imperialists were army officers whose careers were dependent on an imperial arena. Operating beyond the control of their political 'masters' in Paris, colonial commanders in Africa were able to dictate the pace and direction of French expansion to a remarkable degree.[7]

The fortunes of the SGP recovered under the Second Empire. By 1870, membership had risen to 645.[8] The society's ethos fitted well with the imperialist aspirations of the new Napoleonic age. The speculative boom produced by the reconstruction of Paris spawned a new, self-confident and internationally minded bourgeoisie, fertile recruiting ground for a geographical society. In 1860, the Marquis de Chasseloup-Laubat, Napoléon III's Naval Minister in the mid-1850s and the Colonial Minister between 1858 and 1860, became President of the SGP and was joined three years later by a new Secretary-General, Charles Maunoir, previously at the Ministry of War. Under their guidance the SGP began to attract a younger membership which would meet on Friday evenings for lavish, self-congratulatory dinners at a Parisian restaurant, La Petite Vache.[9]

After 1870, both geography and imperialism became increasingly central to French intellectual and political debate.[10] The turmoil of 1870–71 (defeat in the Franco-Prussian War, the loss of Alsace-Lorraine and the class war of the Paris Commune) produced widespread dismay and fractious attempts to identify the moral, psychological and political roots of France's decline. Educational reform, an area of perennial political dispute, loomed large in these debates. Many were convinced that 1870–71 had been the result of an educational system which had failed to inculcate the kind of patriotism and devotion to duty which Prussian soldiers had displayed to such effect. Ignorance of geography seemed particularly damaging for obvious military reasons. According to the editor of *La République Française*:

> During this calamitous war, we relied on generals who didn't know whether the Rhine flowed north–south or south–north. Senior officers were heard to inquire the distance from Metz to the former eastern frontier. Goethe was once heard to say, 'the French are a highly spirited people, but really, they know little geography'. We merely smiled at this whimsy, not realising that the day would come when the Germans would add the roar of cannon to the mocking voice of their great writer.[11]

Geography – together with its sister subject, history – was allocated a central role in revised programmes of patriotic civic education. The words of the historian Ernest Lavisse at a school prize-day in Nouvion-en-Thiérache (Aisne) in August 1905 perfectly encapsulate the republican image of geography and history in civic instruction. Lavisse boomed:

> To create our nation, nature, from whom we had previously claimed our share of land and sky, had to contribute for several centuries; then came politics, iron, and fire, and finally mind and heart, each with its contribution ... [O]ur fatherland is ... not merely a territory; it is a human structure, begun centuries ago, which we are continuing, which you will continue. The long struggle of our fathers, the memory of their actions and their thoughts, the monuments of their genius, our language, our type of mind, our way of understanding life, all that – with the rich beauty of our land, with the mildness of our sky, with the poetic variety of our landscape, our mists in the north and our southern sunlight, our superb mountains and our beautiful plains, our green seas, and our blue ocean – that is your rich inheritance. It is our country, the daughter of our spirit ... Suppose, then, you say to me: 'It is an accident that brought me into the world in France ... First of all, I am born a man. I wish to belong to humanity. It is humanity that I wish to serve.' I will answer: 'Humanity, that does not exist as yet; it is a great and beautiful idea; it is not a fact. You must have a fixed place in which to act, and I defy you to serve humanity otherwise than through the medium of the fatherland.[12]

The pressure of educational change gave a boost to geography at all levels in the educational system. By the end of the century, a dozen new university chairs had been established, including one in Colonial Geography at the Sorbonne, occupied in 1892 by Marcel Dubois.[13] Sensing the mood of the times, the SGP announced in 1871 that it would seek to promote geography in schools and universities, something it had previously avoided because educational questions were so intensely politicised. Membership immediately increased, rising from 600 in 1870 to 1,353 in 1875 when the SGP hosted the Congrès International des Sciences Géographiques. Three years later, the SGP occupied spacious new headquarters at 184 Boulevard Saint-Germain. Most new members came from the liberal professions, banking, commerce and trade. Several influential publishers, newspaper proprietors and journalists from periodicals such as the *Revue des Deux Mondes* and the *Journal des Débats* also joined.[14]

It was this liberal, patriotic constituency which began the clamour for a more assertive policy of imperial expansion as a complementary route to national rejuvenation.[15] Gradually, the centre of gravity of the French imperial impulse began to shift from the colonial periphery to the metropolitan core and from the army to a small but influential group within the French bourgeoisie. The arguments in favour of imperial expansion were complex but centred on several oft-stated assertions about the economic, strategic and cultural benefits of empire. The raw materials and resources of a large, flourishing empire would be essential, it was claimed, for France's future as an industrial and trading nation. Faced with a sluggish domestic population growth, it was also argued that French military power would be dependent on the bottomless reservoir of colonial 'man-power'.[16] The cultural idea of empire as a means of spreading the inestimable benefits of French language and civilisation amongst the less fortunate peoples of Africa and Asia was also regularly cited, particularly in view of the rising global hegemony of the English language.[17] The most insistent claim was that French political prestige depended on possession of a large overseas empire. A non-European empire was a political and strategic necessity if France was to remain a great power and overcome its loss of territory and pre-eminence in Europe. In this sense, French imperialism was 'the highest stage not of French capitalism but of French nationalism'.[18]

The quest for colonies did not go unchallenged and many, across all shades of political opinion, dismissed overseas expansion as a wasteful diversion from France's historic destiny in Europe.[19] However, a few prominent politicians, notably Léon Gambetta and Jules Ferry, became committed imperialists during the early Third Republic, though not without cost to their careers.[20] By the 1890s, an identifiable and highly effective colonial pressure group had emerged, the so-called Parti Colonial. At its head was Eugène Étienne, the *député* for Oran who later served as both Interior and War Minister.[21] The Parti Colonial was not a formal political party and had no consistent ideological perspective. It comprised fifty or sixty clubs and societies, some seeking to promote

French interests in particular regions (by 1914, these included Comité de l'Afrique française, the Comité de l'Asie française, the Comité de l'Océanie française, the Comité de l'Orient, the Comité de Madagascar and the Comité du Maroc) and others championing imperialism on moral, religious, commercial or ideological grounds (for example, the Comité Républicain aux colonies, La France colonisatrice, the Société de colonisation française, the Sociéte antiesclavage de France, the Ligue maritime française and the Société des études coloniales et maritimes).[22]

Although disparate and small (less than 10,000 adherents), the Parti Colonial had powerful representatives in the Chamber of Deputies, the Senate, the civil and diplomatic services and the universities. Its key members – operating close to, but rarely within, the French government – were able to play a decisive role not only in promoting the imperial cause but in formulating official policy. This was made possible by the instability and constitutional weakness of the Third Republic. Most governments survived for months rather than years and ministers were rarely in position long enough to learn their brief let alone impose their own ideas. Hard-pressed cabinets tended to concentrate on domestic and European issues and were notoriously uninterested in colonial affairs. Colonial policy was thus left to civil servants and advisers with an interest in their empire; in short, to the leading members of the Parti Colonial. Prominent, unelected colonialists were able to wield an influence over policy out of all proportion to their limited public support. It is no exaggeration to claim that France's huge overseas empire was carved out by a handful of imperialists operating firstly in the colonial army and subsequently in a small, unrepresentative pressure group whose remarkable successes were achieved because of – rather than despite – public indifference and political impotence.[23]

The SGP was one of the most important organisations in the Parti Colonial. Andrew and Kanya-Forstner refer to it as 'the elder statesman of the colonialist movement' and show that over half of the leading two hundred colonialist personalities were members. Twenty-six of the forty-five largest colonial societies were chaired by SGP members.[24] As membership grew, however, different schools of thought emerged about the regions into which French expansion should be concentrated. There was also debate about the most desirable forms of imperial rule and about the role which the SGP should play. Although committed to the empire, Maunoir clung to his scientific credentials and was reluctant to become embroiled in political controversy. This brought him into conflict with more virulent imperialist factions. When the Society was attacked in an 1894 editorial in La Politique Coloniale for its 'faint-hearted' commitment to the imperial cause, Maunoir was unrepentant:

'the SGP refuses to play ball with colonialists and politicians and sticks firmly to the scientific terrain', he wrote. One of his last acts in the SGP was to re-affirm its original constitution which explicitly forbade involvement with 'questions connected to politics'.[25]

More aggressive imperialists in the SGP, particularly those from the business community, wanted overt political action to promote imperial trade.[26] As early as 1873, a Commission de la Géographie Commerciale was established in the SGP to liaise with Chambres Syndicales and to lobby political leaders. Frustrated by Maunoir's caution, the Commission set itself apart as a separate organisation, the Société de Géographie Commerciale de Paris (SGCP), in 1876. The two societies had overlapping membership and were closely linked but the SGCP focused on economic questions and was more overtly imperialist, eventually establishing overseas branches in Tunis (1896), Hanoi (1902) and Constantinople (1904).[27]

Similar developments took place in the larger provincial cities, particularly those like Bordeaux, Lyons and Marseilles with commercial links to the empire. In 1874, a group of geographical enthusiasts and businessmen from Bordeaux, led by Pierre Foncin and Eugène Azam, established a Société de Géographie Commerciale de Bordeaux.[28] Almost at the same time, a similar society was established in Lyons and, in 1876, two more were founded in Marseilles and Montpellier. By 1909, there were twenty-seven geographical societies in France and a further four in French Algeria. Several called themselves commercial geographical societies including those at Bordeaux, Nantes, Le Havre and St Nazaire. The Marseilles society styled itself the Sociéte de Géographie et d'Études Coloniales de Marseille. These provincial societies, whose clientele was drawn mainly from local business communities, became vigorous imperialist pressure groups, part of a broader development which J. F. Laffey refers to as 'municipal imperialism'.[29] Meanwhile, the membership of the SGP rose to a peak of 2,500 in the mid-1880s and declined slightly thereafter. Overall, membership of French geographical societies increased to more than 16,500 in 1901, about one-third of the global membership of such organisations.[30]

The SGP and French imperial war aims

In 1914, French imperialists in the SGP could look with satisfaction at their atlas maps. Apart from the swaths of British red, the blue of France was easily the most prominent colour. The French empire covered 10.5 million square kilometres, principally in Africa (9.5 million) and Indochina (0.8 million). The constitutional relationship between the empire and la mère patrie was complex. Algeria was technically not a colony at all but an integral part of metropolitan France though only its white population had full local, regional and national political rights. Non-Europeans in Algeria (about 90 per cent of the population) were largely disenfranchised and administered by an unelected Governor-General answerable to the Ministry of War. Tunisia and French Morocco were protectorates governed by traditional monarchs supervised by the Ministry of Foreign Affairs at the Quai d'Orsay. The rest of the empire was directed from the Colonial Ministry, established in 1894 on the Rue Oudinot. It was responsible for the long-established and partly self-governing island colonies in the Caribbean (such as Guadeloupe and Martinique), the Indian Ocean (such as Réunion) and the Pacific (including Tahiti, New Caledonia, the Wallis Islands and the Anglo-French condominium in the New Hebrides). There were four huge colonial federations; French West Africa (most of modern Mauritania, Senegal, Guinea, Mali, Ivory Coast, Upper Volta, Benin and Niger); French Equatorial Africa (most of modern Chad, Central African Republic, Congo and Gabon); French Indochina (most of modern Vietnam, Laos, Cambodia and the leased Chinese coastal enclave of Kwang-Chow-Wan); and Madagascar and the Comoro Islands. These sprawling territories were run by Governors-General in Dakar, Brazzaville, Hanoi and Tananarive (Antananarivo).[31] The strategic east African port of Djibouti and its environs (French Somaliland), the Indian enclaves (such as Pondichéry, Mahé and Chandernagor), the islands of Saint Pierre and Miquelon off the Newfoundland coast and French Guiana constituted the final parts of the Rue Oudinot's imperial responsibilities. All together, the French empire was inhabited by 46 million people.[32]

French geography was also at the peak of its prestige in 1914. Apart from the numerous geographical societies, French academic geographers, led by Paul Vidal de la Blanche, had an unrivalled reputation.[33] The SGP, still the largest French geographical society with around 2,000 members, was presided over by Prince Roland Bonaparte, the nephew of Napoléon III and the Society's most generous benefactor. The Secretary-General was Baron Étienne Hulot, a committed imperialist who had succeeded the ailing Maunoir in 1897. The Central Committee was chaired by Charles Lallemand, the Director of the Service de Nivellement Général de la France, and included General Robert Bourgeois, the Director of the Service Géographique de l'Armée;[34] Auguste Pavie, the famed explorer of Indochina; Paul Labbé, the Secretary General of the SGCP; and Franz Schrader, the map and atlas compiler. Leading academics such as Lucien Gallois from the Sorbonne and Paul Pelliot and Georges Blondel from the Collège de France were also Central Committee members as was Louis Raveneau in his capacity as editor of the *Annales de Géographie*, the influential journal co-founded in 1891 by Vidal de la Blache and Dubuis.[35]

In the last days of July 1914, thousands of French geographers assembled for their annual conference at Brive where they enjoyed the usual range of lectures, field excursions and lavish dinners.[36] Less than a week later, many of the delegates were preparing for the grim business of war. Ten members of the Central Commission and fully half the membership of the SGP were conscripted, including Prince Roland Bonaparte who was seconded to Bourgeois's Service Géographique de l'Armée.[37] On 7 August, the Central Commission met in emergency sesion with Vice-President Franz Schrader in the chair. In view of the loss of key personnel and the possibility that Paris would fall, the SGP suspended its normal activities and ceased publication of the house journal, which since 1900 had been called simply *La Géographie*. It was agreed symbolically to sever links with the German geographical movement and to rescind membership of those based in enemy countries. This ban was subsequently extended to geographers from neutral countries who expressed sympathy with Germany, the most notable 'casualty' being the Swedish explorer Sven Hedin.[38] One of the leading women members, the formidable Comtesse Roederer, proposed that the society's headquarters be opcned as a crèche where the working wives of departed soldiers could leave their young children. For the first three months of the war, the society's splendid Second Empire salons reverberated to the whoops and cries of infants playing carelessly beneath the brooding portraits of Bougainville and de Lapérouse.[39]

After the Battle of the Marne, the immediate threat to Paris abated. The Central Committee re-convened on 9 September and again on 13 November. At the first meeting, the society's resources and expertise were offered to the Ministry of War. A full programme was to be maintained and publication of *La Géographie* continued in the belief that a patriotic journal would have propaganda value. The society then set about raising money for the war effort. The 21,000 francs already raised by the Comité d'assistance to help retired colonial soldiers and *fonctionnaires* were donated to the Ministry of War for ambulances and medical provisions. Throughout the war, the SGP spent much of its time and effort collecting money for medical provisions, particularly for injured colonial troops and the labour corps from Africa and Indochina.[40]

Monthly public *séances* re-commenced on 27 November 1914. Any lingering apolitical objectivity and internationalism had long since been abandoned. The tone was set by the opening address, on the role of geography and science in war, by the ironically named Charles Lallemand, the Society's acting President. After a diatribe against the willing involvement of German scientists in the preparation for war, Lallemand outlined the geopolitical changes which France should demand, and which the SGP should promote, once Germany was defeated. These included the return of an expanded Alsace-Lorraine; the expansion of Belgium to the Rhine; the extension of Serbia to include Bosnia-Herzogovina; the re-establishment of Poland and Armenia under Russian protection; the break-up of the Austro-Hungarian and Turkish empires; the return to France of Congolese territory ceded to German Cameroons in 1911;[41] the transfer of 'the greater part' of the German colonies in Africa to Britain; and the removal of German influence in Asia and the Pacific. That such demands could realistically be advanced when a large portion of France was under German occupation was, to say the least, optimistic. Lallemand insisted, however, that the war offered unprecedented opportunities to change the global geopolitical order to enhance the position of France and, by implication, the 'civilised' world at the expense of Germany. Geographers, acknowledged experts in the physical and human characteristics of the world's regions, had a sacred duty to keep French public opinion and the political leadership informed about the geopolitical implications of the war. 'Never, in so short a space of time', he argued, 'will geopolitical changes of this magnitude be possible again'.[42]

Of the 29 meetings held at the SGP during the war, 27 concentrated on geopolitical questions raised by the war although strict censorship laws precluded full publication of these debates and effectively prevented any serious discussion of military questions.[43] With the aid of invited politicians and dignitaries, the SGP sought to establish itself as a major centre of geopolitical discussion and progaganda. Attention initially focused on the most desirable political landscape for Europe after the war.[44] The Society also kept up its attacks on German geographers for supporting their country's war effort. Geographers in neutral countries were bombarded with anti-German SGP leaflets.[45]

Imperialists in the SGP were soon advancing their own agenda. On 22 January 1915, Hulot gave a lecture on the war in the colonies. Anxious to distance the SGP from Lallemand's generous proposal that most of German Africa should pass to Britain after the war, Hulot's survey stressed the extent of Allied successes on the colonial fronts and the opportunities this presented for *all* Allied imperial powers.[46] Asia was deemed unlikely to offer much scope for French expansion. The small, isolated and defenceless German possessions here had all fallen to France's Allies within weeks. The Chinese coastal enclave of Kiau Tschau and the German Pacific islands (the Carolines, the Palau, the Marianas and the Marshalls) were safely in Australian, New Zealand or Japanese hands. German Samoa had been under New Zealand occupation since the end of August and the Australians held Kaiser Wilhelmland (the north-east corner of New Guinea) and the Bismarck Archipelago. In Africa, the situation was more promising. German Togoland had been invaded and overrun by French and British troops from neighbouring Dahomey and Gold Coast in the opening days of the war.[47] Elsewhere, the contest was less one-sided but eventual success seemed likely. In German Cameroons, Anglo-French forces were pushing back German resistance in the mountainous interior and South African troops under Generals Smuts and Botha seemed destined to win the battle for German South West Africa (modern Namibia). In German East Africa (roughly modern Tanzania), Anglo-Belgian forces helped by South Africans, Indians and Portuguese were also engaged in extraordinary running battles with the troops of General von Lettow-Vorbeck.[48] The imminent collapse of Germany's African empire therefore offered the best opportunities to expand France's imperial interests.

After Hulot's address, colonial questions began to feature more prominently in SGP *séances*. The clear objective was to scotch any idea that conquered colonial territory should be returned to Germany after the cessation of hostilities. This was justified on both moral and geopolitical grounds. Morally, Germany was deemed to have forfeited the right to an overseas empire because the unfortunate inhabitants of its colonies had been subjected to regimes which were dismissed as unspeakably barbaric and inhuman, a 'fact' confirmed by the testimonies of 'liberated' indigenous peoples.[49] Geopolitically, Germany's vast imperial ambitions, particularly in Africa, were deemed to be in direct conflict with the legitimate, long-standing rights and responsibilities of the Allies, especially France.

In fact, Germany's colonial war aims, though extensive, were never promoted as aggressively as the European demands. While Chancellor Theobald von Bethmann Hollweg's war-aims memorandum of September 1914 contained a demand for a German central African empire comprising the existing colonies plus large concessions from Portugal, Belgium, France and Britain, this was drawn up by Colonial Minister Wilhelm Solf and added to the memorandum as something of an afterthought. Their impressive scale was partly due to Solf's erroneous assumption that there would be no German territorial claims in Europe. The colonial scheme was only one element in a much larger programme and although revised and even extended in 1917 and 1918, it did not figure prominently in Germany's wartime propaganda. The assumption in Berlin was that colonial settlements would realise themselves once victory was achieved in Europe.[50]

Hulot was certainly aware of this but realised that the cause of French imperial expansion was best served by emphasising the scale of

Germany's imperial ambition. If it could be demonstrated that Germany's desire for a large overseas empire was so great that all-out war in Europe was deemed a price worth paying, then the case for intensive development of France's own imperial resources and for French imperial expansion at Germany's expense became compelling. Fortunately for French colonialists, German geographers provided more than enough ammunition to sustain this argument. Hulot was able to seize on pre-war German colonial propaganda and the Solf demands in Africa to provoke horrified speculation in the SGP at the prospect of a vast German *Mittelafrika* comprising a continuous and broadening belt of territory stretching east to west across the continent from German East Africa to German South-West Africa and into French and British West Africa.[51] The description of Germany's *Weltpolitik* in 1915 by the doyen of Heidelberg geographers, Alfred Hettner, also provided excellent copy:

> The English regard themselves as a chosen people and believe themselves entitled to rule the seas and the world; but we deny them this right and make ourselves their equals. The Russians [and the French] . . . claim a right to extend their empires even further, but we challenge their right, too, with our own . . . We want to be the educators of the world, to carry our culture out into the world. The German Ideal shall heal the world . . . Germany . . . cannot live from the crumbs which fall from World Powers' table; she must be a World Power herself, the peer of the others.[52]

Albrecht Penck, the much-respected Professor of Geography and Rector of the University of Berlin from 1917, likewise obliged by producing a remarkable article on geography and war which was mined for incriminating quotations such as 'Der jetzige Krieg wird uns einen Platz an der Sonne sichern' ('This war will secure for us a place in the sun') and 'Wissen ist Macht, geographisches Wissen ist Weltmacht' (knowledge is power, geographical knowledge is world power').[53] In the absence of direct German discussion about an overseas empire, these general assertions would more than suffice.

Eventually, the relative absence of colonial demands in German wartime propaganda was itself used as evidence of evil Teutonic intentions in Africa and Asia. The argument was fully elaborated by Jan Smuts in an address to the RGS on 28 January 1918, though similar views had been expressed in the SGP. Insisting on the influence of the German colonial movement and the power of the German Colonial Ministry (which had helpfully produced a new map of *Mittelafrika* in 1917), Smuts suggested that 'German colonial aims are not really colonial, but are entirely dominated by far-reaching conceptions of world politics. Not colonies, but military power and strategic positions for exercising world power in future, are her real aims. Her ultimate objective in Africa is the establishment of a great Central African Empire . . . based on the writings of great German publicists, professors, and high colonial authorities'. This vast empire, the source of Germany's future economic and military power, would be disastrous for the Allies but catastrophic for Africa: 'The civilization of the African natives and the economic development of the dark continent must be subordinated to the most far-reaching schemes of German world power and world conquest . . . the African native must play his part in the new slavery.'[54] French colonialists now had another, highly emotive argument in favour of imperial expansion to add to their familiar list: France *had* to expand overseas to annex German colonies in order to frustrate German ambition.

Having dispensed with German colonial pretensions, the next step was to develop intellectually and morally justified claims to former German lands without offending the ambitions and sensibilities of the Allies. This was a difficult though increasingly urgent task as provisional territorial arrangements were already being drawn up by Allied military authorities. Togoland was divided as early as 31 August 1914 into French and British sectors, the latter including Lomé, the few kilometres of railway and the destroyed German radio mast at Kamina (see Figures 9.1a and 9.1b).[55] Although acceptable temporarily, French imperialists were quietly insistent that wartime arrangements would need to be re-negotiated once peace was established. Whatever changes occurred in the global colonial order, the long-term interests of the French empire had to remain paramount.

British fears that France would eventually claim most, if not all, of Togoland and Cameroons despite the numerical inferiority of French troops in the region revived old rivalries. According to one Foreign Office official writing on 1 December 1914: 'It is a curious thing that we looked with indifference on the occupation of Duala [the capital and main port of Cameroons] by our declared enemy, but are up in arms at the idea of its passing to our ally.'[56] In their eagerness to pre-empt French demands, leading British imperialists set out their own territorial claims. A mere two days after Hulot's February 1915 address on Germany's African ambitions, the influential British explorer and colonial administrator Sir Harry H. Johnston outlined his own version of events before the RGS. Unlike Hulot, Johnston ended with a set of specific British claims. Using three coloured maps, subsequently published in the *Geographical Journal*, Johnston contrasted the political geography of Africa in July 1914 with his nightmare image of the continent's likely political configuration in 1916 had Germany been victorious in Europe. His third and final map showed a new African

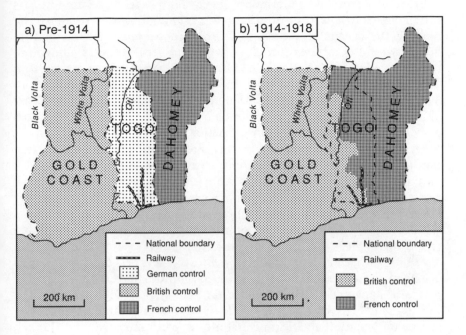

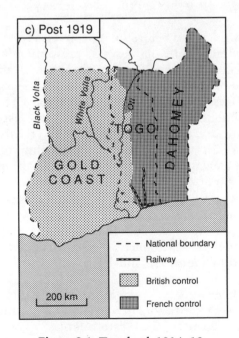

Figure 9.1 Togoland, 1914–18

[233]

political structure after Germany's eventual defeat. According to this Britain and South Africa would gain the vast majority of German Africa, making the dream of a Cape-to-Cairo British empire into a reality. France would be offered 'much of Togoland', the strips of Congolese territory ceded to Germany in 1911 but only the southern and eastern portion of Cameroons. The north and west of Cameroons, including the Sanagá river and Duala, would become part of British Nigeria not least because of the 'civilizing' role which British Baptist missionaries had played in the region before being 'expelled' by the Germans in 1885.[57]

Although clearly concerned by this, French imperialists were reluctant to become embroiled in an unseemly wartime dispute with their principal ally. SGP publications were therefore restricted to impassioned statements about the usefulness of existing colonies to the war effort, replete with oblique references to France's legitimate ambitions within vaguely defined spheres of influence. These lectures invariably ended with pleas for more investment and a greater emphasis on developing imperial resources after the war.[58] Lest this be interpreted as a sign of acquiescence, however, an anonymous letter signed by 'Un Africain' appeared in the Bulletin de la SGCP which, escaping the censor's blue pen, offered a spirited rejection of Johnston's 'premature' claims on behalf of the 'Lion Britannique' while insisting that French reticence on this matter was motivated solely by a temporary need to maintain allied unity.[59]

SGP lectures were given before small, converted audiences and their impact on policy and public opinion was probably minimal. However, the political influence of the SGP, already strong in the Quai d'Orsay and the Rue Oudinot, had expanded since 1914, particularly in the all-important Ministry of War. The routeway was General Bourgeois's Service Géographique de l'Armée. Bourgeois was responsible for producing all the military maps needed by the huge French army including the highly secret cartes de mobilisation indicating troop and artillery deployments. In December 1914, after prompting from Hulot, he persuaded the Minister of War that a Commission de Géographie should be established, linked to the Service, to produce detailed geograpical surveys of the théatres d'opération for use by the military authorities. Bourgeois recruited Paul Vidal de la Blache, Albert Demangeon, Lucien Gallois, Emmanuel de Martonne, Emmanuel de Margerie and Louis Raveneau.[60] Throughout 1915 and 1916, this small band of scholars produced forty confidential Notices descriptives et statistiques, ranging in size from ninety to six hundred pages, on the actual and potential theatres of the European war, including nine surveys of the French départements along the Western Front.[61] The Ministry of War also made full use of Pierre Camena d'Almeida, Professor of Geography at the University of Bordeaux, who developed techniques to estimate German troop movements based on detailed studies of the German army.[62]

The Commission de Géographie raised the profile of geographers within an important ministry but its European focus was of little use to ardent colonialists.[63] Moreover, its intellectual scope and political influence was limited by the military requirements of the army. By mid-1915, Hulot was determined to turn the SGP into an important 'think-tank' which would debate France's post-war territorial claims (both in Europe and in the colonial world), offer informed proposals to ministers and influence state policy. When Aristide Briand became Prime Minister in October 1915, the society was provided with a new opportunity.[64] During the first year of the war, under the premiership of René Viviani, French war aims were still vague. The only consistent demand was a call for the return of the 'lost' provinces of Alsace-Lorraine as an atonement for German aggression though even here, the details were unclear. By autumn 1915, the war had claimed nearly a million French casualties and the need for coherent and detailed war aims became urgent, not least because of rumours about a possible negotiated peace between Germany and Russia. Briand wanted to re-assert cabinet authority over Marshall Joffre's army and was eager to specify clear political objectives. A wide-ranging process of consultation began, focused mainly on Europe.[65] With a cabinet eager for policy options, the SGP was well positioned to promote its own, distinctly imperialist, agenda.

A central character linking the SGP with the cabinet was Louis Marin, the conservative, pro-imperialist *député* for Meurthe-et-Moselle.[66] Marin had been pressurising the Colonial Minister, Gaston Doumergue, to establish a committee to co-ordinate French imperial war aims. He was particularly concerned that the empire should not be bargained away by those preoccupied with a European settlement. Marin decided the SGP was the ideal focus for such a committee for three reasons. Firstly, it was still possible to present the SGP as an objective, independent scientific society, despite its obvious patriotic and imperialist commitments and close connections with government. Its leading members, particularly those in academic life, were inter-nationally renowned scholars whose involvement in developing French war aims would lend intellectual gravitas. Secondly, the SGP's interests were global and synthetic, bringing together information from the sciences and the humanities. It was therefore the ideal adjudicator between rival demands. Thirdly, the SGP possessed its own headquarters with a large library and map collection.[67]

Encouraged by Doumergue, Marin approached the SGP's central committee on 9 February 1916. Hulot agreed to establish 4 committees (*sous-commissions*) to consider France's war aims with respect to the Franco-German border, central Europe, Africa, and Asia-Oceania. Each committee was to comprise a core of about 20 people plus co-opted experts. Their task was to commission reports, collect evidence and decide policy alternatives in each area. Of the 61 people involved on these committees, 45 were members of the SGP and 25 – the inner core – were drawn from its central committee.[68]

Both European committees were chaired initially by Franz Schrader and subsequently by Charles Lallemand. The Franco-German border committee, with 25 members, met 6 times between 16 February and 10 July. The central European committee, with 20 members, met on just 3 occasions between 1 March and 17 May. Both came together on 20 October in an unsuccessful attempt to develop a general European programme. Apart from Schrader, Lallemand and Hulot, several academics were involved on both these committees including Lucien Gallois, Louis Raveneau, Emmanuel de Margerie, Georges Blondel, Jean Brunhes and Christian Schefer.[69] The 'colonial' committees were equally busy. The Asia-Oceania committee, chaired by Émile Senart (the President of the Société Asiatique), involved 25 people and met 7 times between 23 February and 12 July.[70] The smaller African commission, chaired by Schrader and comprising 15 members, met just twice in February and June.[71] Those involved on all 4 committees included Lallemand, de Margerie and Schrader but only the indefatigable Hulot and Raveneau attended every *séance*.

The SGP and the political geography of Africa

The African committee began work on 18 February 1916. Its dominant personality was Auguste Terrier, the Secretary-General of the Comité de l'Afrique Française and a close friend of Hulot.[72] According to Terrier, the government required a general African policy based on four important questions, all predicated on a German defeat in Europe. Firstly, how should German colonies be divided amongst the Allies and which territorial claims should France advance? Secondly, was a general post-war re-distribution of African territory between the Allies necessary and how could this be achieved to the benefit of France? Thirdly, how could a new French African empire become economically more profitable? Fourthly, in what ways should pre-war international accords in Africa be revised?

Almost before it began, the African committee was overtaken by events. With Togoland provisionally divided, the next area of potential French expansion was Cameroons. The end of effective German resistance in Cameroons in February 1916 coincided with the resolution of secret Anglo-French talks on the Ottoman empire (see below). The main French and British negotiators – François Georges-Picot and Sir Mark Sykes – briefly considered west-central Africa at the end of their Ottoman discussions. Colonial Minister Doumergue, realising his negotiating position was weakened by the relatively small number of French troops in Cameroons, wanted to avoid discussing partition until the war was over and advocated an Anglo-French condominium in the meantime. The more acquisitive British diplomats in the Admirality and Colonial Offices, led by Sir Frederick (later Lord) Lugard (the Governor-General of Nigeria) and Viscount Harcourt (the former Colonial Secretary), wanted a quick and definitive partition. Like Johnston, they felt Britain should retain north-western Cameroons, including the port at Duala. Concessions to France should be offered only in return for territory elsewhere such as French Somaliland or the French share of the New Hebrides.[73]

Doumergue and Picot decided to stake an optimistic claim to the whole of pre-1911 Cameroons excepting a narrow strip of land along the Nigerian border. Unknown to them, the British Foreign Secretary, Sir Edward Grey, decided on 22 February (the day after the Germans began their assault on Verdun) that pre-1911 Cameroons should be offered to France to prevent suspicion that Britain sought colonial aggrandizement while the French army was bled white on the Western Front. An angry British Colonial Office persuaded Grey to keep his intentions secret until the French position was announced. To Picot's surprise, his ambitious opening proposal was accepted by the British negotiators after a token debate and formally agreed on 4 March. In this way, Britain gained 10 per cent of a country which the Foreign Office had been prepared to surrender in its entirety (see Figure 9.2b and 9.2c).[74] Elsewhere in German Africa, French hopes faded. Doumergue's offer of troops for the East African campaign was turned down (to the relief of the French Ministry of War), undermining French territorial ambitions in the region.[75] The SGP leadership were aware of these developments though censorship restricted published discussion to brief reports on the military conquest of Cameroons.[76]

On 14 June, the African committee re-convened to discuss its findings. Terrier reviewed Anglo-French discussions about Cameroons. 'Thanks to the impression produced in Britain by the defence of Verdun', he stated, 'the British have given us all that we asked . . . This partition is extremely advantageous and is more than even the most optimistic colonialist hoped for.' Elsewhere, France could expect nothing 'despite . . . hopes for east African territory as a link to our possessions in the Indian Ocean'. Given this fact, 'French colonialists must at

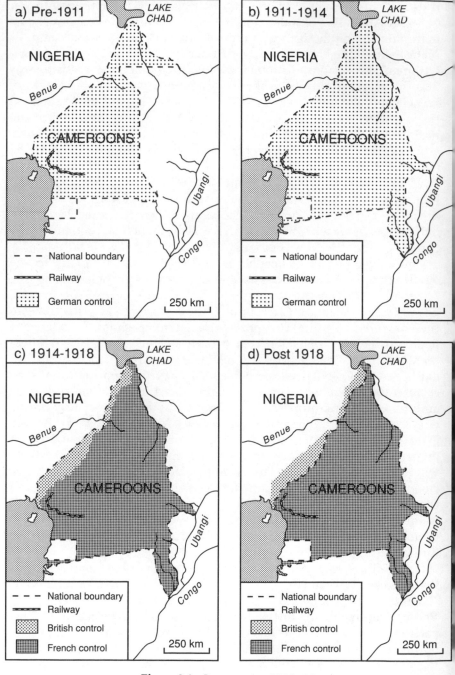

Figure 9.2 Cameroons, 1911–18

least underline the enormous value of British gains in south-west and east Africa with a view to future negotiations, without at this stage openly abandoning all demands in east Africa.' Such realism was in marked contrast to Terrier's ambitions for a re-structured colonial Africa. To replace the chaotic patchwork of colonial enclaves, post-war negotiations should seek 'cohesion and harmony'. Logically, France should have a single, uninterrupted north and west African empire, including the British 'coastal enclaves' of Gambia, Sierra Leone, the Gold Coast and British Togoland. Nigeria was too enormous a prize even for Terrier who contented himself with demands for a modified northern border with Niger and Chad. To counter British attachment to these long-established 'enclaves', France should offer assurances about British commercial interests and, more importantly, provide territorial compensations elsewhere. His preferred option was to cede French possessions in India and their share of the New Hebrides. On the other Allied colonial powers, Terrier was less specific. The 'vast ambitions' of Italian colonialists in north and east Africa, as expressed in 'brutal articles' in the colonial press, should be strongly resisted by France and Britain. Particular care was also needed with Morocco. Any discussion of Spain's role should be avoided 'if it leads to questions about the status of Tangiers'.[77] Portugal's African empire was, Terrier felt, in serious jeopardy; if it collapsed, Portuguese Guinea and the Portuguese islands should be claimed by France. Terrier concluded by insisting that France's African demands must be part of an agreed worldwide colonial policy, a kind of blueprint for the future colonial order and the basis of government policy in future negotiations. 'The SGP is perfectly placed to take charge of this', he insisted, 'because of its age and tradition, the valuable services it has rendered the geographical and colonial cause, its prestige abroad and its wide-ranging, global interests.'[78]

Augustin Bernard, the doyen of French North African geographers, also reported on the revision of international agreements. On the 1906 Algeciras accord, he agreed that 'It is vital to French interests that the question of Tangiers should be avoided and that the existing situation – an international zone with a preponderant role for France – must be maintained.' However, the details of this and other pre-war agreements would need to be revised to protect and enhance French commerce. It was intolerable, claimed Bernard, that Dahomey had effectively become a German rather than a French colony. In 1913, 46 per cent of its trade was with Germany and just 23 per cent with France. Like Terrier, Bernard felt the smaller British colonies in west Africa should be ceded to France. In negotiations with Britain. 'Let us cite ... the need for French logic and simplification and the desire of a geographer to see African territories distributed in a more rational manner.' African civilisation was still in an embryonic stage and territorial changes were both feasible and desirable: 'One hears of a US representative in Lisbon who declared on his arrival, with an innocence characteristic of American diplomats, that he could see no reason why Portugal and Spain were separate states. This, of course, ignores the traditions and secular antagonisms which created the older nations of Europe. In Africa, the re-distribution of territory is still possible and often desirable.' A committed Africanist, Bernard concluded his report by drawing on the ideas of Onésime Reclus about the need to concentrate French imperial interests in Africa and gradually abandon the far-flung Asian possessions.[79] For Bernard, this implied carefully weighing any potential gains in the Middle East against their impact on France's African ambitions. 'African questions are actually linked to the questions raised by the division of the Turkish empire . . . [Our policies should be based] on the principle of concentration and in opposition to a policy of dispersion . . . Given the situation of our merchant navy, the weakness of our birth-rate and the formidable loss of men and capital which the Great War will produce, a global imperial policy is no longer feasible.'[80] Having established its agenda, most members of the African committee turned their attention to Asia and the Middle East, in effect merging the two colonial committees from mid-June 1916.

The SGP and the division of the Ottoman empire

The Asia-Oceania committee first met on 23 February 1916. Its six subsequent meetings examined five questions in detail: French India, the New Hebrides, Mauritius, China and, most importantly, the Middle East. On 8 March, the committee debated a long report from Guillaume Capus on the status of French India. This bluntly concluded that, with the possible exception of Pondichéry, the Indian enclaves should be offered to Britain in exchange for concessions elsewhere.[81] This argument was reinforced on 29 March by reports from Henri Froidevaux and Jules Harmand, a former Consul-General in Calcutta.[82] Both agreed these scattered outposts had been retained for sentimental reasons only. Harmand, in particular, was unmoved: 'Our persistence [in India] is a humiliation, a permanent reminder of past defeat'. Further debate on 10 and 24 May confirmed this view. The only dissenting voice was that of Charles Humbert, the Senator for the Meuse, who noted that plans to cede French India 'were based on the [unproven] proposition that the British have the same mentality as ourselves, that is to say an abstract desire to remove enclaves which disrupt the geographical unity of their empire'. Even Humbert felt that the only possession worth holding was an enlarged Pondichéry.[83]

On 5 July, two surveys were presented by Paul Carié and Henri Froidevaux about Mauritius and the New Hebrides.[84] The former had been occupied by the British during the Napoleonic wars and the latter had been governed by an unsatisfactory Anglo-French condominium since 1906. Carié concentrated on the Mauritians' 'devotion' to France and ended with a fresh claim to the island while Froidevaux reluctantly concluded that, unless Britain agreed to withdraw, France should regard the New Hebrides as collateral against gains elsewhere. These suggestions were supported on 12 July by an astute report from Paul Bertin on France's role in the Pacific where British, Japanese and American ambitions jostled with those of France and with the emerging sub-imperialism of Australia and New Zealand.[85] These reports, which formed the basis of SGP resolutions forwarded to the government, had a clear message, consistent with the African committee's conclusions. Insofar as circumstances allowed, France should seek to consolidate imperial gains in and around the African continent. To this end, concessions should be offered in Asia and the Pacific in return not only for Mauritius, which would become part of an off-shore 'African' empire based on Madagascar, but on the understanding that French claims in Africa would be favourably received.

This raised the question of France's future in the Middle East, the subject of a long-awaited report delivered on 28 June by Robert de Caix.[86] Allied policy towards the Middle East was extremely confused. Prior to 1914, both Britain and France had attempted to maintain the crumbling Ottoman empire as a bulwark against the southward expansion of Russia. Turkey's entry into the war in November 1914 made the Ottoman empire a target for Allied strategists and imperialists. By 1915, with stalemate on the Western Front, the idea of a sea-borne offensive against Constantinople through the Straits of the Dardanelles began to preoccupy the British, particularly Winston Churchill and civil servants in the Admiralty. The disastrous Gallipoli campaign, which ground on pointlessly and murderously through 1915, served only to intensify Allied intrigues in the region. At the secret Treaty of London, signed on 26 April 1915, Italy entered the war on the Allied side in return for vague promises of an 'equitable share' in the Ottoman empire. The following summer and autumn, Britain ambiguously promised the Arab Hashemites an independent state under British protection covering Syria, Transjordan, Palestine, northern Iraq and most of Arabia in return for their support against the Turks (leaving just the Persian Gulf and Mesopotamia under direct British rule).[87]

The Sykes–Picot talks, which began at Grey's suggestion, were partly to clarify matters.[88] The first round opened on 23 November 1915. Picot, the lone French delegate armed with a brief he had drafted himself, had been French consul in Beirut since March 1914 and expected to encounter Lord Kitchener, the War Minister and leader of the so-called 'Egyptian party' whose desire to protect the Suez route to India effectively determined British policy in the Middle East. In the event, he met a group of senior diplomats to whom he rehearsed French claims to *la Syrie intégrale*, a vaguely defined zone limited in the north by the Taurus mountains beyond Adana (in modern-day Turkey) and in the south by the Egyptian border. *La Syrie intégrale* was to expand from the Mediterranean a varying but unspecified distance into the desert interior. This directly clashed with British ambitions in Palestine and with the proposed independent Arab state. Further discussions were held on 21 December 1915 when Sir Mark Sykes, Tory MP for Hull and Kitchener's devoutly Catholic adviser on Oriental matters, emerged as the principal British negotiator. Thereafter, almost daily meetings took place and gradually an agreement was reached, signed on 3 January 1916 and passed by both cabinets the following month subject to an Arab rising and Russian ratification. This provisionally divided the entire region into separate zones under the direct control of Britain, France, Russia and Italy. The 'independent' Arab state was further divided into 'spheres of influence' under British or French military protection and economic-political domination. Palestine was to become an international zone with the exception of the port towns of Acre and Haifa which were reluctantly ceded to Britain. Following Russian modification of the original plans, the Sykes-Picot accords were ratified in mid-May (see Figure 9.3).

The SGP's Asian committee was well-informed about these secret negotiations largely because Robert de Caix was Picot's closest confidant.[89] De Caix was one of trio of senior colonialist diplomats at the Ministry of Foreign Affairs who effectively ran the Comité de l'Asie Française, the others being Philippe Bertholet, Briand's powerful *chef de cabinet*, and Pierre de Margerie, the Ministry's *directeur politique* and a cousin of geographer Emmanuel.[90] These men had worked hard to expand French influence in the pre-war Ottoman Middle East in the face of stiff competition, especially from Germany. After Turkey's entry into the war, several members of the Comité de l'Asie Française, including Picot, unsuccessfully advocated sending a French expeditionary force to the Lebanon and Syria to protect French interests and pave the way for post-war French hegemony.[91] The predicted collapse of Turkish authority and the likely scramble for control of the region made it necessary to devise carefully thought out territorial claims for a future French zone. *La Syrie intégrale* was the preferred design of de Caix and the Comité de l'Asie française.

Despite the Sykes-Picot agreement, de Caix wanted to keep his plan

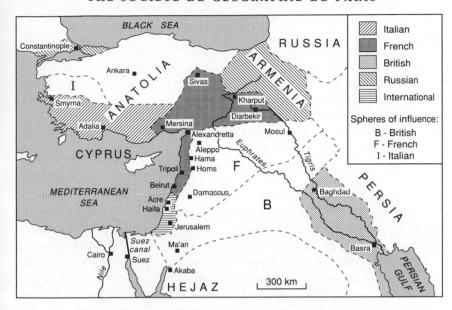

Figure 9.3 Wartime division of the Ottoman empire, 1915–17

alive as an option for future negotiations. The arguments in favour of *la Syrie intégrale* were based largely on cultural, historical and religious rationales. Greater Syria was the land of the Crusades where Frankish blood had flowed for centuries in defence of Christendom. Catholic France, *la fille la plus ainée de l'eglise*, was therefore the most historically legitimate Great Power guardian of the Holy Land. Italy was too weak for the task and control by Protestant Britain would alienate European Catholics. By custom, France had offered protection to the Christian and Maronite communities of the Levant, an arrangement formalised in 1860. French was also the dominant language, though it was under pressure from English, Italian and German. There were over 500 French-run religious schools in the Levant with more than 100,000 pupils (10 per cent of all those educated in the Ottoman empire). Even anti-clerical republican colonialists accepted the legitimacy of these claims. As Léon Gambetta memorably remarked, 'Anti-clericalism is not an article for export.'[92] Economically, too, French finance was extremely important not only in the Levant but throughout the Ottoman empire. While its own empire was starved of capital and trade, French investment in the Ottoman empire before 1914 was more than three times that of Germany, the next largest investor. France also held 60 per cent of the Ottoman public debt and supplied 45 per cent of foreign capital to the private sector.[93]

De Caix's report re-stated familiar arguments against Turkey's 'unsustainable' authority and alluded to the long-term ambitions of Britain, Italy and Russia. France, he insisted, must not 'abandon to others a terrain marked by seven centuries of French effort'. In an oblique reference to the arguments of Augustin Bernard and Onésime Reclus that post-war France would lack the capacity to sustain a non-African empire, de Caix countered that Syria would require no costly military 'pacification' nor a complex and expensive administration. In light of the Sykes-Picot accords, de Caix was anxious to dismiss the idea of an independent Arab state. There was no tradition of self-rule in the area and Arab nationalism was a mere chimera: 'how in this mosaic of disparate and often hostile communities can one imagine an autonomous Syria?' This was supported by Emmanuel de Martonne in a report to the same *séance*: 'Ethnographically, Syria presents a mixture of races and religions and it is impossible to speak of a Syrian nationality or to pretend that there is a general sentiment in favour of complete autonomy. Syria for the Syrians is a meaningless idea.'[94]

Two days later, at the meeting of the SGP's Central Commission, a proposal was passed for submission to the government insisting that 'geographically, ethnically and historically' the separation of Palestine from Syria was untenable. The entire region should be seen as a single territory to which France had historic rights. The 'irrational' internationalisation of Palestine was condemned as 'a source of endless discussion and perhaps conflict between those who are today Allies'. Showing a remarkable ability to delude itself about its own role, the SGP further insisted that to accept British claims would be 'a concession not to the real aspirations of the British people, but to the narrow and suspicious minds of colonial specialists who have no real interests in the Syrian question'.[95]

On 17 July, a deputation comprising Hulot, Marin, Schrader, de Margerie and de Martonne was received by Briand to discuss SGP resolutions on Africa, Asia and, most importantly, the Middle East. Briand was welcoming but cautious. He acknowledged the SGP's African and Asian policy and accepted that pre-war and wartime agreements would be re-negotiated. On the Middle East, he pointed out that Britain would certainly claim its share of the Ottoman spoils and that France would have to accommodate this. In the meantime, he insisted that publication of the SGP's recommendations would not be in the national interest.[96] By way of compensation, and as part of a concerted attempt to head off growing socialist opposition to the war, Briand agreed to provide government funds to support a series of Sunday afternoon open lectures in Paris by senior members of the SGP 'to clarify public opinion about Germany's war aims, exorbitant ambitions and the danger they present not only for France but for the entire world'. These lectures, which had begun on 5 June and continued until 24 April 1917, were revealingly concentrated in the poorer, working-class districts to the north, north east and south of the city around Montmartre, La Villette and the Place d'Italie (the ninth, tenth, thirteenth and eighteenth *arrondissements*), precisely the areas where socialist, anti-war attitudes were most widespread.[97]

By the end of the year, Hulot could only hope that the SGP's proposals had filtered through to Briand's cabinet and influenced its policy. On 18 December 1916, US President Woodrow Wilson requested all belligerent nations to define their war aims. Here was the perfect opportunity for a clear public declaration of French territorial claims in Europe and overseas. Unfortunately, despite the SGP's efforts, French demands were vague and hedged around with provisos and caveats. Most disturbing for French colonialists, Asquith's government fell in the same month, replaced by Lloyd George's coalition cabinet including two leading Francophobe imperialists, Lords Curzon and Milner. The new British Prime Minister was himself also famously ill disposed towards French territorial claims, especially in the Middle East. On 26 December 1916, Senart gloomily informed Hulot that the SGP's work was destined to have little impact: 'I must confess . . . I find these researches, alas, too academic. I certainly don't underestimate their interest, but so far as their practical application in the near future is concerned, I am a little discouraged.'[98]

The SGP and the end of the war

The early months of 1917 were a turning point in the war. The collapse of the Tsarist regime in February and the subsequent withdrawal of Russia from the war removed the main ally for France's Middle Eastern ambitions. The Bolshevik habit of publishing the texts of earlier secret treaties also caused chaos, sending French, British and Italian diplomats scurrying back to their maps and charts. On 20 March, Briand was replaced by Alexandre Ribot as Prime Minister amidst growing, mainly left-wing, advocacy of a compromise peace. For the colonialists in the SGP, this was particularly worrying as a negotiated peace would probably involve surrendering part of the Fench overseas empire to Germany in return for Alsace-Lorraine.[99] The United States' entry into the war on 6 April was also a mixed blessing. While American troops would provide much-needed support on the Western Front, where German armies from the eastern theatre were now amassing, their involvement meant Wilson's anti-imperialism and idealistic commitment to national self-determination would influence future negotiations, to the obvious detriment of France's imperial vision.[100]

In these circumstances, political pressure for an official war aims committee intensified and the SGP was no longer able to play its central, co-ordinating role. Official attention focused on Europe. The SGP's two European committees, which had fizzled out in October 1916, had been re-launched by Briand as the Comité d'Études comprising a group of distinguished academics chaired by the historian Ernest Lavisse and with geographers Paul Vidal de la Blache and Emmanuel de Martonne as Vice-Chair and Secretary. The Comité d'Études met for the first time on 17 February and re-convened each week until 2 June 1919 in a smoke-filled room in the Institut de Géographie at the Sorbonne. At its second meeting, on 23 February, Lavisse defined the Comité's geographical brief as Europe and Turkey-in-Asia. By the end of the war, however, only the exhaustive historical and geographical research on France's Rhineland policy and on the claims to an enlarged Alsace-Lorraine, including the Saar, had been fully completed.[101]

The colonial lobby refused to give up its campaign, particularly in view of the claims advanced by colonialists in other Allied countries. Untrammelled by draconian censorship laws and apparently with the support of their Foreign Minister, Baron Giorgio Sonnino, Italian colonialists had been pressing for huge slices of territory along the Adriatic, in Anatolia, the Middle East, Arabia, and northern and eastern Africa. These were examined in a long-delayed report by Camille Fidel, Secretary-General of the Société des Études Coloniales et Maritimes and Secretary of the Ligue Coloniale Française. Fidel's 97-page document was originally commissioned by the SGP's African committee in February 1916 but was not completed until the following October.[102] Like most colonialists, Fidel knew from experience that the French imperial cause was best served by enflaming public opinion about the imperial ambitions of other states. French governments had proved extremely reluctant to promote their imperial claims unless national pride appeared to be at stake. Fidel's objective was to frighten political leaders into action. From his perspective, it was immaterial whether the threat came from an enemy state such as Germany of 'friendly' allies such as Britain or Italy. Using dramatic quotations from Italian colonialists such as Guiseppe Piazza, the director of the Rivista Coloniale, and from newspaper articles in L'Idea Nazionale, Fidel painted a disturbing picture of Italian plans: 'Italian public opinion is at present obsessed by, and well-informed about, post-war colonial problems whilst French public opinion wallows in ignorance.' The objective of the Italian government was to dominate the eastern Mediterranean and extend its African empire as far south as Lake Chad and Abyssinia thus changing the Triple Alliance into a Quadruple Alliance. France would be the principal victim, losing vast tracts of north and central Africa as well as French Somaliland. At the end of his long discourse, Fidel recommended that, 'as an extreme concession', France should consider allowing Italy the southern part of Anatolia limited by a line from Smyrna (Izmir) in the east to Adana in the west provided Italy supported French claims to *la Syrie intégrale*, including the port of Alexandretta. Other minor concessions might also be possible in the Sahara along the Algerian–Libyan border. Elsewhere Fidel called on the government forcefully to reject Italian claims.[103]

Fidel's alarming scenario prompted the SGP to re-convene its Asian and African committees on 30 March. Mindful of Briand's publication ban, the 15-man joint committee (chaired by Lallemand and including, for the first time, the ageing Eugène Étienne) recommended that the report and its associated proposals be passed to the Ministry of Foreign Affairs, the Colonial Ministry and leading colonialist politicians in the Chamber of Deputies and the Senate.[104] An audience was also demanded with the newly installed Prime Minister, Alexandre Ribot, and provisionally fixed for 12 May but subsequently postponed. Marin once again assumed a leading role in the campaign. On 5 May, he wrote to Hulot suggesting a high-powered meeting of representatives from all the leading French colonial societies in the SGP to agree a common policy document. This would, he felt, force the government into action: 'I believe we will obtain, either through Ribot or other personalities, a favourable response to our plans for *la Syrie intégrale* despite the poor reception it was originally given . . . I think the SGP would be rendering an important service which would earn it, in the future, an immense debt of gratitude. Confidentially, I can inform you that the new government is very poorly informed because Foreign Affairs do not have accurate documents on all colonial questions.'[105] The time was ripe for another collective *démarche*. Hulot agreed and on 11 May a Groupe d'études pour l'examen des questions relative au règlement de paix met in the SGP. This group, made up of familiar names from the Asian and African committees, re-asserted the same territorial objectives already agreed on.[106] On 23 May, a delegation representing this new group, including Hulot, Marin, Senart, Schrader, Lallemand, Labbé, de Caix, Terrier and Fidel, was received by Ribot. Like Briand, Ribot assured his guests that their views had been heard but was clearly concerned that premature colonial demands would cause tension in the alliance.[107]

Towards the end of 1917, the SGP's colonial campaign finally began to influence ministerial decisions as politicians and diplomats at last took responsibility for policy making. Under the short-lived Colonial Ministry of René Besnard (12 September to 16 November 1917), a

Commission de documentation was created to prepare the ground for a post-war colonial settlement. The following February, four months after the Balfour Declaration on a Jewish homeland in Palestine added the final nail in the coffin of the Sykes–Picot accords, Henri Simon (who succeeded Besnard in the Rue Oudinot after Georges Clemenceau assumed the Premiership in November 1917), established a more impressive Commission d'étude des questions coloniales posées par la guerre comprising senior politicians and diplomats such as Étienne, Doumergue, Berthelot and Pierre de Margerie. Both these government committees drew extensively on the documents and resolutions put forward over the preceding eighteen months by the SGP in their rear-guard efforts to maintain French territorial claims.[108]

Étienne was evidently reluctant to leave matters in the hands of a government committee, even one on which he sat. In the spring of 1918, on the eve of the last-ditch German attempt to break the deadlock on the Western Front, Étienne launched a parallel, unofficial attempt to obtain an agreed programme of colonial objectives. Étienne had suggested to Hulot that the SGP host a general colonial committee at almost exactly the same time as Marin the previous May.[109] As President of the Ligue Coloniale Française, he had also established his own, rather ineffective colonial war-aims committee as early as February 1916, just as the SGP began its deliberations. With support from Simon, Étienne persuaded 29 leading colonialists, representing 39 colonial societies from Paris, the provinces and North Africa, to meet in the SGP on 9 March to plan the final phase of their campaign. This spawned a smaller committee, including Étienne, de Caix, Fidel, Froidevaux, Schefer, Terrier and Marin, which was to draft an agreed document.[110] Before anything could be achieved, however, all thoughts were diverted to the Western Front where the German spring offensive and the Allied counter-offensive finally brought the carnage to a close. In the midst of this, both de la Blache and Hulot died, the former suddenly on 5 April, the latter after an illness on 28 June. Although de la Blache had kept a discrete distance from the SGP's colonial campaign, his death deprived the geographical movement of its intellectual focus while the passing of Hulot removed the SGP's principal organisational talent.

Conclusion

As the politicians and diplomats took over, the role of all but the most politically experienced members of the SGP began to decline. Over the last months of the war, and at the post-war peace negotiations in Paris, the SGP's colonial proposals and blue-prints were modified and often transformed during acrimonious discussions at the highest level between Clemenceau, Lloyd George, Wilson and the other political leaders. French fears that American idealism, mixed with a growing mood of socialist anti-imperialism in Europe, would be major stumbling blocks to the uncomplicated annexation of Ottoman and German lands were proved justified. The compromise between imperialists and anti-imperialists was the system of mandates, first mooted by Jan Smuts. According to this, the former Ottoman and German territories would be administered by one of the allied colonial powers under the auspices of the new League of Nations, thus involving elements of both inter-national and national-colonial control. Mandates were not colonies, the rhetoric went, but were to be 'held in sacred trust for the good of civilisation'. Most observers – both imperialists and anti-imperialists – justifiably suspected that mandates, particularly those in the tropics, would quickly be absorbed into the existing colonial empires. For J. A. Hobson, mandates were merely a 'veil for the annexation of enemy countries and the division of the spoils'. According to Wm Roger Louis, after 1918 'imperialism wore an increasingly elaborate fig-leaf'.[111]

The British empire gained most from the peace process. The wartime architects of British imperial policy in the Foreign, Colonial and India Offices were, like their Victorian predecessors, preoccupied with the need to protect the trade routes to India. Their objective was to secure and extend British rule around the rim of the Indian Ocean and British claims in the Middle East and east Africa were dictated by that broad geopolitical aim. Despite wartime agreements, Britain was granted the Middle Eastern mandates for Palestine, Transjordan and Iraq together with most of German East Africa, a modified share in Togoland and most of the unexpected strip of Cameroons negotiated in 1916 (see Figures 9.1c, 9.2d, 9.3 and 9.4).[112] By the early 1920s, newly published atlases showed British red in two wide and continuous belts from Palestine eastwards to the Persian Gulf, India, Burma and Australasia; and from the Suez Canal southwards across east and central Africa to the Cape. The great British imperial dream of an African empire stretching from Cairo to the Cape was, at last, a reality.[113]

Many observers, in Britain and elsewhere, considered this huge empire to be strategically unsustainable.[114] Partly in response to this fear (which pre-dated the 1914–18 war), imperial responsibilities were progressively shared between the British government in London, the Dominion governments in Canada, South Africa, Australia and New Zealand (which themselves became sub-imperial powers) and the increasingly independent authorities in Egypt and Iraq. Although it expanded territorially after 1918, British imperial power became more diffuse and the amount of territory governed directly from Whitehall declined sharply. By 1939, the empire administrated from London

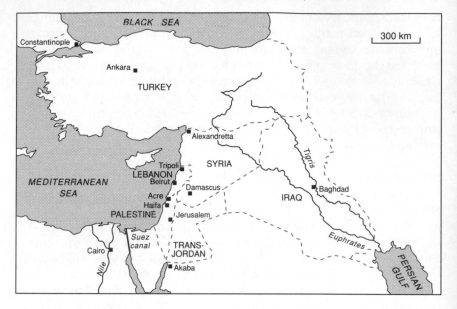

Figure 9.4 Post-war division of the Ottoman empire

covered less than 14 million square kilometres, though this still represented nearly 450 million people.

The French empire also expanded as a result of the war though its gains fell far short of the SGP's wartime ambitions. France assumed the mandates from Lebanon and Syria (the vestiges of *la Syrie intégrale*) as well as most of Togoland and the bulk of Cameroons (see Figures 9.1c, 9.2d, 9.3 and 9.4). In total, this amounted to more than three-quarters of a million square kilometres inhabited by six million souls. The other pawns in the complex wartime debates – the Indian enclaves, the New Hebrides, Mauritius, Djibouti and French Somaliland – were all unaffected. Yet while the British shed some of their imperial responsibilities in, and to, the Dominions, the French clung tenaciously to their overseas possessions. By 1939, the French empire covered nearly 12 million square kilometres and had a population of almost 69 million people.[115]

The war had a discernible impact on the rhetoric, if not the reality, of French imperialism. In the wave of euphoria that followed the armistice, the colonial lobby was quick to emphasise the debt France owed to the people and resources of the empire. Never again should France allow its empire to stagnate, deprived of investment and commerce from the mother country. Victory in 1918 had only been achieved through the belated recognition of the empire's enormous potential value. A new

developmental ethos swept through colonialist circles during the 1920s
– the so-called *mise-en-valeur* school of imperial thought – associated
with Camille Fidel, Henri Simon, Albert Sarraut and, especially,
Edmond du Vivier de Streel. Sadly, the imperial investment and
development they called for was rarely forthcoming.[116]

The foregoing discussion raises two related questions. Firstly, precisely what impact did the French geographical movement have on
imperial policy during and immediately after the war? There is no
simple answer to this for it is never easy (or sensible) to draw a rigid
division between the state (the preserve of executive politicians and
policy makers) and the academy (the haven of intellectuals and opinion-
formers). Insofar as it is possible to identify two such arenas, the line
between them is always extremely ill defined, particularly in times of
national crisis. The locus of political and intellectual power in a modern
state is, by definition, extremely diffuse and key individuals are able to
move so effortlessly between academic and political environments that
separating the two becomes highly problematic. During the Great War,
identifying a clear professional distinction between French geographers
on the one hand and French politicians or diplomats on the other is far
from easy. French colonialist geographers did not so much influence
their political leaders as assume the mantle of their responsibility in
certain areas of state policy.

The conspicuous executive role of professional academics in
formulating state policy during the 1914–18 war illustrates the complex
inter-weaving of science and politics in France which has a long, and not
always honourable, tradition.[117] As respected and influential intellectuals, the geographers were able to negotiate as a co-ordinated group
directly with their political leaders and were invited to identify policy
objectives. After the war, too, French geographers – led by Emmanuel de
Martonne, Emmanuel de Margerie, Albert Demangeon, Lucien Gallois
and Jean Brunhes – were recruited as high-ranking experts on the Service
Géographique Française, not only to advise the French delegation to the
Peace Conference, but actually to delimit the borders of the new
European geopolitical order. The revised frontiers of Rumania, for
example, owed a great deal to work carried out by de Martonne first on
the French Comité d'Études and subsequently on the Rumanian
Territorial Commission at the Peace Conference.[118] Other nations also
used their leading geographers in a similar, elevated fashion. Isaiah
Bowman, the President of the American Geographical Society in New
York, was a dominant figure on the US government's mammoth House
Inquiry into the geopolitical problems of the post-war world.[119] The
British, by contrast, tended to rely on the technical expertise of their
civil servants and on the negotiating skills of their politicians.[120]

French geographers were not reluctantly press-ganged into the service of the state. An overwhelmingly patriotic, republican group, they genuinely and understandably believed their country to be involved in a just war and, like their colleagues in other countries, they willingly offered their skills and expertise with few thoughts for the larger moral questions raised by their involvement. If, as a result of their war efforts, they could promote a broader ideological and intellectual agenda, then so much the better.[121] As the war progressed, French geographers became a well organised and influential pressure group, working behind the scenes to support their vision of an expanded colonial France. Ideally positioned to exploit wartime secret diplomacy, the geographers became precisely the sort of lobbyists that E. D. Morel contemptuously referred to as 'the flotsam and jetsam which crowd the diplomatic corridors'.[122] Yet despite their frenetic activity, huge ambitions and significant influence within political circles, the impact of French geographers on the post-war geopolitical order was limited by the highly personalised nature of international relations and diplomatic discussion. Most of the more important decisions were taken by a very small group of political leaders. In their private discussions, these men were often obliged to take decisions which contradicted years of painstaking geographical and historical research in support of a particular policy or territorial claim.

The second important question raised by this account is whether the wartime activities of French geographers influenced the subsequent development of the discipline? At one level, the outbreak of the Great War had an immediate and ostensibly detrimental impact on the subject. The free flow of ideas produced by international conferences and the international movement of academics all but ceased during the dark days of 1914–18.[123] Yet this did not mean that intellectual activity was somehow suspended while the war continued. On the contrary, professional geographers in France – and elsewhere – were busier than ever. The war, and the new world order which it created, spawned new sets of geographical questions and new kinds of geographical research.[124]

Perhaps the principal distinguishing characteristic of post-war geography was its heightened sense of politics. Pre-war geography had already become, by its very nature, an imperial science and, as such, an intensely political discipline. Indeed, several of the most influential geographers of the pre-1914 generation were pioneers in the study of political geography and global geopolitics, most notably Sir Halford Mackinder in Britain and Friedrich Ratzel in Germany.[125] However, the majority of geographers would have insisted that their pre-war work was objective and apolitical. The fairly obvious ideological and political implications of their activities were rarely acknowledged. During the war, and for several years afterwards, the politics which had occupied the unexamined core of geographical inquiry was forced into the open. The role French geographers played as political propagandists and political advisers during and after the war influenced the kinds of questions that they asked about the new world order. After 1918, political geography became one of the major pillars of geographical research and teaching not only in France but across Europe and America. Throughout the early 1920s, the pages of virtually every French geographical journal were filled with detailed and well informed analyses of the new states and of the broader geopolitical and economic order. The same can be said of other geographical journals around the world. The crowning achievement of this episode was undoubtedly Isaiah Bowman's *The New World: problems of political geography* (1922), probably the most widely read geography book of the inter-war period.[126] In those countries which felt most aggrieved by the peace accords, Germany and Italy, post-war political geography acquired disturbing and openly partisan characteristics. Under Karl Haushofer and Ernesto Massi, academic geopolitics in Germany and Italy became notoriously associated with the frightening ambitions of the Nazi and Fascist regimes in those countries.[127] The excesses of geopolitical research before and during World War Two tainted the entire sub-discipline. After 1945, teaching and research in political geography and geopolitics were actively discouraged througout Europe and North America and it is only since the 1980s that political geography has re-emerged to skake off its unsavoury past.[128]

It is often claimed that the Great War transformed the moral, cultural and intellectual climates of the nations that were involved. Certainly, the values and outlooks of those who witnessed at first hand the carnage of the battlefields seem to have been utterly transformed. The war, it is said, ushered in a new and disturbing era of twentieth-century modernity.[129] For academics who fought on the home fronts with pens and typewriters rather than howitzers and machine guns, the war also had a profound intellectual impact. The business of teaching and research does not take place in a political vacuum and is always, in some degree, determined by political context. The events of 1914–18 created not only a new global geography but, to some extent, a new discipline of geography which sought new ways to conceptualise the modern world.

Notes

1 For a broader analysis of the connections between geography, geopolitics and war, see Yves Lacoste, *La géographie, ça sert, d'abord, à faire la guerre*, Paris, 1976.
2 On the SGP, see Alfred Fierro, *La Société de Géographie de Paris (1826–1946)*, Geneva and Paris, 1983 and the definitive work of Dominique Lejeune, particularly

'La Société de Géographie de Paris dans le mouvement social de la première moitié du XIXe. siècle (1821–1864)' and 'La Société de Géographie de Paris dans le mouvement intellectuel du XIXe. siècle (1821–1864)', *Actes du 104e. Congrès National des Sociétés Savantes: Bordeaux 1979*, Paris, 1980, pp. 27–41 and pp. 43–56; 'La Société de Géographie de Paris: un aspect de l'histoire sociale fançaise', *Revue d'Histoire Moderne et Contemporaine*, 29 (1982), pp. 141–63; Les Sociétés de Géographie en France dans le mouvement social et intellectuel du XIXe, siècle', unpublished Doctorat d'État ès Lettres, Université de Paris X – Nanterre, 1987, 4 vols; *Les sociétés de géographie en France et l'expansion coloniale au XIXe. siècle*, Paris, 1993.

3 Fierro, *La Société*, p. 287.

4 SGP prizes were offered, *inter alia*, for the triangulation of the Alps (1,500 francs in 1823), for a statistical investigation of road and canal commerce between Paris and Le Havre (600 francs in 1824), for a geographical survey of the Russian language (500 francs in 1824) and to the first European to return from Timbuctoo (9,000 francs in 1827). Fierro, *La Société*, p. 247.

5 Maxine F. Taylor, 'Nascent expansionism in the Geographical Society of Paris, 1821–1848', *Proceedings, Western Society for French History*, 7 (1979), pp. 229–38; Christian Schefer, *La politique coloniale de la Monarchie de Juillet: l'Algérie et l'évolution de la colonisation française*, Paris, 1947.

6 M. R. Buheiry, 'Anti-colonial sentiment in France during the July Monarchy: the Algerian case', Unpublished Ph.D. thesis, University of Princeton, 1974.

7 C. M. Andrew and A. S. Kanya-Forstner, 'Centre and periphery in the making of the second French empire, 1815–1930', *Journal of Imperial and Commonwealth History*, 16 (1988), pp. 9–34; A. S. Kanya-Forstner, *The Conquest of the Western Sudan: a study in French military imperialism*, Cambridge, 1969.

8 Fierro, *La Société*, p. 287. The RGS Fellowship was over 2,300 in 1870.

9 Fierro, *La Société*, pp. 51–87.

10 Michael Heffernan, 'The science of empire: the French geographical movement and the forms of French imperialism, 1870–1920' and Olivier Soubeyran, 'Imperialism and colonialism versus disciplinarity in French geography', both in Anne Godlewska and Neil Smith (eds), *Geography and Empire*, Oxford, 1994, pp. 92–114 and pp. 244–64; Vincent Berdoulay, *La formation de l'école française de géographie (1870–1914)*, Paris, 1981, pp. 45–75; D. V. MacKay, 'Colonialism and the French geographical movement', *Geographical Review*, 33 (1943), pp. 214–32; Agnes Murphy, *The Ideology of French Imperialism, 1871–1881*, Washington DC, 1948, esp. pp. 1–40; William H. Schneider, 'Geographical reform and municipal imperialism in France, 1870–80', in John M. MacKenzie (ed.), *Imperialism and the Natural World*, Manchester, 1990, pp. 90–117. See also Brian Hudson, 'The new geography and the new imperialism', *Antipode*, 9 (1977), pp. 12–19; Felix Driver, 'Geography's empire: histories of geographical knowledge', *Environment and Planning D: Society and Space*, 10 (1992), pp. 23–40; Thomas Richards, *The Imperial Archive: knowledge and the fantasy of empire*, London, 1993.

11 *La République Français*, 27 November 1871, p. 3. Partially quoted in Mona and Jacques Ozouf, 'Le tour de France par deux enfants', in Pierre Nora (ed.), *Les lieux de mémoire: vol. II – La République*, Paris, 1986, pp. 291–312 on p. 318.

12 Ernest Lavisse, 'The fatherland', in Ferdinand Buisson and Frederic Ernest Farrington (eds), *French Educational Ideals of Today: an anthology of the molders [sic] of French educational thought of the present*, London, 1919, pp. 92, 94, 99. See also Jacques and Mona Ozouf, 'Le thème du patriotisme dans les manuels primaires', *Le Mouvement Social*, 49 (1964), pp. 3–32; Numa Broc, 'Histoire de la géographie et nationalisme en France sous la IIIe. République (1871–1914)', *L'Information Historique*, 32 (1970), pp. 21–6.

13 See Numa Broc, 'L'Établissement de la géographie en France: diffusion, institutions, projets (1870–1890)', *Annales de Géographie*, 83, 459 (1974), pp. 545–68; Numa Broc, 'Nationalisme, colonialisme et géographie: Marcel Dubois (1856–1916)', *Annales de Géographie*, 87, 481 (1978), pp. 326–33; Berdoulay, *La formation*, pp. 77–108, esp. pp. 91–106. See also George Weisz, *The Emergence of the Modern Universities in*

France, 1863–1914, Princeton, 1983; Harry W. Paul, *From Knowledge to Power: the rise of the science empire in France, 1860–1939*, Cambridge, 1985.

14 See Fierro, *La Société*, pp. 75–8 and 287; George Kish (ed.), *Bibliography of International Geographical Congresses 1871–1976*, Boston, 1979; Mechtild Rössler, 'La géographie aux congrès internationaux: échanges scientifiques et conflits politiques', *Relations Internationales*, 62 (1990), pp. 183–8.

15 Useful accounts of French imperialism after 1870 include Henri Brunschwig, *French Colonialism, 1871–1914: myths and realities*, London, 1966; Jean Ganiage, *L'Expansion coloniale de la France sous la Troisième République 1871–1914*, Paris, 1968; Raoul Girardet, *L'Idée coloniale de la France de 1871 à 1962*, Paris, 1972; Jacques Marseille, *Empire coloniale et capitalisme français: histoire d'un divorce*, Paris, 1984.

16 See, for example, Charles Mangin, *La force noire*, Paris, 1911.

17 Raymond F. Betts, *Assimilation and Association in French Colonial Theory 1890–1914*, New York, 1961; John Chipman, *French Power in Africa*, Oxford, 1989; L.-J. Calvet, 'Le colonialisme linguistique de la France', *Les Temps Modernes*, 324–6 (1973), p. 72–89; M. D. Lewis, ' "One hundred million Frenchmen": the assimilation theory in French colonial policy', *Comparative Studies in Society and History*, 4 (1961–2), pp. 129–53.

18 Christopher Andrew and A. S. Kanya-Forstner, *France Overseas: the Great War and the climax of French imperial expansion*, London, 1981, p. 26. See also J. Bouvier,'Les traits majeurs de l'impérialisme français avant 1914', *Le Mouvement Social*, 86 (1974), pp. 99–128.

19 See Charles-Robert Ageron, *L'Anticolonialisme en France de 1871 à 1914*, Paris, 1973; Charles-Robert Ageron, *France coloniale ou Parti Colonial?*, Paris, 1978.

20 Jules Ferry's administration collapsed in 1885 after fierce opposition, led by Georges Clemenceau, to his expansive colonial policies in Indochina. See Charles-Robert Ageron, 'Gambetta et la reprise de l'expansion coloniale', *Revue Française d'Histoire d'Outre-Mer*, 59, 215 (1972), pp. 165–204; Charles-Robert Ageron, 'Jules Ferry et la colonisation', in François Furet (ed.), *Jules Ferry: fondateur de la République*, Paris, 1986, pp. 191–206.

21 Herward Sieberg, *Eugène Étienne und die Französische Kolonialpolitik*, Cologne, 1968.

22 C. M. Andrew and A. S. Kanya-Forstner, 'The French "Colonial Party": its composition, aims and influence, 1885–1914', *The Historical Journal*, 14, 1 (1971), pp. 90–128; C. M. Andrew and A. S. Kanya-Forstner, 'The *Groupe Colonial* in the French Chamber of Deputies 1892–1932', *The Historical Journal*, 17, 4 (1974), pp. 837–66; C. M. Andrew and A. S. Kanya-Forstner, 'Gabriel Hanotaux, the Colonial Party and the Fashoda strategy', *Journal of Imperial and Commonwealth History*, 3 (1974), pp. 55–104; C. M. Andrew, P. Grupp and A. S. Kanya-Forstner, 'Le mouvement colonial français et ses principales personnalités (1890–1914)', *Revue Française d'Histoire d'Outre-Mer*, 62 (1975), pp. 640–73; C. M. Andrew and A. S. Kanya-Forstner, 'French business and the French colonialists', *The Historical Journal*, 19, 4 (1976), pp. 981–1,000; L. Abrams and D. J. Miller, 'Who were the French colonialists? A reassessment of the *Parti Colonial*, 1890–1914', *The Historical Journal*, 19, 3 (1976), pp. 685–725; S. M. Persell, *The French Colonial Lobby 1889–1938*, Stanford, 1983.

23 Andrew and Kanya-Forstner, 'Centre and periphery'; Andrew and Kanya-Forstner, *France Overseas*, pp. 9–32. On popular interest in the empire, see William H. Schneider, *An Empire for the Masses: the French popular image of Africa, 1870–1900*, Westport, 1982; Thomas G. August, *The Selling of Empire: British and French imperialist propaganda 1890–1940*, Connecticut, 1985.

24 Andrew and Kanya-Forstner, *France Overseas*, p. 102; Andrew, Grupp and Kanya-Forstner, 'Le mouvement colonial', pp. 658–673.

25 On 16 November 1895. See Fierro, *La Société*, pp. 232–3.

26 French colonial trade was underdeveloped. In 1914, just 10 per cent of the France's external trade and an even smaller percentage of overseas investment were directed towards the empire. A quarter of French overseas investment went to the Russian

empire. With a merchant navy one-tenth the size of Britain's, France was rarely able to capture even half the trade of its own empire. See Andrew and Kanya-Forstner, 'French business' and Andrew and Kanya-Forstner, *France Overseas*, pp. 14–17.

27 The statutes of the SGCP are in the Archives Nationales (henceforth AN) F^{17} 13034 though the Society's archives have not survived. The SGCP also established several branch sections in provincial French cities.

28 In 1883, Foncin co-founded the Alliance Française, the semi-official organisation dedicated to spreading French language and culture around the world. Azam, a professor of medicine, is discussed in Michael S. Roth, 'Remembering forgetting: "Maladies de la mémoire" in nineteenth-century France', *Representations*, 26 (1989), pp. 49–68, esp. pp. 61–5.

29 See J. F. Laffey, 'Municipal imperialism in nineteenth-century France', *Historical Reflections/Reflexions Historiques*, 1, 1 (1974), pp. 81–114; J. F. Laffey, 'Roots of French imperialism in the nineteenth century: the case of Lyon', *French Historical Studies*, 6, 1 (1969), pp. 78–92; J. F. Laffey, 'The Lyon Chamber of Commerce and Indochina during the Third Republic', *Canadian Journal of History*, 10 (1975), pp. 325–48; J. F. Laffey, 'Municipal imperialism in decline: the Lyon Chamber of Commerce, 1925–1938', *French Historical Studies*, 9, 2 (1975), pp. 329–53; Schneider, 'Geographical reform'. The most detailed source on the provincial societies is Lejeune, *Les Sociétés de Géographie en France*, 4 vols, 1987.

30 The RGS Fellowship was over 3,500 in 1890. Fierro, *La Société*, p. 287; Kollm, 'Geographische', pp. 411–13; Schneider, 'Geographical reform', p. 92.

31 Vietnam was divided into Amman-Tonkin in the north and Cochin China in the south. Cambodia and Amman-Tonkin were officially protectorates.

32 Franz Ansprenger, *The Dissolution of the Colonial Empires*, London, 1989, pp. 305–6; Andrew and Kanya-Forstner, *France Overseas*, pp. 18–23; William B. Cohen, *Rulers of Empire: the French colonial service in Africa*, Stanford, 1971.

33 Vidal de la Blache had retired from his chair at the Sorbonne by 1914 but continued to write prolifically and to teach at the *École Libre des Sciences Politiques* in Paris. There is a large literature on his life and works: see, as recent examples, Berdoulay, *La formation*, pp. 141–227; Paul Claval (ed.), *Autour de Vidal de la Blache*, Paris, 1992; André-Louis Sanguin, *Vidal de la Blache, 1845–1918: un génie de la géographie*, Paris, 1993; Howard F. Andrews, 'The early life of Paul Vidal de la Blache and the making of modern geography', *Transactions, Institute of British Geographers*, N.S., 11 (1986), pp. 174–82; Howard F. Andrews, 'Les premiers cours de géographie de Paul Vidal de la Blache', *Annales de Géographie*, 95, 529 (1986), pp. 341–67.

34 Eighteen per cent of SGP members came from the armed forces, the largest single group. Fierro, *La Société*, p. 275.

35 *Archives de la Société de Géographie de Paris, Salles des Cartes et Plans, Bibliothèque Nationale* (henceforth BN-SGP) 41/4122, an anonymous typed document entitled *La Société de Géographie pendant la guerre 1914–1918* (p. 12). See Alfred Fierro, *Inventaire des manuscripts de la Société de Géographie*, Paris, 1984. Many of the characters in this chapter are subjects of short, factual bio-bibliographies by various authors in the annual volumes of T. W. Freeman (ed.), *Geographers: bio-bibliographical studies*, London, various dates.

36 The 1914 conference was organised by the Brive section of SGCP. See *Bulletin de la SGCP*, xxxvi, 5 (1914), pp. 317–23.

37 BN-SGP 41/4122.

38 The correspondence between Hedin and Charles Le Myre de Vilers, former President of the SGP (1905–8), is in *La Géographie*, xxx (1914–15), pp. 157–60.

39 BN-SGP 41/4122; Baron Hulot, 'La Société de Géographie pendant la guerre', *La Géographie*, xxx (1914–15), pp. 84–8. Women represented less than 5 per cent of the SGP membership in 1913.

40 BN-SGP 41/4122 and Hulot, 'La Société', p. 85. A Comité d'assistance aux troupes noires was established on 10 April 1915 followed by a Comité d'assistance aux travailleurs indochinoises on 8 January 1916. *La Géographie*, xxxi (1917), pp. 75–7. The society's capital resources were around 2 million francs in 1914. Average annual

expenditure and income were about 70,000 francs. Fierro, *La Société*, p. 285; G. Odier, 'Bilan de la Société de Géographie au 31 décembre 1915', *La Géographie*, xxxi (1917), p. 160.

41 Two strips of territory were ceded from French Equatorial Africa to German Cameroons in November 1911, extending the latter's eastern and southern borders and giving access to the Congo river. This was in return for Germany's reluctant agreement to allow France a free hand in Morocco following unsuccessful attempts in 1905 and 1911 to establish a German colonial presence on the Moroccan coast. Territory was also exchanged between the two countries in north-eastern Cameroons to straighten the border along the Logone river. See Figures 9.2a and 9.2b and Wm Roger Louis, *Great Britain and Germany's Lost Colonies 1914–1919*, Oxford, 1967, p. 14.

42 Charles Lallemand, 'Discours pronouncé à la séance de rentrée', *La Géographie*, xxx (1914–15), pp. 81–3.

43 See, however, General Malterre, 'Les variations des fronts de guerre et situation générale actuelle', *La Géographie*, xxxi (1917), pp. 140–51 (presented 25 February 1916). The unexpurgated 58-page typed text of his lecture, complete with contemptuous asides about British pre-war diplomacy, is available in BN-SGP 31/3841. The words 'supprimé par la Censure' occasionally appeared across certain passages of *La Géographie*. See, for example, Abbé Wetterlé, 'L'Alsace-Lorraine de demain, *La Geographie*, xxxi (1917), pp. 59–62, esp. p. 60 (presented 21 December 1915).

44 Louis Leger, 'Les Slaves d'Autriche-Hongrie', *La Géographie* xxx (1914–15), pp. 161–81 (presented on 19 December 1914); Georges Bienaimé, 'La Pologne', *La Géographie*, xxx (1914–15), pp. 337–58 (presented on 22 January 1915); Pierre Nothumb, 'La Belgique d'autrefois et la Belgique d'aujourd'hui', *La Géographie*, xxx (1914–15), pp. 478–87 (presented 22 January 1915); Emmanuel de Martonne, 'La Roumanie et son rôle dans l'Europe orientale', *La Géographie*, xxx (1914–15), pp. 241–50 (presented on 7 March 1915); Paul Labbé, 'La Serbie fidèle', *La Géographie*, xxxi (1917), pp. 131–5 (presented 25 February 1916); Édouard Blanc, 'La Russie et ses ressources militaires', *La Géographie*, xxxi (1917), pp. 230–3 (presented 9 June 1916); Henri Lichtenberger, 'Le projet allemand d'une Europe central', *La Géographie*, xxxi (1917), pp. 310–12 (presented 3 November 1916); René Henry, 'La monarchie habsbourgeoise: sa composition ethnographique et sa formation historique – point de vue français et européen', *La Géographie*, xxxi (1917), pp. 381–6 (presented 12 January 1917); Édouard de Keyser, 'Les affinités de la France et de la Belgique, leur rapprochement après la guerre', *La Géographie*, xxxii (1918), pp. 136–42 (presented 7 December 1917).

45 Charles Rabot, 'La géographie en Allemagne pendant la guerre', *La Géographie*, xxx (1914–15), pp. 267–76. The black propaganda campaign occasionallly produced angry exchanges. Emmanuel de Martonne tangled with Professor Hein, a geologist from Zurich University after the latter returned an SGP leaflet on German 'atrocities' in Belgium. A furious De Martonne wrote to Hein on 4 April 1917: 'these outrages require more than a shrug of the shoulders . . . This attitude . . . is beneath you, a professor from a country which, like Belgium, was neutral and has luckily been able to remain so. It is Germany which will carry, for ever, the responsibility for having unleashed the most appalling conflagration in history . . . I have only one message to the people you admire so much despite their crimes: *Gott strafe Deutschland!*' BN-SGP 16bis/2795. Lieutenant-Colonel de Gennes took the prevailing anti-German feeling to extremes by objecting to the use of the 'Germanic' word 'hinterland' in an SGP report reproduced in the *Journal Officiel* on 20 June 1915. BN-SGP 9bis/2316.

46 Baron Hulot, 'La guerre aux colonies', *La Géographie*, xxx (1914–15), pp. 226–8.

47 Togoland was thus the unlikely site of the first shot fired by a 'British' soldier in the Great War, the man in question being a black African RSM named Alhaji Grunschi. See Michael Crowder, 'The 1914–1918 European war and West Africa', in J. F. A. Ajayi and Michael Crowder (eds), *History of West Africa*, Vol. II, London, 1974, pp. 484–513, esp. pp. 484–91; F. J. Moberly, *Military Operations: Togoland and the*

Cameroons 1914–1916, London, 1931; Melvin E. Page (ed.), *Africa and the First World War*, London, 1987.

48 The East African campaign was to continue as a half-forgotten, private squabble until after the 1918 armistice. On the German overseas empire, see Lewis H. Gann and Peter Duignan, *The Rulers of German Africa 1884–1914*, Stanford, 1977; Arthur J. Knoll and Lewis H. Gann (eds), *Germans in the Tropics: essays in German colonial history*, New York, 1987; Woodruff D. Smith, *The German Colonial Empire*, Chapel Hill, 1978.

49 Paradoxically, French and British criticisms had previously been aimed not at the colonial administration of Germany, which was often held in jealous awe, but at that of their wartime allies, Portugal and Belgium. For a powerful attack on Belgium's mismanagement of the Congo, see E. D. Morel, *Red Rubber: the rubber slave trade on the Congo*, London, 1906. Morel was the founder of the Congo Reform Association in 1904.

50 Significantly, the existing German empire was unprofitable. In 1913, only 0.6 per cent of German exports went to the colonies which in turn supplied just 0.5 per cent of the country's imports. The empire cost Germany £6 million per annum. See Louis, *Great Britain*, p. 23; Fritz Fischer, *Germany's Aims in the First World War*, London, 1967, pp. 102–4, 586–91, esp. p. 596; H. W. Koch (ed.), *The Origins of the First World War: great power rivalry and German war aims*, 2nd edn, London, 1984, pp. 319–42.

51 Baron Hulot, 'Les projets de l'Allemagne en Afrique avant la guerre', *La Géographie*, xxx (1914–15), pp. 298–300 and Robert Chauvelot, 'L'Allemagne dans le Pacific', *La Géographie*, xxx (1914–15), pp. 302–6 (both presented 22 February 1915). These arguments were certainly not limited to the geographical literature or to the salons of the SGP: see, for example, Eugène Gallois, 'Les agissements de l'Allemagne dans le monde', *Bulletin de la SGCP*, xxxvii, 7–9 (1915, pp. 231–40; Jacques Dompierre, 'Les ambitions coloniales de l'Allemagne', *Revue des Deux Mondes*, 86 (1916), pp. 193–4, 197–8, 204–5, 206–7, 214–15; Anon, 'L'Annexionisme sans annexion: programme du pangermanisme économique', *Bulletin de la SGCP*, xxxix, 7–10 (1917), pp. 232–56; F. Pierre-Alype, *La provocation allemande aux colonies*, Paris, 1915.

52 Alfred Hettner, 'Die Ziele unserer Weltpolitik', in *Der Deutsche Krieg*, 64 (1915), p. 28, quoted in Fischer, *Germany's Aims*, pp. 159–60. *Der Deutsche Krieg (The German War)* were a series of propaganda pamphlets produced by a group of academics and businessmen, close to Bethmann Hollweg and including both Max and Alfred Weber. Hettner became highly, if temporarily, politicised during the war: see *Englands Weltherrschaft und der Krieg*, Liepzig, 1915; 'Das Britische und das Russische Reich', *Geographische Zeitschrift*, 25 (1916), pp. 353–71; *Der Friede und die deutsche Zukunft*, Stuttgart, 1917; 'Deutschlands territoriale Neugestaltung', *Geographische Zeitschrift*, 25 (1917), pp. 57–72.

53 Albrecht Penck, 'Der Krieg und das Studium der Geographie', *Zeitschrift der Gesellschaft für Erdkunde zu Berlin*, (1916), p. 159–76, 222–48, esp. p. 227. The latter phrase is often unfairly attributed to Penck. It was the rallying cry of *Verbandes deutscher Schulgeographen*, an association of ultra-nationalist German geography schoolmasters, real-life Kantoreks straight from the pages of Erich Maria Remarque. Penck, who was extremely critical of the organisation, was simply quoting their slogan.

54 General the Rt Hon. J. C. Smuts, 'East Africa', *The Geographical Journal*, li, 3 (1918), pp. 129–49. This lecture received enormous publicity and was reported at length in the *Times* on 29 January, 1918. Louis, *Great Britain*, suggests it was 'probably the most important speech about German colonies made during the entire war' (p. 101). See also Georges Blondel, 'L'Allemagne, ses ressources et ses ambitions', *La Géographie*, xxx (1914–15), pp. 89–102.

55 Lord Eustace Percy, a Foreign Office official, claimed Anglo-French arrangements in Togoland were made 'with a sole view to efficiency, and with no political *arrière-pensée*'. Wm Roger Louis, *Great Britain*, p. 57; Andrews and Kanya-Forstner, *France Overseas*, p. 61.

56 Louis, *Great Britain*, p. 13–14.
57 Sir Harry H. Johnston, 'Political geography of Africa before and after the war', *The Geographical Journal*, xlv, 4 (1915), pp. 273–301, esp. pp. 283–4. This article (a summary of which was translated by the Comité de l'Afrique Française as 'L'Afrique de demain', *L'Afrique Française*, 25, 6–7 (1915), pp. 159–60) also received widespread attention and was generally interpreted as an accurate assessment of British imperial policy. The New York Times called it 'the most important unofficial document that has crossed the Atlantic since the beginning of the war'. See Roland Oliver, *Sir Harry Johnston and the Scramble for Africa*, London, 1957; H. H. Johnston 'The German colonies', *The Edinburgh Review*, 220 (1914), pp. 298–312.
58 Dr d'Anfreville de la Salle, 'Le Maroc pendant la guerre', *La Géographie*, xxx (1914–15), pp. 311–19 (presented 2 April 1915); Auguste Chevalier, 'Les ressources de l'Indochine et leur utilisation pendant la guerre', *La Géographie*, xxxi (1917), pp. 455–64 (presented 8 December 1917); J. Dybowski, 'Nos colonies pendant la guerre', *La Géographie*, xxxi (1917), pp. 378–80 (presented 8 December 1917); Charles Alluaud, 'L'Afrique orientale anglaise et allemande avant et pendant la guerre', *La Géographie*, xxxi (1917), pp. 223–4 (presented 9 June 1916); Charles Robot, 'La participation de la France d'outre-mer à la guerre', *La Géographie*, xxxi (1917), pp. 123–6 (presented 3 November 1916); Baron Hulot, 'Coup d'oeil sur les fronts anglais', *La Géographie*, xxxi (1917), pp. 464–6 (presented 16 March 1917); A. Gruvel, 'Les ressources maritimes et fluviales des colonies françaises et leur utilisation pendant et après la guerre', *La Géographie*, xxxii (1918), pp. 66–8 (presented 12 November 1917).
59 *Bulletin de la SGCP*, xxxvii, 4–6 (1915), pp. 186–7.
60 The papers of the Commission de Géographie are in the *Archives de la Guerre*, Vincennes (AG) 9.N.110. The Parisian geographers were later joined by B. Auerbach of the University of Nancy, Jules Sion of the University of Montpellier, and Antoine Vacher of the University of Lille.
61 Service Géographique de l'Armée *Rapport sur les travaux exécutées du 1er aout 1914 au 31 décember 1919: historique du Service Géographique de l'Armée pendant la guerre*, Paris, 1936, pp. 189–289, esp. pp. 208–17; Ministère de la Defense Nationale et de la Guerre, *Le Service Géographique de l'Armée: son histoire, son organisation, ses travaux*, Paris, 1938, pp. 79–86; Arthur Levy, *Les coulisses de la guerre: le Service Géographique de l'Armée 1914–1918*, Paris, 1936; Anon, 'Le Service Géographique de l'Armée et la cartographie de guerre', *La Géographie*, xxxii (1918–19), pp. 463–84; Berdoulay, *La formation*, p. 32.
62 Berdoulay, *La formation*, p. 32.
63 Only one publication, Emmanuel de Martonne's *Notice sur la Syrie* (1916), was concerned with a non-European zone.
64 Georges Suarez, *Briand: sa vie, son œuvre avec son journal et de nombreaux documents inédits. Tome IV – Le pilote dans la tourmente, 1916–1918*, Paris, 1940; Bernard Oudin, *Aristide Briand – la paix: une idée neuve en Europe*, Paris, 1987.
65 Pierre Renouvin, 'Les buts de guerre du gouvernement français, 1914–1918', *Revue Historique*, 90 (1966), pp. 1–38; Douglas Johnson, 'French war aims and the crisis of the Third Republic', in Barry Hunt and Adrian Preston (eds), *War Aims and Strategic Policy in the Great War*, London, 1977, pp. 41–54; David Stevenson, *French War Aims against Germany 1914–1919*, Oxford, 1982; David Stevenson, *The First World War and international politics*, Oxford, 1986, esp. pp. 87–138.
66 Marin was a member of the SGP, the SGCP, the Comité de l'Afrique Française, the Comité de l'Asie Française and the Comité de l'Orient. See Herman Lebovics, *True France: the wars over cultural identity, 1900–1945*, Ithaca, 1992, p. 12–50.
67 Marin's idea of a general war-aims committee probably sprang from his efforts within the SGCP which was promoting itself as an wartime economic 'think-tank'. On 4 December 1915, the SGCP's *Conseil* (which included both Hulot and Bonaparte) established a Commission d'Études Économiques under the Presidency of Raoul Peret, a former Minister of Commerce. Its self-appointed task was the production of a 'programme des réformes économiques à l'occasion de la guerre et de ses suites', a

general economic plan to allow the nation both to win the war and to enhance its post-war economic performance, mainly (though by no means exclusively) at the expense of Germany. Seven small sub-committees were established to study national policy alternatives on transport, credit and banking, overseas trade, private capital and investment, colonial expansion and development, 'defense économique' and general financial questions. These committees, each with around six members, met up to eight times each between December 1915 and June 1917 and produced interminable policy resolutions of mind-boggling tedium which, sadly for them, seem to have progressed no further than the pages of the relatively obscure *Bulletin de la SGCP*. See *Bulletin de la SGCP*, xxxviii, 1–3 (1916) entire number; xxxviii, 4–6 (1916), pp. 214–312; xxxviii, 10–12 (1916), pp. 472–612; xxxix, 1–3 (1917), pp. 55–82; xxxix, 4–6 (1917), pp. 182–301; xxxix, 7–10 (1917), pp. 282–4. Marin chaired the sub-committees on the colonies and on general financial questions. It is worth noting, however, that Marin's recommendations on colonial matters were deemed sufficiently sensitive to be censured in their entirety. See *Bulletin de la SGCP*, xxxix, 4–6 (1917), pp. 179–81. The *Bulletin* also carried dozens of detailed wartime articles on specific economic issues.

68 BN-SGP 9, 16bis, 24 and 41; Fierro, *La Société*, pp. 104–5; Christopher M. Andrew and A. S. Kanya-Forstner, 'The French Colonial Party and French colonial war aims, 1914–1918', *The Historical Journal*, 17, 1 (1974), pp. 79–106; Andrew and Kanya-Forstner, *France Overseas*, pp. 33–54; Peter Grupp, 'Le "parti colonial" français pendant la Première Guerre Mondiale: deux tentatives de programme commun', *Cahiers d'Études Africaines*, 54, xiv–2 (1974), pp. 377–91, esp. pp. 388–91. Doumergue was, of course, subsequently to become President of the French Republic.

69 BN-SGP 9/2284–2287 (Franco-German border); BN-SGP 9/2278–2282 (Europe).

70 BN-SGP 9/2269–2274. Members were Émile Senart, Louis Marin, Guillaume Capus, Henri Froidevaux, Louis Raveneau, Baron Hulot, Édouard Anthoine, Henri Cordier, Paul Labbé, Charles Lallemand, Paul Bertin, Robert de Caix, Jules Girard, Jules Harmand, Emmanuel de Margerie, Edmond Perrier, Emmanuel de Martonne, André Salles, Christian Schefer, Maurice Spronck, Commandant Barré, Paul Carié, Gustave Regelsperger, Franz Schrader and Auguste Terrier.

71 BN-SGP 9/2268. Members were Schrader, Perrier, Charles Alluaud, Édouard Blanc, Augustin Bernard, Guillaume Capus, de Margerie, Raveneau, Joseph Renaud, Terrier, Hulot, Lallemand, Barré, Georges Blondel, Froidevaux, General Lacroix and Schrader.

72 Hulot to Auguste Terrier, 14 February 1916 and 15 June 1916, *Fonds Auguste Terrier, Institut de France*, ms. 5901, *feuilles* 109 and 115–18.

73 On Anglo-French tension over Cameroons, see Akinjide Osuntokun, *Nigeria in the First World War*, London, 1979, pp. 206–36, 270–88; Akinjide Osuntokun, 'La France et la redistribution des territoires du Cameroun, 1914–1916', *Afrika Zamani*, 12–13 (1981), pp. 36–52; Lovett Elango, 'The Anglo-French "Condominium" in Cameroon 1914–1916: the myth and the reality', *International Journal of African Historical Studies*, 18 (1985), pp. 657–73; Marc Michel, 'Le Cameroun allemand aurait-il pu rester unifié? Français et Britanniques dans la conquête du Cameroun (1914–1916)', *Guerres Mondiales et Conflits Contemporains* 168 (1992), pp. 13–29; Peter J. Yearwood, 'Great Britain and the repartition of Africa 1914–1919', *Journal of Imperial and Commonwealth History*, 18 (1990), pp. 316–41. Harcourt wrote a secret Colonial Office memorandum of 25 March 1915, bluntly entitled 'The spoils', which was the first serious attempt to define Britain's colonial war aims. Louis, *Great Britain*, p. 59–60.

74 Louis, *Great Britain*, pp. 58–62; Andrew and Kanya-Forstner, *France Overseas*, pp. 97–9.

75 Charles Alluaud, 'La frontière Anglo-Allemande dans l'Afrique orientale', *Annales de Géographie*, 25, 133 (1916), pp. 206–17.

76 Baron Hulot, 'La conquête du Cameroun', *La Géographie*, xxxi (1917), pp. 136–40 (presented 24 March 1916).

77 Tangiers was a source of endless irritation. Although nominally part of the Sultan's empire and hence under French protection, it was given special status under the terms of both the Entente Cordiale of 1904 and the Algeciras Act of 1906. Britain, France and Spain all suspected each other of seeking to control this strategically-important port.

78 BN-SGP 9/2268.

79 Onésime Reclus, *Lâchons l'Asie, prenons l'Afrique. Où renaître? et comment durer?*, Paris, 1904. See also his *France, Algérie et colonies*, Paris, 1880 and *L'Allemagne en morceaux: paix draconienne*, Paris, 1914.

80 BN-SGP 9/2268; Christopher M. Andrew and A. S. Kanya-Forstner, 'France and the repartition of Africa', *Dalhousie Review*, 57 (1977), p. 475–93; Christopher M. Andrew and A. S. Kanya-Forstner, 'France, Africa, and the First World War', *Journal of African History*, 19, 1 (1978), pp. 11–23; André Kaspi, 'French war aims in Africa, 1914–1919', in Prosser Gifford and Wm. Roger Louis (ed.), *France and Britain in Africa: imperial rivalry and colonial rule*, New Haven, 1971, pp. 369–96; Andrew and Kanya-Forstner, *France Overseas*, pp. 103–4.

81 BN-SGP 9/2269.

82 BN-SGP 9/2272–3.

83 BN-SGP 9/2271.

84 BN-SGP 9/2275.

85 BN-SGP 9/2275. Other reports included a survey of German commercial interests in China by Guillaume Capus on 12 July: BN-SGP 16bis/2788.

86 BN-SGP 9/2274.

87 Elie Kedourie, *In the Anglo-Arab Labyrinth: the MacMahon–Husayn correspondence and its interpretations, 1914–1939*, Cambridge, 1976; David Fromkin, *A Peace to End All Peace: the fall of the Ottoman empire and the creation of the modern Middle East*, New York, 1989, pp. 173–86; Jeremy Wilson, *Lawrence of Arabia: the authorised biography of T. E. Lawrence*, London, 1986, pp. 151–678.

88 Fromkin, *A Peace*, pp. 188–99; Andrew and Kanya-Forstner, *France Overseas*, pp. 87–102; Elie Kedourie, *England and the Middle East: the destruction of the Ottoman Empire, 1914–1921*, Hassocks, 1978 (first published 1956).

89 Senior members had been urging the SGP leadership to act in support of a French Syria for several months. BN-SGP 9bis/2322, letter to Hulot from M. Marfond. Other members put forward their own unusual solutions to the region's geopolitical problems. The extraordinary Vicomte de Breuil, a member of the SGP from 1874, noisily proclaimed through late 1915 and early 1916 that local tribesmen in the central Arabian desert wanted their own independent kingdom after the war where he, de Breuil, would become, by popular demand, the new monarch thus securing France's future in the region. An embarrassed Hulot was obliged to write a long report to the *Ministère des affaires étrangères* on 26 October 1915 dismissing de Breuil's claims and distancing the SGP from his one-man campaign. BN-SGP 9bis/2297.

90 Bernard Auffray, *Pierre de Margerie (1861–1942) et la vie diplomatique de son temps*, Paris, 1976.

91 Andrew and Kanya-Forstner, *France Overseas*, pp. 67–74.

92 Andrew and Kanya-Forstner, *France Overseas*, p. 41; Mathew Burrows, 'Mission civilisatrice': French cultural policy in the Middle East, 1860–1914', *The Historical Journal*, 29, 1 (1986), pp. 109–35; Mathew Burrows, 'Laïcité and the Levant: French cultural policy at the time of the Separation, 1900–1914', unpublished Ph.D. thesis, University of Cambridge, 1983.

93 J. Thobie, *Intérêts et impérialisme français dans l'Empire Ottoman*, Paris, 1978; Andrew and Kanya-Forstner, 'French business'; Andrew and Kanya-Forstner, *France Overseas*, pp. 40–54; Sevtek Pamuk, *The Ottoman Empire and European Capitalism, 1820–1913: trade, investment and production*, Cambridge, 1987; William I. Shorrock, *French Imperialism in the Middle East*, London, 1976.

94 BN-SGP 9/2274 and BN-SGP 16bis/2787–8.

95 BN-SGP 9/2275.

96 Hulot to Senart, 23 July 1917, BN-SGP 16bis/2782. The only published record were

brief notes in *La Géographie*, xxxi (1917), pp. 151–2 and 305.

97 BN-SGP 41/4122; Baron Hulot, 'Conférences sur l'Allemagne dans les vingt arrondissements de Paris', *La Géographie*, xxxi (1917), pp. 302–4, 386–93. Lectures were published as *Les appétits allemands. Tome I: Les ambitions de l'Allemagne en Europe. Tome II: Les rêves d'hégémonie mondiale*, Paris, 1918. See *La Géographie*, xxxii (1918), pp. 72–3.

98 BN-SGP 9/2269; Andrew and Kanya-Forstner, *France Overseas*, p. 114.

99 Censorship hampered the publication of anti-war tracts in France, though see Jean Lhomme, *En 1916, une Europe renovée: la charte des nations*, Paris, 1915. In Britain, the anti-war Union of Democratic Control produced several pamphlets denouncing secret diplomacy and Allied wartime imperialism and promoting free trade and the internationalisation of the colonial world. See, for example, H. N. Brailsford, *The War of Steel and Gold*, London, 1914; J. A. Hobson, *Towards International Government*, London, 1915; E. D. Morel, *Truth and the War*, London, 1916; E. D. Morel, *Africa and the Peace of Europe*, London, 1917. See Marvin Swartz, *The Union of Democratic Control in British Politics during the First World War*, Oxford, 1971.

100 To court US support for its views of the post-war order, the SGP organised a Franco-American *séance* on 6 November 1917 in the presence of the American Ambassador to France. See Henri Gourd, 'L'évolution du sentiment américain vis-à-vis de la France et ses conséquences actuelles', *La Géographie*, xxxii (1918), pp. 125–35. Gourd was the President of the French Chamber of Commerce in New York.

101 *Travaux du Comité d'Études. Tome I: L'Alsace-Lorraine et la frontière du Nord-Est. Tome II: Questions Européenes – Belgique, Slesvig, Tchécoslovaquie, Pologne et Russie, Questions Adriatiques, Yougoslave, Roumanie, Turquie d'Éurope et d'Asie*. On Briand's personal involvement, see Andrew and Kanya-Forstner, *France Overseas*, pp. 86–7.

102 BN-SGP 9/2269 entitled *Le point de vue italien dans les problèmes coloniaux d'après-guerre: exposé et discussion. Rapport (avec carte) presenté au nom des groupements coloniaux et géographiques de France*; Fidel to Hulot, 6 August, 1916, BN-SGP 16bis/2782. Other works by Fidel include *L'Allemagne d'outre-mer: grandeur et décadence*, Paris, 1915.

103 BN-SGP 9/2269 and, in handwritten form, 24/3712. See also M. Carazzi, *La Societa Geografica Italiana e l'esporazione coloniale in Africa, 1867–1900*, Florence, 1972; Robert L. Hess, 'Italy and Africa: colonial ambitions in the First World War', *Journal of African History*, iv, 1 (1963), pp. 105–26. On another area of Franco-Italian tension, see D. Grange, 'Un sujet de tension dans les rapports franco-italiens: les questions balkaniques (août 1916–août 1917)', in P. Guillen (ed.), *La France et l'Italie pendant la première guerre mondiale*, Grenoble, pp. 403–26.

104 Hulot to Fidel, 10 October 1916, BN-SGP 16bis/2782; Grupp, 'Le "parti colonial" français', pp. 383–4.

105 Marin to Hulot, 5 May 1917, BN-SGP 24/3711.

106 BN-SGP 24/3711. Names are listed in Grupp, 'Le "parti colonial" français', p. 384.

107 BN-SGP 24/3711; BN-SGP 9/2269; and Andrew and Kanya-Forstner, *France Overseas*, pp. 121–2.

108 Andrew and Kanya-Forstner, *France Overseas*, pp. 144–9.

109 BN-SGP 24/3711; 9/2269; Hulot to Terrier, 17 May 1917, *Fonds Auguste Terrier, Institut de France*, ms. 5901, *feuille* 117.

110 Grupp, 'Le "parti colonial" français', pp. 385–7; Andrew and Kanya-Forstner, *France Overseas*, pp. 149–51; and more generally James J. Cooke, 'Eugène Étienne and the acquisition of the Togo and Cameroons 1916–1918', *Proceedings, Western Society for French History*, 11 (1983), pp. 296–303.

111 Wm Roger Louis, *Great Britain*, pp. 135 and 160. For the American perspective, see Wm Roger Louis, 'The United States and the African peace settlement of 1919: the pilgrimage of George Louis Beer', *Journal of African History*, 4, 3 (1963), pp. 413–33. The mandates idea was laid out in Jan Smuts, *The League of Nations: a practical suggestion*, London, 1918. Three kinds of mandates were agreed on – A, B and C –

depending on the intensity of sovereignty claimed by the mandated power and the 'stage of civilisation' of the indigenous inhabitants.

112 In German East Africa, the Ruanda-Urundi district was ceded, after difficult negotiations, to Belgium. South Africa acquired the mandate for German South West Africa. To their dismay, the smaller Allied colonial powers in Africa, Italy and Portugal, received only small concessions in the form of boundary alterations. Japan gained control of all German Pacific islands north of the equator, German Samoa passed to New Zealand and the Bismark Archipelago and Kaiser Wilhelmland were mandated to Australia.

113 A classic account of British imperial thought is Ronald E. Robinson and Jack Gallagher (with Alice Denny), *Africa and the Victorians: the official mind of imperialism*, London, 1981 (first published 1960). See also Wm Roger Louis, *Imperialism: the Robinson and Gallagher controversy*, London, 1976; P. J. Cain and A. G. Hopkins, *British Imperialism: innovation and expansion, 1688–1914*, London, 1993, pp. 3–52.

114 Albert Demangeon, *The British Empire: a study in colonial geography*, London, 1925 (originally published as *L'empire britannique*, Paris, 1923). On this work, see Paul Claval, 'Playing with mirrors: the British empire according to Albert Demangeon', in Godlewska and Smith, *Geography and Empire*, pp. 228–43.

115 Ansprenger, *The Dissolution*, pp. 305–6. For the final phase of wartime and post-war colonial negotiations, see Andrew and Kanya-Forstner, *France Overseas*, pp. 137–236.

116 Camille Fidel, *La paix coloniale française*, Paris, 1918; David D. R. Heisser, 'The impact of the Great War on French imperialism, 1914–24', Unpublished Ph.D. thesis, University of North Carolina, 1972.

117 Maurice Crosland, *Science under Control: the French Academy of Sciences 1795–1914*, Cambridge, 1992.

118 Sherman David Spector, *Rumania at the Peace Conference: a study of the diplomacy of Ioan I.C. Brătianu*, New York, 1962, pp. 56, 100–1.

119 Lawrence E. Gelfand, *The Inquiry: American preparations for peace, 1917–1919*, New Haven, 1963; Edward Mandell House and Charles Seymour, *What Really Happened at Paris: the study of the Peace Conference, 1918–1919*, New York, 1921; Geoffrey J. Martin, *The Life and Thought of Isaiah Bowman*, Hamden, Connecticut, 1980, pp. 81–97. Other leading geographers at the Peace Conference included Douglas W. Johnson, Professor of Geography at Columbia University; Mark Jefferson, Professor of Geography at Michigan State Normal School (both on the American Delegation); Jovan Cvijic, Professor of Geography at Belgrade and perhaps the most prominent academic member of the Yugoslav Delegation; and Eugeniusz Romer, Professor of Geography at Lwòw, who was equally important to the Polish Delegation. On Cvijic, see Ivo J. Lederer, *Yugoslavia at the Paris Peace Conference: a study in frontiermaking*, New Haven, 1963, pp. 126–8, 174, 223, 310.

120 The small Geographical Section assisting the British delegation was headed by Colonel Sir Walter Coote Hedley and included Major O. E. Wynne and Captain Alan G. Ogilvie, the last named subsequently Professor of Geography at Edinburgh University. See Alan G. Ogilvie, *Some Aspects of Boundary Settlement at the Peace Conference*, New York, 1922; George G. Chisholm, 'Geography at the Congress of Paris, 1919', *Geographical Journal*, 55, 1 (1920), pp. 309–12.

121 For general discussions of the role of intellectuals during the war, see Julien Benda, *La trahison des clercs*, Paris, 1927; Stuart Wilson, *War and the Image of Germany: British academics 1914–1918*, Edinburgh, 1988; Carol S. Gouber, *Mars and Minerva: World War I and the uses of higher learning in America*, Baton Rouge, 1976; Klaus Schwabe, *Wissenschaft und Kriegsmoral: die deutschen Hochschullehrer und die politischen Grundfragen des Ersten Weltkrieges*, Göttingen, 1969.

122 Morel, *Truth*, p. 37.

123 German universities had attracted some of the best British, French and American geographers before 1914 who flocked to sit at the feet of Penck and Hettner and their illustrious predecessors like Friedrich Ratzel (Leipzig) and Ferdinand von Richthofen

(Berlin) (uncle of Manfred von Richthofen, the feared 'Red Baron' flying ace of the Great War). See Samuel van Valkenburg, 'The German school of geography', in Griffith Taylor (ed.), *Geography in the Twentieth Century*, New York, 1951, pp. 91–115; Wallace, *War and the Image of Germany*, pp. 1–42 and 227–8.

124 For other disciplines, see Roy M. MacLeod and E. Kay Andrews, 'Scientific advice in the war at sea, 1915–1917', *Journal of Contemporary History*, 6 (1971), pp. 3–40; Ulrich Trumpener, 'The road to Ypres: the beginning of gas warfare in World War I', *Journal of Modern History*, 47, 3 (1975), pp. 460–505; Gilbert F. Whittemore, Jr, 'World War I, poison gas research, and the ideals of American chemists', *Social Studies of Science*, 5 (1975), pp. 135–63; Michael Pattison, 'Scientists, inventors and the military in Britain, 1915–19: the Munitions Inventions Department', *Social Studies of Science*, 13 (1983), pp. 521–68; L. F. Haber, *The Poisonous Cloud: chemical warfare in the First World War*, London, 1986; Guy Hartcup, *The War of Inventions: scientific developments, 1914–1918*, London, 1988. On the impact of an earlier conflict, see Maurice Crosland, 'Science and the Franco-Prussian war', *Social Studies of Science*, 6 (1976), pp. 185–214.

125 There is a vast literature on both men. For book-length studies, see Brian W. Blouet, *Halford Mackinder: a biography*, College Station, Texas, 1987; W. H. Parker, *Mackinder: geography as an aid to statecraft*, Oxford, 1982; Harriette Wanklin, *Friedrich Ratzel: a biographical memoir and bibliography*, Cambridge, 1961.

126 T. W. Freeman, *A Hundred Years of Geography*, London, 1961, pp. 205–25. On inter-war French geography, see Numa Broc, 'Homo geographicus: radioscopie des géographes français de l'entre-deux-guerres (1918–1939)', *Annales de Géographie*, 102, 571 (1993), pp. 225–54.

127 There is a rapidly expanding literature on this. For a book-length treatment on German geopolitics, see Michel Korinman, *Quand l'Allemagne pensait le monde: grandeur et décadence d'une géopolitique*, Paris, 1990. The Italian story is the subject of David Atkinson, 'Geography and geopolitics in Fascist Italy', unpublished Ph.D. thesis, University of Loughborough, in preparation. A parallel study of German historians is offered by Michael Burleigh, *Germany Turns Eastwards: a study of Ostforschung in the Third Reich*, Cambridge, 1988.

128 The establishment of the journal *Political Geography Quarterly* (now *Political Geography*) in 1982 has provided a major focus for a new generation of critical political geography in the Anglo-American world.

129 Paul Fussell, *The Great War and Modern Memory*, London, 1975; Modris Eksteins, *Rites of Spring: the Great War and the birth of the modern age*, London, 1989; Samuel Hynes, *A War Imagined: the First World War and English culture*, London, 1990; J. A. Winter, 'Catastrophe and culture: recent trends in the historiography of the First World War', *Journal of Modern History*, 64, 3 (1992), pp. 525–32. For a novel interpretation of 1914–18 as historical turning point, see Daniel Pick, *War Machine: the rationalisation of slaughter in the modern age*, New Haven, 1993.

Acknowledgements

I am grateful to John Agnew, Felix Driver, A. S. Kanya-Forstner and my co-editors for helpful comments.

CHAPTER TEN

Geopolitics, cartography and geographical knowledge: envisioning Africa from Fascist Italy

David Atkinson

Introduction

From the foundation of a trading base at Assab on the Eritrean Red Sea coast in 1882 until the defeat of Italian troops in North Africa in 1943, the Kingdom of Italy maintained a territorial presence in Africa. Italy's nineteenth-century African possessions comprised the East African territories of Eritrea and Somaliland. In 1911 Italy occupied the Ottoman colonies of Tripolitania and Cyrenaica in modern-day Libya. Finally, the Fascist regime invaded Abyssinia in 1935 and, after the fall of Addis Ababa in May 1936, proclaimed an Italian empire. This chapter will consider some of the ways in which this colonial project was abetted by Italian geographers and geographical institutions.

I will concentrate especially upon the Italian geopolitical movement of the late Fascist period. An outgrowth of the wider European politicisation of geography after World War One and a self-consciously Italian version of the international interest in 'geopolitics', the journal *Geopolitica* (1939–42) claimed a privileged and synoptic perspective upon the world. I want to consider how this geopolitical vision, as articulated through the journal's cartography, represented Africa in the last years of the Italian colonial presence. Despite its claims to an innovative perspective, *Geopolitica*'s portrayal of Africa continued a long-standing tradition of Italian geography whereby its individuals and institutions participated in, supported and legitimated Italian imperialism on the continent.

Italian imperialism and Italian geography

When the Bersaglieri of the Italian army breached the Aurelian walls of Rome on 20 September 1870, the unification of Italy was finally complete. Italian geography sprang from a number of institutions, organisations and individuals in the new state and the label 'geography' meant many different things to many different groups. Academic geography emerged after Vittorio Emmanuele II, the Savoyard king of Piedmont, became the first monarch of a united Italy and the 1859 Piedmontese educational laws of Count Gabrio Casati were extended to the new territories. These Legge Casati established geography as a discipline in Italian universities. However, the chairs that were created received little institutional support and were often isolated within other departments. Their occupants enjoyed few opportunities to develop their discipline. It was only with the arrival of Giuseppe Dalla Vedova, educated in the German tradition of geography in Vienna, that Italian geography attained a higher academic status. In Padua and in Rome (from 1875), he propagated German positivism in his efforts to establish the scientific credentials of geography. Together with other German-speaking Italian geographers, father and son Giovanni and Olinto Marinelli who occupied the chair at Florence consecutively from 1893 to 1926, Dalla Vedova promoted the distinctive role of geography as the science of the earth's surface and its associated human activities.[1]

While academic geographers were attempting to establish their subject's intellectual standing, 'geographical' activity was far more successfully pursued by other organisations and interest groups. Official bodies included the Istituto Geografico Militare (IGM). First established as the Ufficio Tecnico dello Stato Maggiore Italiano in Florence in 1861 and then as the Istituto Topografico Militare in 1872, it was destined to become the major cartographical agency of Italy and to play a leading role in the collection of geographical information about the Italian Kingdom. Less overtly tied to the state were the geographical and exploration societies. Chief amongst these was the Società Geografica Italiana (SGI) which was established in the then capital of Florence by 1,254 founder members on 12 May 1867. Like the Royal Geographical Society of London or the Société Géographie de Paris, the SGI became the dominant geographical institution in Italy and one that was closely associated with government circles. Other Italian cities and interest groups would found similar, although less influential bodies. In 1879, Milanese merchants and industrialists established the Società d'esplorazione commerciale (later re-named the Società d'esplorazione geografica e commerciale). In Naples the local elite lent their support to a Società Africana d'Italia from 1882.[2] All of these bodies laid claim to the label 'geography' and all of them placed their version of 'geographical science' at the service of Italian imperialism.

As the prestigious *Enciclopedia Italiana* pointed out in 1928: 'After the political unification of Italy, the geographical knowledge of the country benefited above all from the work of public offices and governmental organisations.'³ Massimo d'Azeglio, the Risorgimento hero, noted famously upon the fall of Rome that, now the Italian state was completed, the pressing task was to make Italians. 'Liberal' Italy inherited a people whose first loyalties were local and regional and whose modern political history was one of foreign domination. These local affiliations – the sense of *campanilismo* – hindered efforts to foster national unity and a national identity. As a consequence, understanding the new country's physical extent, and the nature and character of its territory seemed an urgent necessity in the nation-building project and in attempts to build popular affection for Italy.

The collection of geographical information and the results of geographical survey were part of the nation-building undertaken in the infant state. From 1879 the IGM began to publish its *Carta topografica del Regno d'Italia* at the scale of 1:100,000. As the first systematic attempt to map the entire peninsula and the first detailed topographic portrait of the new state, this series had both a practical and a symbolic importance.[4] The maps provided a scientific definition of the extent and character of Italy, but also furnished Italians in their provinces with a much needed impression of their new country.[5] In this same period other aspects of Italy's physical and human geography were collected by state-sponsored organisations. A nationwide cadastral survey was carried out while the Istituto Idrografico della Reale Marina surveyed Italy's coasts and marine basins. From 1884 the country's geology was also mapped at 1:100,000 by the Reale Ufficio Geologico. The Reale Ufficio Centrale di Meteorologia e Geofisica collated national climatological data and the Ministero dei Lavori Pubblici examined all of Italy's inland lakes and rivers. The Reale Istituto Centrale di Statistica took the first national population census in 1861 and at subsequent 10-yearly intervals (with the exception of 1891). It also gathered and published information on other aspects of the fledgling state, including economic activity. It appears that much effort was invested to produce an accurate and comprehensive picture of the state's territory and to acquire the range of information that contributed to what the Italian Encyclopaedia labelled 'geographical knowledge'.[6] Although I haven't the scope here to explore more fully the role of geographical writings, descriptions and survey in the late nineteenth-century creation of an Italian national identity, I can briefly note the role that such geographical knowledge can play in the creation, definition, and articulation of the Italian nation-state.[7] The compilation of 'facts' about Italy and the drafting of this information within the sciences of cartography and geography then lent such accounts a certain authority to shape the way in which the country became known.

The requirement for a comprehensive geographical understanding of

Italy was also one of the stated concerns of the SGI upon its foundation. The first president of the society was Christoforo Negri, an ardent Italian expansionist. As Lucio Gambi demonstrates, although the professional geographers of the SGI wished to sustain their commitment to studies of the geography of Italy these ambitions were overwhelmed by an all-pervasive colonial and expansionist lobby which dominated the institution.[8] Those who might be labelled 'professional geographers' totalled only a fraction of the SGI and the military men, diplomats, industrialists, politicians and aristocrats who constituted the vast majority of the membership, created a society that reflected the interests of Liberal Italy's ruling elite.[9] The Milanese society would attract a more commercially oriented membership than the SGI and there were fewer diplomats and parliamentarians to be found in other provincial societies. Nevertheless, all of these organisations boasted a similar social-profile, and their memberships ensured that they concentrated upon overseas exploration. In the SGI a tradition of African exploration would be sustained and celebrated.[10] And as with the geographical societies of London, Paris or Berlin, such overseas interests were seldom fuelled by purely scientific motives.

The political role of the SGI was evident from the start. In 1870 the SGI sponsored its first overseas expedition and under its patronage three Italians travelled to northern Eritrea. Ostensibly engaged upon scientific exploration and the collection of botanical and geological samples, the expedition also had an ulterior aim which was to monitor a group of Italians who, led by an Italian missionary, had settled in Eritrea's Sciotel valley.[11] With the Suez canal newly completed in 1869, and with France and Britain vying for influence in the Sudan and East Africa, the expedition expressed the Italian colonial lobby's interests and ambitions in the region. It also signified the SGI's close relations with Italy's colonial cause and the beginning of the society's long involvement in Africa.[12]

Towards the close of the century, the Italian Geographical Society continued its exploration of territories that were subject to the imperial ambitions of its military, political and diplomatic constituencies. From the mid-1870s the *Bollettino della Reale Società Geografica Italiana* reported expeditions to East Africa, Morocco and Tunisia. Although the banner of science and notions of trading and commercial links featured prominently in the public profile of such expeditions, the journeys were supported by a society enmeshed in the overseas ambitions of Italy's political class.[13] Having relocated to Rome when it became the country's capital in 1871, the SGI found itself ever more in the vanguard of a colonial lobby comprising politicians, diplomats, landowners and military men. Unsurprisingly, the SGI reflected the colonial interests of these government circles and their ambitions in regions of Africa to which they felt that Italy was entitled.

Whilst a lingering Risorgimento commitment to national liberty and self-determination fuelled some anti-colonial sentiments, many who had been raised on Garibaldi's adventures and the fiery nationalism of Mazzini argued vigorously that Italy should take part in the 'Scramble for Africa' to fulfil its destiny as a great European power. Apart from the traditional nationalism of the Roman establishment and the SGI, a new nationalism found voice in the northern cities among the emergent middle classes of industrialists, merchants and financiers. Reflecting these interests, a parliamentary deputy, Captain Manfredo Camperio, began publishing a Milan-based journal, L'Esploratore, from 1877. Concentrating upon economic and commercial issues and travel, it was this journal which prompted the 1879 foundation of the Società d'esplorazione commerciale.[14] In 1880 the society established a trading post at Benghazi in Cyrenaica and a year later dispatched two expeditions to the region. Caperio sought to develop commercial opportunities and Giuseppe and Angela Haimann, whilst supposedly gathering 'scientific' and 'artistic' data, reported upon the agricultural and trading potential of Cyrenaica – creating an impression that would fuel the demands of Italian nationalists some thirty years later.[15] Elsewhere, still more Italians set out under the banner of scientific or commercial exploration to the Levant, the Balkans, the Aegean, and above all to Africa: to the regions coveted by the financiers, industrialists and merchants who funded the geographical and exploration societies of Italy.

Italy's expansionary project suffered a major blow in 1896, when an invading Italian army was routed by Abyssinian forces at Adowa. So debilitating and humiliating was this defeat that for fifteen years Italy withdrew from any further territorial aggrandisement, save for the consolidation of extant possessions. During this hiatus, which lasted until 1911, some of Italy's academic geographers tried to reassert the scientific credentials of geography within the SGI and other societies that were weakened by their involvement with the failed colonial campaign. Guiseppe Dalla Vedova became president of the SGI in 1900. The wider anti-colonial backlash in Italy after the debacle of Adowa also found some expression within academic geography. Anti-colonial sentiments surfaced when the Società di Studi Geografici was established in Florence in 1895 and similar anti-colonialism was expressed by Cesare Battisti and Renato Biasutti, the professional geographers who founded the Genoa-based journal La Cultura Geografica in 1899.[16] Yet such opposition was hindered by the limited numbers of academic geographers involved, by their isolation within the universities, and by their similar minority status within the geographical societies. With the passage of time the colonial lobby became less muted and again began attempts to influence parliamentary and public opinion.

In 1913, the SGI and the colonial administration of Italian Somaliland funded Giuseppe Stefanini and Guido Paoli to conduct geographical and resource surveys, to construct a geological map, and to investigate the possibilities for agricultural development in this existing Italian colony. Exploration was again paving the way for colonial projects.[17] Meanwhile, in response to pressure from the nationalist lobby (many of whom were active within the SGI or other geographical societies), Prime Minister Giovanni Giolitti had authorised the 1911 Italian invasion of the Ottoman territories of Cyrenaica and Tripolitania. At the 1912 Peace of Ouchy, control of these regions was ceded to Italy and colonial expansion was placed back on the national agenda. The professions of scientific intent with which geographical societies had sought to mask the political implications of their activities in the nineteenth century were partially discarded. For the first time, as Gambi tells us, 'geographers were directly involved in the process of colonisation itself'.[18] Geographical expertise provided intellectual justification for the extension of Italian control. Furthermore, geographical lectures and writings were used to introduce Italy's new territories to the Italian people; a trend that was to become common over the next thirty years.[19] However, Rome's initial control over the Libyan territories was rudimentary and whilst the 1915–18 war and its aftermath shook Italy, Italian authority became restricted to a few coastal strongholds. The rise of Fascism in 1922 and the consolidation of Mussolini's regime after 1925 accompanied the gradual extension of Italian authority into the Libyan interior, the *reconquista*.[20] At the same time the discipline of geography became more established in the universities and geographical institutions enjoyed increasing governmental favour.[21] It was within this milieu, the *reconquista* of Libya, that the work of geographers played a central role in the appropriation and 'capture' of Italy's 'Fourth Shore'.[22] Their efforts did not go unacknowledged. In 1928 Luigi Rusca praised the work of the tenth Italian Geography Congress (September 1927) as essential for the development of what he called a 'colonial consciousness' in the Italian people.[23] Nicola Vachelli, the Director of the IGM claimed that:

> The formation of a colonial science . . . requires, beyond scientific analysis, the accurate and profound knowledge of the geographical environments and of the societies inhabiting them . . . geographical science . . . [when], understood in its true sense, is physical, political, and economic all at once.[24]

It was this sense of geographical knowledge as broad ranging, encompassing economic, political and physical elements simultaneously, that persuaded some in Fascist Italy to consider geographical survey – with its catholic reach, self-proclaimed comprehensiveness and scientific status – as the essential foundation of a colonial domain.

As a result, Libya was the focus of a great deal of geographical attention during Italian attempts to domesticate the colony. The SGI organised several major projects in the region. As Corrado Zoli, SGI president from 1933 to 1944, an ex-colonial governor and a future patron of *Geopolitica*, noted in 1937, once the *reconquista* of the colony by Italian forces was complete by the spring of 1931, the SGI began, 'proceeding with the scientific exploration of the most remote and least known regions of the vast north-African dominion'.[25] One geographer involved with this project was Emilio Scarin. In 1934 he joined an SGI-sponsored expedition to the Fezzan along with Elio Migliorini, and eventually published several studies upon the borders, settlement, regions and environment of Libya.[26] His initial findings were included in the first set of reports from this SGI project which were published in a 700-page tome in 1937. Besides Scarin's work on Libyan settlement patterns, the SGI's report upon 'The Italian Sahara' ranged from histories of the region, through accounts of the geology, morphology, climate, hydrology and vegetation of Libya to reports upon ethnography, demography, resources and communications. All of these factors, measured and collected by Italian 'science', were brought together and published in Italy's 'geographical survey' of the colony.[27] As Fabio Lando notes, the geographers and colonial authorities worked together to such an extent that the territory became 'an authentic laboratory' of geographical science.[28] Italy was also eager to present her colony to an international audience. At the 1934 Esposition du Sahara in Paris, the Italian exhibit boasted a similarly comprehensive range of sciences – all of which were dedicated to describing, analysing, understanding, and ultimately, to taming the desert through science.[29] Both of the aforementioned accounts of Italian Libya also reserved space for the role of cartography in the 'conquest' of the territory.

Italian cartography's links with imperialism stretched back to the mercantile empires of Venice and Genoa. In the Fascist era cartography was again mobilised to support Italian expansion. As with other European nations engaged upon territorial acquisition in Africa, Italy needed maps for immediate practical and military purposes. Yet cartography was also essential for the post-war organisation of conquered regions and for the imposition of Italian authority. Contemporary accounts relate how Italy was known for the systematic manner in which Italian territories were swiftly mapped.[30] Just as cartography

[271]

and geographical survey were used to uncover and define knowledge of the Kingdom of Italy upon unification, so too did Italians use geographical writings, descriptions and surveys to understand and to take symbolic possession of their overseas territories. As chief artefact and repository of geographical knowledge, the map was fundamental to this process. The mapping of Libya, Eritrea, Italian Somaliland and eventually Ethiopia started as soon as military authority was established in these areas.[31] The aim was to survey the entire territory, and to translate these new lands into maps. Once enshrined within 'accurate' and 'objective' cartographical images, these lands were named, demarcated and represented as Italian. By the early 1940s almost all of the vast territory of Libya had been mapped by the IGM at 1:400,000 and such was the requirement for colonial-surveyors and cartographers that extensive courses in 'Colonial surveying' had been established by 1937.[32]

Admittedly, as much of the Libyan interior was 'unknown' (to Western societies) at this time, the vigorous exploration of the area during Italian suzerainty could lay claim to an enlightenment inheritance of opening up these darkest spaces of the Dark Continent. Undoubtedly, such motives played a part in the exploratory activity of the era. However, although 'Geography' encompassed a broad and varied range of interests in post-Risorgimento Italy, as wc have seen, geographical involvement in Africa during this period was seldom merely 'scientific' or divorced from the wider requirements of the colonising power. Whilst Italy needed practical geographical knowledge to establish hegemony over its territory, and whilst the country recognised the ability of geographical and cartographical survey to take symbolic possession of these lands, geography remained an imperial science.

Geography in imperial Italy

From the balcony of Palazzo Venezia above the crowds in central Rome, Mussolini proclaimed the foundation of empire on 9 May 1936. With the conquest of Abyssinia, an 'Italian East Africa' of Eritrea, Ethiopia, and Italian Somaliland was established. To the north, Cyrenaica and Tripolitania were integrated with the Fezzan to form Italian 'Libia' in 1938 and the coastal territories were then included as part of Metropolitan Italy proper.[33] During this 'Imperial phase' of Fascism, Africa played a very significant role in the cultural life of the nation and of the academy. In the bombastic atmosphere of the period, appeals to the geographical sciences to legitimate Italian expeditions and colonialism overseas were far less necessary than they had been. The right of Italy to its 'reborn' empire seemed self-evident. Nevertheless, the various geographical discourses of Fascist Italy maintained their high profile within

Italian imperialism. In 1939, for example, the IGM published an explicit account of its own role in the preparatory work for the invasion of Abyssinia, in the military exercise itself, and in the subsequent imposition of Italian authority. There was little sense that geography had neglected its 'objective' scientific status.[34]

Indeed, many geographers became closely involved with the numerous colonialist groups and societies that emerged in this period. One which was particularly close to the regime's politics was the Istituto Coloniale Fascista (ICF), later the Istituto dell'Africa Italiana.[35] A national movement with branches in many cities, the ICF attracted the support of several academic geographers who later became involved in *Geopolitica*. The ICF was not primarily concerned with the exploration of proposed colonial territory or in establishing Italian authority over territory already conquered. Its interests revolved around the optimal development of Italian colonies. At home it concentrated on disseminating knowledge of, and information about, the empire to the Italian people, through lectures, newspapers, publications and public exhibitions. Introducing the empire to Italians began to occupy geographers and geographical institutions increasingly after 1936.

The regime actively supported geographical science throughout this period. There was talk of a national geographical academy and Italy's first national atlas was published in 1940.[36] This was prepared by the Touring Club Italiano (TCI) whose own popular series of travel guides had been surrendered to the direct influence of the regime in 1935.[37] One result was the appearance in 1938 of almost half a million copies of the TCI's *Guida dell'Africa Orientale Italiana* that were distributed to Italian households free of charge. Its 640 pages encompassed descriptions of all parts of Italy's East African possessions and sought, in the words of the preface, to contribute to the understanding and development of the empire. Inside the front cover, a map of Europe and Africa highlighted the territorial extent of Italy and its empire – straddling the Mediterranean and extending across the two continents.[38] In the same year Italy's schoolchildren were reading *L'impero d'Italia* by Luigi Filippo De Magistris, Professor of Economic Geography at Genoa and Gian Cesare Pico, Royal Director of Education.[39] This introduced the history and geography of Italy's empire and the role of Fascism in reviving Italian imperialism to all pupils in the fifth year of elementary school. Attempts to expand the Italian people's 'geographical awareness', to use the phrase popular in imperial Italy, were well under way.

In the imperial capital of Rome, the SGI also became involved with this project. In the early 1940s it produced a series of reports upon the regions of Africa, the Balkans, and the Mediterranean that were either coveted by Italy or had been recently included in the Italian empire. This

Paesi d'Attualità series was retitled *Paesi dell'ordine nuovo* in 1942 with the news that Giuseppe Bottai, the Fascist education minister had further increased the SGI's government subsidy to fund this series for the next three years.[40] In addition, the society's *Bollettino* was published under a new cover and put on general sale as *I paesi del mondo*. The intention was to give the publication a wider appeal.[41] Again Bottai lent his support to this venture. Widely considered one of the intellectuals of the regime, Bottai had a coherent vision of future Fascist society and had embarked upon a broad cultural project to bring this about. Part of his programme involved encouraging the *coscienza geografica* of the Italian people.[42] To this end the first article in the *Bollettino della Reale Società Geografica Italiana* and *I paesi del mondo* for 1939 was a programmatic article by Bottai entitled 'Mete ai Geografi' ('Aims for geographers'). Here he stressed that the importance of geographical science in the modern era was its capacity to observe, understand, and acquire a scientifically thorough knowledge of the world. This importance was heightened now that Italy had attained the 'imperial level' and needed to disseminate a *coscienza geografica* amongst its populace. Bottai went on to confirm the future of academic geography in the Fascist state, arguing that the value of a unified and holistic geography, restored to its place at the centre of the modern nation, lay in its ability to unify human understanding of the world.[43] The favour that Bottai extended to geography, to its capacity to understand the world and then to introduce this world to Italians, was appreciated by geographers at the time. He called them to a conference in January 1941 at which he confirmed his support for university-level geography and for a national geographical academy. One reported approvingly upon 'his courageous programme for the renewal of geographical studies and of the national geographical conscience, that he began with the foundation of *Geopolitica*'.[44]

Geopolitica: *political geography in Fascist Italy*

In January 1939 Italy's first geopolitical journal was launched. *Geopolitica*, subtitled 'A monthly review of political, economic, social and colonial geography', was in print between 1939 and December 1942. On its cover were the names of its three co-founders: Giuseppe Bottai and two academic geographers, Giorgio Roletto and Ernesto Massi. According to an article that Massi wrote for a 1939 edition of the German *Zeitschrift für Geopolitik*, Italian geopolitics originated in 1930 when, as a graduate student, he discovered the work of Karl Haushofer, the leading figure in the German *Geopolitik* movement. Massi introduced these ideas to his Professor, Giorgio Roletto, who held the chair of geography at the University of Trieste. Roletto had been appointed with a remit to examine the economic and political issues of this frontier city. Together with his student he began to develop this new version of political geography from an Italian perspective.[45] Both men were convinced that geographical analysis was of genuine value in the description and explanation of the 1930s global order and both of them considered political geography to extend this explanatory grasp. Massi was especially interested in the distinctive geopolitical tradition that was emerging in Europe in this period. Throughout the 1930s he had followed the development of these ideas, and by 1939, with the foundation of *Geopolitica*, he was prepared to argue that the geopolitical perspective had a distinctive utility above and beyond orthodox political geography.[46]

From the mid-1930s Massi worked in Lombardy, at the University of Pavia and at the Catholic University of the Sacred Heart in Milan. He became Cultural Director of the local branch of the ICF and through the organisation's newspaper, *Impero Italiano*, promoted the insights of geography and geopolitics to the proper understanding of Italy's colonial situation.[47] In August 1937 *Impero Italiano* reported that a cartographic office had been established at the ICF's Milanese base. This was in response to a requirement for maps that could represent the synthesis of physical, economic, and political factors that were a feature of the numerous lectures and courses provided by the institute and organised in part by Massi.[48] The first maps produced by the office were integral to a pamphlet by Massi that dealt with the distribution and control of global resources such as rubber and cotton after World War One. He concluded that the extant distribution of colonies and particularly French and British dominance in Africa, was unfair. A more equal distribution, with resources controlled and exploited by nations such as Italy, would be of benefit to all.[49]

The maps that complemented his arguments were drafted by Mario Morandi, a shadowy figure who had worked with Massi and the ICF in Lombardy from around 1936, regularly contributing maps to *Impero Italiano*, to the ICF's monograph series, and to the work of Giorgio Roletto and others who were later involved with *Geopolitica*. Morandi was already developing his distinctive mapping style when Massi was introduced to Giuseppe Bottai by Agostino Gemelli, Rector of the Catholic University of Milan.[50] Given Bottai's support for geography, when Roletto and Massi approached Bottai with a proposal for an Italian geopolitical journal, Bottai not only arranged funding but also, by naming himself as a co-founder of the publication, lent it the considerable advantage of his reputation.[51] Bottai also penned the opening pages of *Geopolitica*, producing another programmatic statement. Geography was again invoked as a discipline of central importance to the modern state and Bottai lamented the lack of exact and clear geographical knowledge with which Italy had faced a radically restructured world in 1919 after the Great War. Yet more particularly, he continued, Italy had lacked a sense of the crucial political aspect of geography. 'All in all, [Geography] has to acquire what I have been calling a political con-science', he wrote.[52] Bottai intended *Geopolitica* to provide Italian geography with what he called a 'political-conscience' that would com-plement other more orthodox parts of the geographical discourse, such as *I paesi del mondo* or the *Paesi d'Attualità* series that he was also promoting. Roletto and Massi also wrote an introduction to their new journal: 'Per una geopolitica italiana'. They too insisted that Italian geography had much to offer the modern state, primarily because of the unique ability of geography to synthesise the increased complexity of foreign policy problems and the intimate connections between their political, social and economic elements.[53] According to Roletto and Massi, their science enjoyed a unique and privileged synthesising per-spective: the geopolitical 'way of seeing'. It was this kind of holistic geopolitical survey, and its assumed analytical and explanatory potential, that became one of the main tenets of *Geopolitica*. A new, aggressively stylised form of cartography was seen as an appropriate and powerful medium for this vision.

Mario Morandi's collaborations with Massi continued into the 1940s and he became closely involved with *Geopolitica*: initially as one of the publication's two official cartographers and later as co-editor, standing in for Massi in 1941 whilst the latter went away on active service.[54] Morandi's main contribution to *Geopolitica* was an extensive series of geopolitical maps. The major editorial emphasis upon cartography is evidenced by their frequency within the journal, by the attention that was paid to their production and presentation and by their consistently high standard. All of this is evident in Morandi's most striking images: a series of over forty double-page maps, one in each monthly issue, that attempted to represent graphically an Italian geopolitical perspective. The *Sintesi Geopolitiche* series was almost always independent of any related text, the maps constituting a purely cartographical statement.[55] The series ranged across the globe and through a number of themes and historical periods. As some *Sintesi Geopolitiche* mapped Africa, they elucidate something of the geopolitical interpretation of the African situation and Italian geopolitical ambitions towards the continent.

The Sintesi Geopolitiche *and its vision of Africa*

The possession and control of African territory was crucial to the Fascist regime in several ways. Fascism made much of its colonial settlement programme, 'Demographic Colonisation', whereby Italy's traditional drain of emigrants was to be re-directed to Italian territories rather than 'lost' to the country. Some 110,000 Italians were established in Libya by the early 1940s and optimists hoped that up to five million colonists would emigrate to the Ethiopian uplands. Italian forces were to expand across Africa, settling emigrants from the mother-country in their wake.[56] Pro-colonialists also stressed the value of Italy's African possessions as a resource base.[57] Few of the raw matcrials required by a modern state could be found on the Italian mainland and this problem was brought into sharper relief by League of Nations sanctions which were prompted by the invasion of Abyssinia. In self-sufficient 'Autarkic Italy', Africa was seen as the potential source of the resources Italy needed.[58] Moreover, Mussolini's long-term goals included making Italy a great power and the undoubted master of the Mediterranean. The conquest of the Libyan 'Fourth Shore' and its incorporation into metropolitan Italy enclosed the central Mediterranean between Italian provinces and acted as a bridge into Africa.[59] Equally, some Italians considered the possibility of linking the East African empire to Libya, by-passing British-controlled Suez and creating a greater Italy extending from the Alps to the Indian ocean. Consequently, Italy's African territories possessed some strategic advantage. Finally, her international status as a colonial power and the attendant prestige was of great significance to Fascist Italy.[60]

Unsurprisingly, given *Geopolitica*'s closeness to the regime, the *Geopolitical Synthesis* maps that represented Africa reflected this context. The conscious rejection of cartographical ethics in favour of a subjective and partial vision is, of course, a mainstay of propaganda mapping.[61] Throughout the 1930s a tradition of geopolitical cartography also emerged from the European geopolitical discourses within which Roletto and Massi deliberately situated *Geopolitica*.[62] Morandi was clearly aware of such techniques, and was prepared to amend or exclude features from his maps, to mis-represent Africa, in order to support the arguments of *Geopolitica*. The absence of actual geographical detail and variation and its substitution by a fabricated and partial representation of a particular place can also be characteristic of geopolitical writing. As certain work upon such discourses has pointed out, geopolitical writings regularly mis-represent places, regions and countries. This ability is founded largely upon the authority that attends such 'scientific' texts.[63] Furthermore, there is also a history, especially in colonial contexts, of supposedly accurate and 'scientific' cartography mis-representing territory through the similar, if unintended, omission of actual variation and detail. In such cases, whether through ignorance of the region or indifference, territory, places and populations, can be represented upon maps as little more than uniform space. A colonising nation may then have few qualms over invading regions that appear on maps to be little more than vacant territory.[64] In this tradition of mapping, as well as in propaganda mapping, representations rely upon the social credibility of maps, upon their aura of accuracy and truth. The cartography of *Geopolitica* exhibited all of these traditions by presenting Africa as a tract of largely vacant space. Seldom did Morandi include elements of human geography upon his maps, and although he represented a basic physical geography of rivers, lakes and desert more frequently, this never reflected the full geographical complexity of the continent or the variation to be found across its regions. In representing Africa as a level playing field and resource base for the colonial powers *Geopolitica* created an image of Africa that suited Italy's imperial ambitions. Nevertheless, these maps invoked their accumulated authority as academic and cartographical discourses to construct and make credible their partial visions of Africa. The presentation of the continent as an empty space to be won, controlled and developed was a constant theme of both maps and journal and also played a role in the cultural legitimation of Italian imperialism.

Southern Africa

Figure 10.1 was the seventeenth of the *Sintesi Geopolitiche* series and dates from October 1940. It represents the situation in sub-Saharan Africa and is evidence of the map as a vehicle for, and articulation of, this particular geopolitical vision. This image is one of a small number of 'Sintesi Geopolitiche' accompanied by explanatory notes.[65] These notes – a guide to reading geopolitical maps – reveal that despite the three sub-graphics that deal with demographic statistics and the arrows indicating migratory patterns, population issues are not as central to the argument as they might initially seem. Instead the notes direct attention to the distribution of imperial territory and influence across this terrain, and to the expansion of South Africa in particular.

Morandi berates South African imperialism: 'especially towards Rhodesia and Niassa. South West Africa, in open violation of the Mandate statute, is practically annexed to the Union. In Mozambique too, the interests of British South Africa have assumed concrete form, especially in the economic field'. The aggressive expansionism of South Africa is represented by the serrated 'front' symbol. Attempts to portray

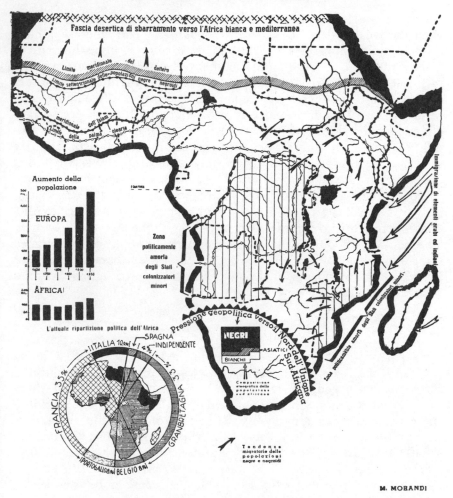

Figure 10.1 *Geopolitical Synthesis 17: Black Africa*

phenomena (such as expansionary pressure) required a new 'vocabulary' of map symbols, the 'front' symbol being one of the most frequent and distinctive.[66] The scale and resources of the British African empire was a common complaint of the Italian geopoliticians, and a recurrent theme of Massi's from the late 1930s. Yet this did not imply solidarity with other smaller colonial powers. Here the only non-British empires indicated are those of Portugal and Belgium – and these are named as 'minor colonial states', their territories labelled 'politically amorphous'. The weakness of these colonial powers is suggested by their constant loss of population by emigration.

Having established the expansionist ambitions of South Africa along with the weakness of the adjacent Belgian and Portuguese colonies, Morandi then points out the location of these colonies upon the Cairo to Cape trans-African route, implying that their strategic location is the rationale behind British pressure. Unusually, the Italian colonial territories are not represented upon the main map. The silence may imply that an Italian presence in sub-Saharan Africa is necessary to counteract the spread of British power. According to a common geopolitical argument, Italy was qualified for this task because the country developed its colonies more productively than competitors, and in this case, better than the 'lesser powers'[67] This echoes a recurrent theme of the journal that, rather than being another 'minor colonial state', Fascist Italy was one of the three foremost European powers in Africa, with a consequent remit to contest the continent with the British and French. The map was not only a lesson in seeing Africa geopolitically, it argued for, and justified, an Italian presence south of the Sahara, appealed for her status as a major regional power, and articulated Italian attitudes towards her colonial rivals.

Northern Africa

Sintesi Geopolitiche number 15 was published in August 1940 with the title 'L'Africa Bianca'.[68] It dealt with North Africa and introduced a further common geopolitical theme: the notion that Europe and Africa formed a single, functional geopolitical unit – 'Eurafrica'. The origins of the 'Eurafrica' idea are unclear. Histories of academic geopolitics usually accredit it to the 'Pan-region' theory of German *Geopolitik*. Whatever its lineage, the notion of 'Eurafrica' found its way into Italy's public realm in this period, whilst within the Italian *Geopolitica* movement it found frequent expression.[69] It also came to be used at the sub-continental scale to infer an extremely close relationship between Italy, a circum-Mediterranean hinterland, and its African colonies both within and beyond this zone.

Morandi's monochromatic style of mapping, developed during his career as a cartographer, is especially evident here. The most striking features on this map are the three vertical arrows that represent the transfer of European imperial authority into Africa by the French, the British and the Italians, the dominant trinity of colonial powers. These arrows, broad, dark, and die-straight, convey visually a sense of strength and permanence: the unbreakable nature of the European presence in Africa and its firm anchoring of the continent to Europe. That the lines terminate in arrowheads lends an impression of dynamism, of a continuing push to the south.[70] In contrast, the northern ends of the arrows are securely rooted in metropolitan France, in Britain's North African capital in Cairo, and in Rome and the Italian colonial capital of Tripoli. Fascism's favoured 'modern' lettering style is also to be found upon the arrows. Common in the public urban spaces of Italian Africa, it signalled the modernity of Fascist Italy and the translation of this superiority onto her colonies. The arrows take no account of the sketchy population and physical geography with which the map represents North Africa. Through their modernity and strength they reduce the region's complex human, geographical, and historical fabric to the simple commodity of space – here represented in the 'Eurafrican' relationship.

The 'Axis' idea that we see here was a familiar one to the Italian geopoliticians.[71] A geopolitical axis represented the spine of a territory or of a politically united block of space, and thus the Italo–African axis signifies the strong cohesion between European and African Italy. Extension of the axis beyond Libya implies an optimism of increased Italian influence beyond the Sahara. This last ambition ties in to the specific nuance that the 'Eurafrica' concept often bore in its Italian incarnation, that of central Africa as a resource base for Italy. In German *Geopolitik* 'Pan-regions' were to be self-sufficient 'Large space economies', consisting of a developed industrial core and a rural, resource-rich periphery.[72] In both German and Italian geopolitics, Africa was to be Europe's hinterland, but in Italy this idea received a special emphasis. This was due to the country's chronic shortage of raw materials and fossil fuels.[73] The 'Pan-region' idea theorised the availability of all necessary resources through the integration of the hinterland, the 'living-space', with the metropole. This prospect ensured the concept some detailed attention in *Geopolitica*, as Roletto and Massi had both produced a good deal of work upon resource location and control throughout the 1930s. Equally, any theory that proposed resource self-sufficiency would gain support in a country where national 'Autarky' had been an ambition of the regime since the mid-1930s.[74]

Figure 10.2 also addresses the colonial settlement issue. Whilst the Italian axis was on the same scale as those of Britain and France, implying that Italy was a colonial power of similar magnitude and prestige, only the Italian expatriate communities are indicated on the map. The proportional circles convey the impression that Italians were by far the most populous of the European minorities in North Africa; thus justifying Italian claims to a significant role in Africa on an equal footing with her colonial rivals.[75] This position is emphasised by the sub-graphic to the right of the main image which supposes a tripartite division of Africa in which the northern and southern sections are well populated by Europeans. Central Africa on the other hand, is comparatively under-populated. Consequently, the substantial migration shown by the curving arrow that connects Italy to East Africa is justified because it does not interfere with the balance of existent colonial settlement. Rather the Italians are colonising a region that previously enjoyed little European attention. Italy's role as a colonial power in North and East Africa is thus simultaneously proposed and legitimated.

African resources and Italian access

The 'Eurafica' theme is reprised in the next map (Figure 10.3) that appeared in December 1941 and was one of the very few in *Geopolitica* that was not drawn by one of the staff-cartographers. It was signed by Alfio Biondo and accompanied his article entitled 'La Transafricana italiana'. It reveals the particular attention to resources that the Italians brought to the 'Pan-region' concept.[76] Biondo too was involved with the ICF in Lombardy, eventually becoming editor of *Impero Italiano*. His article for *Geopolitica* probably originated in these links to Massi and Morandi. Consequently, we might expect Biondo to reproduce the ICF's vigorous promotion of colonialism, of the exploitation of Africa, and to realise the utility of maps in communicating his argument. An earlier published version of the thesis, in *Impero Italiano*, had boasted a similar, yet larger map than Figure 10.3, but before its translation to *Geopolitica* this amateurish map had to be 'cleaned up' to meet the graphical standards of the geopolitical journal. Such evidence of an editorial policy of quality control with regard to maps supports the claim that the journal privileged the cartographical medium and was unwilling to dilute its potential impact by lowering its standards of graphical presentation.[77]

Biondo's arguments are two-fold. He employs shading contrasts to encircle and highlight Italy and its adjacent Mediterranean territories in dark tones. Italy then appears to be centrally positioned to the north of Africa, and it is from this location, and in the same dark tone, that a root-like system descends from Tripoli through the barrier of the Sahara, to branch out into the regions of central Africa. This root system, its strength implied by its angularity and straight lines, proceeds to integrate the communications of the African interior by connecting this network with the major rail lines to the coast. Under this scheme, the Italian integration of Africa's communications would be by virtue of her unique location at the head of the continent combined with her trans-Mediterranean base in Libya. These geographical factors and Italy's ability to transcend the barrier of the Sahara justify Italy's unique role in Africa.

The integration of the continent permits its resource value to be tapped. In this case the opportunities for generating hydro-electric

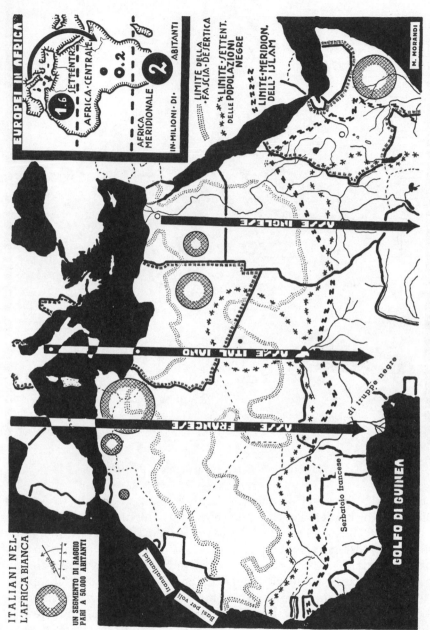

Figure 10.2 Geopolitical Synthesis 15: White Africa

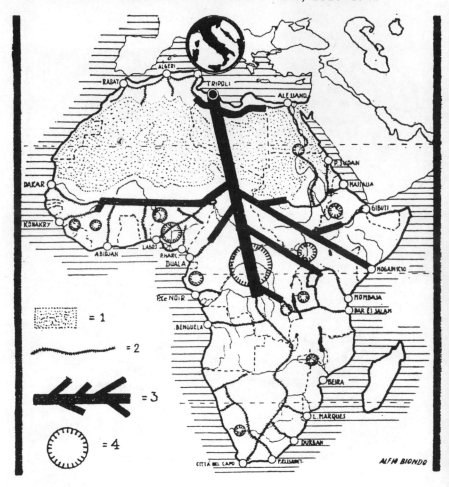

Figure 10.3 'La Transafricana italiana'

power are emphasised with almost all the major potential sites located either on the branches of the 'Transafricana', or upon the railways that it integrates. The largest site of all is in the Belgian Congo at the foot of the main artery from Tripoli. Uplands throughout the Italian peninsula had been subject to the early and extensive development of hydro-electric power in response to the national deficiency in fossil fuels.[78] The extension of such development to Africa and the resultant cheap energy, the surrounding article explains, would power this communications network and allow the spatial integration of central Africa. In turn, Biondo argued, this would facilitate the transportation of the region's riches to Europe, and more specifically to Italy. Although the argument was couched in egalitarian terms whereby the circumvention of the Anglo-French stranglehold upon African resources would be of universal benefit in peacetime, a more nationalist slant surfaced in the argument that the 'Transafricana' also provided a strategic guarantee of adequate raw materials during conflict. Thus the author pursues the notion of a productive African hinterland, integrated and developed by Italy. The map strengthens this agenda by its reductionist geography of Africa as an open resource base waiting to be exploited and through the organic metaphor of the 'Transafricana' as a root system sustaining Italy through contact with these African riches.[79] It also emphasises the north–south axis of this relationship by denying east–west linkages and by framing Africa to either side with substantial black lines, whilst the north and south borders of the map remain unenclosed, suggesting a neat and natural longitudinal segmentation of the globe into 'pan-regional' blocks.

The 'Mare Nostrum' and the Libyan 'Fourth Shore'

Figure 10.4, the third *Sintesi Geopolitiche*, dates from March 1939 and anticipated later editions by encompassing the recurrent themes of *Geopolitica* towards Africa. In addition, it rehearsed the strategic advantages that the Libyan colony lent to its mother country across the Mediterranean basin. As only the third *Sintesi Geopolitiche*, the map was preceded by a reader's guide which directed attention to the 'importance of [our] compatriots distributed throughout the basin'. The 1930s Italian population of the African Mediterranean coast was represented exclusively amongst the foreign communities with the benefit of large proportional symbols.[80] Again this selective mapping gave an exaggerated impression of a Maghreb populated largely by Italians, especially in the central and eastern section. The colonial settlement programme had several rationales. Colonialism was of immediate value to Italy as a source of labour-intensive work for the 1930s unemployed, and as a social safety-valve for the dispossessed and Italy's emigrants. Internationally, it was intended to be a measure of the new Italy's dynamism and to signal the nation's arrival as a great power. Here it is allied to the theme of the Mediterranean as a distinct geopolitical region in which Italy was the natural hegemonic power.

The spatial unity of the region is hinted at by the sub-graphic which depicts the integrated nature of the Mediterranean basin's migratory patterns.[81] Such unity is emphasised on the main map by the use of tonal contrasts. These picture a Mediterranean that is enclosed by the dark tones which represent Italian territory, the 'authoritarian states' allied to Italy, the circles that portray Italian expatriate communities,

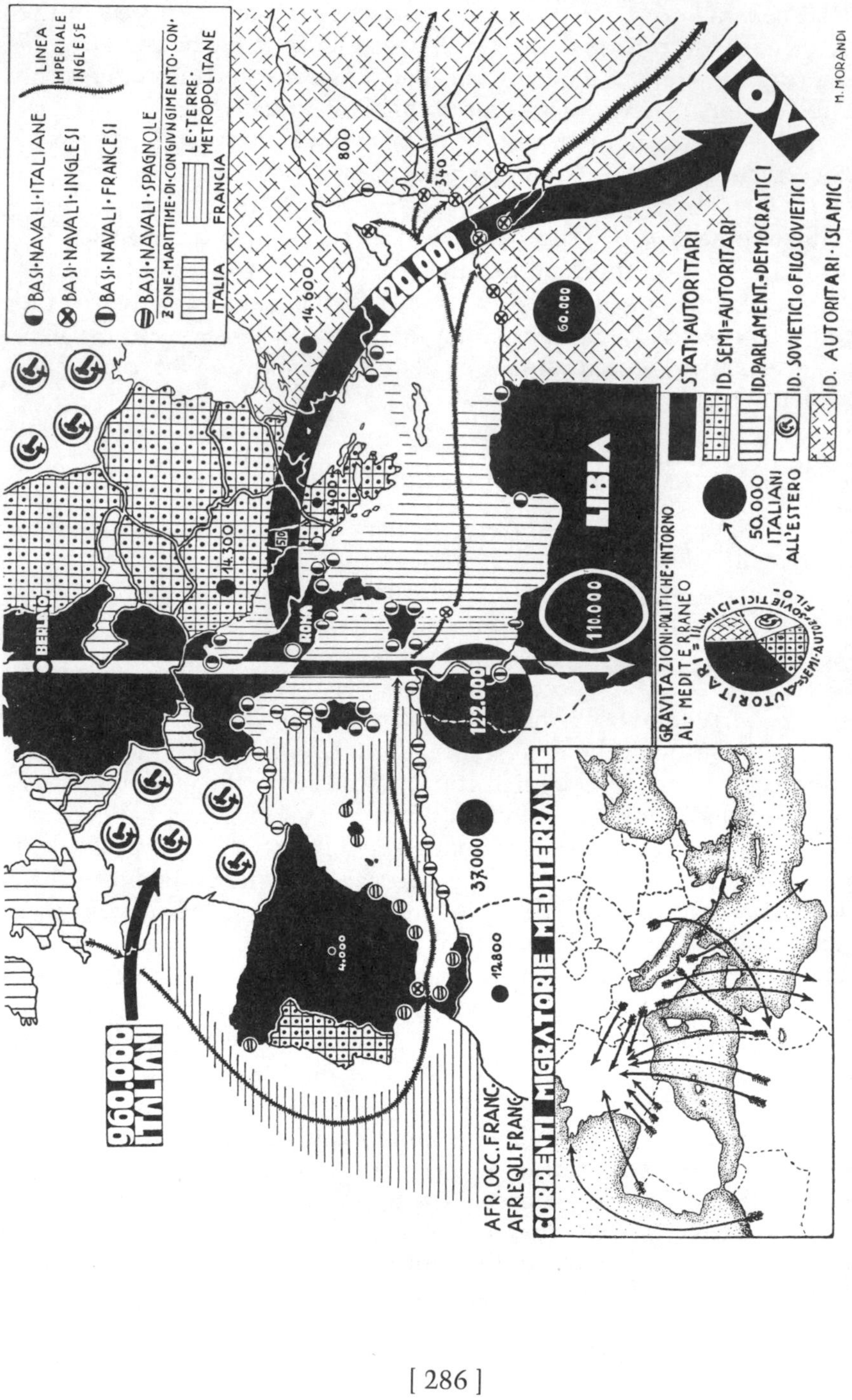

Figure 10.4 *Geopolitical Synthesis 3: The Mediterranean*

and the sweeping arrow that represents the Italian colonisation of East Africa.[82] The map is constructed to persuade the reader visually that the Mediterranean is centred upon Italy and surrounded by an Italian presence; that the region is, in fact, Italy's own natural zone of influence and growth.[83] The trans-Mediterranean Libyan territory plays a crucial role in creating the impression that the basin was an Italian sea.

Libya then plays a further part in the integration and control of Italy's 'Mare Nostrum'. Whilst another vertical arrow reflects the idea of 'Eurafrica' as it anchors Africa to the Rome–Berlin axis, its passage through Libya and the sizeable Italian population in Tunisia facilitates not only the continuation of the Axis presence across the Mediterranean, but it also blocks the British imperial passage to Suez. The inclusion upon the map of naval bases and zones of maritime control furthers this argument and emphasises the strategic utility of Libya. This is because the maritime zones, indicated by horizontal lines for French control and vertical lines for the Italians, are dependent upon the adjacency of metropolitan territory: a criterion that produces Italian hegemony in the central and eastern Mediterranean thanks to the Libyan shore. In contrast, the Royal Naval presence upon what H. J. Mackinder labelled 'The Highway of Empire' is reduced to a slender line that is broken regularly by symbols that represent Italian force. On these terms Italy's African 'Fourth Shore' functions as a buttress to Italy's hegemonic ambitions in the Mediterranean region. As ever, it is also proof of Italy's great power status based on its prestigious African empire.

The East Africa empire

The fifth map is another *Sintesi Geopolitiche* from January 1940 (Number 11. Figure 10.5).[84] As with the great majority of these images, this was published without any text or explanatory notes. Again it confirms the faith that the editors and cartographers placed in the communicative ability of these graphical essays, and their suitability for articulating a geopolitical perspective. Once more concern at the ubiquity of the British empire, its dominions and influence in Africa is expressed in the main map. In this case by the bar of arrows that represent South African pressure in central Africa, and the northern delimitation of the Italian territory by the economic barrier along the Egyptian border. In the Horn of Africa, British Somaliland and French Djibouti, although partially enclosed by Italian arrows, each constitute a hostile presence. The unknown quantity of Arab nationalism stretches across the Arabian peninsula.

The map notes the economic and settlement potential of the

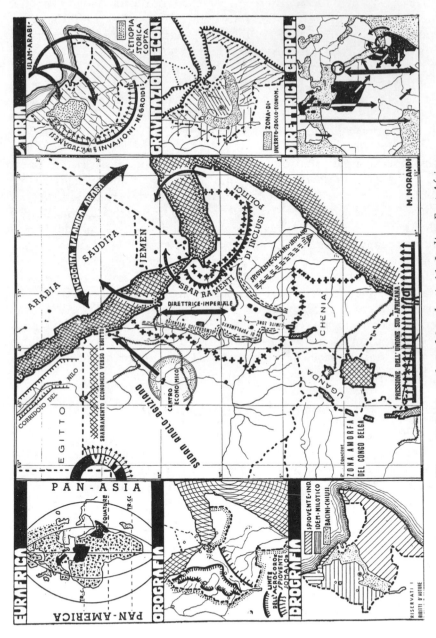

Figure 10.5 *Geopolitical Synthesis 11: Italian East Africa*

Ethiopian highlands. The actual territory of the empire is geographically defined and also bears an arrow representing the north–south geo-political development of Italy from its nineteenth-century Eritrean colony to its 1936 prize of Addis Ababa. Economic aspects in the synthesis include the economic centre of the Anglo-Egyptian Sudan. Enclosed by Italian Libya to the north west, this zone outlets to the Red Sea as the Egyptian economic frontier prevents it from accessing the Nile corridor. This orientation leads us back once more to the Anglo-Italian contest in the Mediterranean. In concert with the small map to the bottom-right of the page, the main image portrays Italian claims to the Red Sea's outlet to the Indian Ocean. Italian control of the Red Sea's access into the Gulf of Aden would make redundant Britain's control of Suez and lessen the Royal Navy's hold over the Mediterranean. Italian ambitions to by-pass Suez via an overland Libya–Ethiopia route were also in response to Britain's hegemony over the Mediterranean. The smaller map on the bottom-right challenges the British Cape to Cairo corridor. It was in the envisioning and comprehension of this kind of scenario that *Geopolitica* claimed a unique advantage.

The six smaller images that flank the main map help the 'synthesis' to provide a more comprehensive coverage. They provide limited thematic narratives that are related to the central image. In this case, they seek to contextualise the geopolitical representation of Italian East Africa. Small maps of the empire's relief and hydrology relate the salient aspects of the local physical geography and in the same way portray the historical pressure on the Ethopian heartland from adjacent Arabic and Sudanese peoples. The economic orientation of the various imperial regions is indicated, based largely upon the principal communications of each area via river or rail to oceanic ports. Finally, two maps situate this locale within its wider geopolitical context. East Africa conse-quently figures prominently within the theorised 'Pan-region' of 'Eurafrica', and at a smaller scale, is a contestant in the African colonial competition, enclosing the mouth of the Red Sea, and in competition with the British empire that separates Italian Libya from the East African empire.

Conclusion: envisioning Africa

These maps constituted a way of seeing: an articulation of the geo-political vision and the construction of a geopolitical world through the holistic synthesis of 'political, economic, social, and colonial geo-graphy' to which *Geopolitica* laid claim. Roletto and Massi hoped to articulate this unique vision and special insight through the journal, and it was in the cartography of *Geopolitica* that their vision found its most immediate and striking expression. I have considered how these maps applied geopolitical analysis and expertise to the situation in Africa. *Geopolitica*'s claims to a privileged, synthetic vision were spurious. The perspective of the *Sintesi Geopolitiche* was partial. *Geopolitica*'s Africa was a continent laid open to imperial competition and the over-riding force of the European powers. Africa's actual geography, its complex histories, cultures, and patterns of human activities, were ignored. Rich and diverse regions were reduced to a simplistic, de-socialised and largely abstract space, to be contested by the European powers. Amongst these nations Italy desired a prime role. The Mediterranean was represented as an Italian zone for the expansion of her trade and population.[85] This 'Mare Nostrum' linked Italy to Africa, where Italians enjoyed a legitimate remit for further settlement, resource-exploitation and expansion.

Geopolitica had a print run of about one thousand copies and was available upon public bookstands until April 1941. Sperling and Kupfer, the journal's Milanese publishing house, boasted a national distribution network and Bottai encouraged all middle and higher educational insti-tutes to subscribe to the journal.[86] However, it is impossible to say how many Italians saw these maps and how influential they were. Roletto and Massi probably found their most sympathetic readership amongst the Fascist organisations with which they had connections. As I noted earlier, the ICF reproduced geopolitical cartography extensively in their own newspapers and their monograph series. Morandi's maps could also be found in the books and articles of other geographers who were linked to *Geopolitica*.[87] Similarly, bodies such as the Milanese Scuola di Mis-tica Fascista, the Instituto Nazionale di Cultura Fascista, and the Istituto per gli Studi di Politica Internazionale, which all flourished in the 1930s had links with *Geopolitica*.[88] Such groups may well have given credence to the journal's claims to expertise and an authoritative geopolitical perspective, to an ability to understand and explain the situation in Africa. If this was the case, pro-Fascist *Geopolitica*'s repre-sentation of Africa may have served to reinforce colonial sentiments within these pro-Fascist organisations. More probably, *Geopolitica*'s legitimation of Italian imperialism functioned as useful intellectual justification in wider Fascist culture. Whilst Roletto and Massi pre-sented *Geopolitica* as possessing a distinct way in which to understand and explain the world, based partly upon the wide-ranging vision of geopolitical survey and partly upon links to the academic discipline of geography, their representation of Africa might have earned unwarranted credibility as a form of 'scientific analysis'.

In this manner, *Geopolitica* – an outgrowth of academic geography – functioned in much the same way as other geographical discourses had in modern Italy. *Geopolitica* used its proclaimed scientific status to survey, propose, and justify Italian imperialism in Africa. In envisioning and representing Africa in this way, *Geopolitica* was continuing a long tradition whereby Italian geography abetted Italian imperialism. Whether by recourse to exploration, cartography, the collation of information and survey, or the dissemination of geographical knowledge, Italian geographers – in their various guises – offered serious intellectual support for Italian activities and interventions in Africa. In the last years of 'Imperial Italy', as *Geopolitica* sought to instil a 'political consciousness' into geography, it is not surprising that it continued to provide geographical legitimation for imperialism. In January of 1943, amidst the increasing chaos of the Italian war effort, *Geopolitica* ceased publication. Four months later Axis forces capitulated in Tunisia. Despite the efforts of perhaps the most overtly politicised journal in the history of Italian geography, the Italian Kingdom lost its last foothold in Africa.

Notes

1 G. Ferro and I. Caraci, *Ai confini dell'orizzonte, Storia delle esplorazione e della geografia*, Milan, 1992, 2nd edition, esp. pp. 187–93; Lucio Gambi, *Una geografia per la storia*, Turin, 1973, esp. pp. 11–14.

2 On geography, in all of its forms in this period, see Gambi, *Una Geografia per la storia*; G. Ferro and I. Caraci, *Ai confini dell'orizzonte*, Milan, 1992; I. Caraci, *La Geografia Italiana tra '800 e'900 (dall'unità à Olinto Marinelli)*, Genoa, 1982; I. Caraci, 'Storia delle Geografia in Italia dal secolo scorso ad oggi', in G. Corna-Pellegrini (ed.) *Aspetti e Problemi di Geografia*, vol. 1, Milan, 1987, pp. 47–94; and F. Fulvi, *Lineamenti di storia della Geografia in Italia*, Florence, 1990. On the IGM see Istituto Geografico Militare, *Il primo centenario dell'Istituto Geografico Militare*, Florence, 1973; for an inter-war view of the SGI see E. De Agostini, *La Reale Società Geografica Italiana e la sua opera dalla fondazione ad oggi (1867–1936)*, Rome, 1937; and for a post-war perspective, M. Carazzi, *La Società Geografica Italiana e l'esplorazione coloniale in Africa, 1867–1900*, Florence, 1972; A. Milanini Kemény, *La Società d'esplorazione Commerciale in Africa e la politica coloniale, (1879–1914)*, Florence, 1973, deals with the Milanese society.

3 *Enciclopedia Italiana*, entry 'Italia', 1928, vol. 19, p. 696.

4 R. Parry and R. Perkins, *World Mapping Today*, London, 1987, p. 461.

5 On such issues see; J. B. Harley, 'Maps, knowledge and power', in D. Cosgrove and S. Daniels, (eds), *The Iconography of Landscape. Essays on the symbolic representation, design, and use of past environments*, Cambridge, 1988, pp. 277–312.

6 *Enciclopedia Italiana*, 'Italia', p. 696.

7 On geographical knowledge and the 'construction' of places and nations see: J. B. Harley, 'Deconstructing the map', in T. J. Barnes and J. S. Duncan (eds), *Writing Worlds: discourse, text and metaphor in the representation of landscape*, London, 1992, pp. 231–47; and M. J. Heffernan, 'An imperial utopia; French surveys of North Africa in the early colonial period', in J. C. Stone (ed.), *Maps and Africa*, Aberdeen, 1994, pp. 81–107.

8 See L. Gambi, 'Geography and imperialism in Italy from the unity of the nation to the "new" Roman Empire', in A. Godlewska and N. Smith (eds), *Geography and Empire*, Oxford, 1994, pp. 74–91, esp. pp. 76–80; and H. Capel, 'Elementos para la comprobacion del modelo: el caso de la Geografia Italiana', in *Institucionalizacion de la geografia y estrategias de la comunidad de los geografos*, Barcelona, 1977, pp. 5–26.

9 Carazzi, *La Società Geografica Italiana*, esp. pp. 3–53.
10 This celebration continued into the Fascist era: C. Zoli, 'Relazione sull'opera svolta della Reale Società Geografica Italiana nel triennio 1935–1937', in *Atti de XIII Congresso Geografico Italiano*, 1937, pp. 91–6, for example, celebrates the SGI as a society without equal in its long and distinguished history of involvement in Africa. Even today, at Villa Celimontana, the SGI headquarters in Rome that was gifted to the society by Vittorio-Emmanuele III in 1926, the paraphernalia of exploration sits alongside African artefacts in glass display cases.
11 S. Rist, 'La Società Geografica Italiana e la spedizione allo Sciotel', *Bollettino della Società Geografica Italiana*, Serie XI, vol. IX (1992), pp. 111–124.
12 Carazzi, *La Società Geografica Italiana*, esp. pp. 54–98; see also A. A. Castagno, 'The development of the expansionist concepts in Italy, 1861–1896', unpublished Ph.D. thesis, University of Columbia, 1956.
13 Gambi, 'Geography and imperialism'.
14 Milanini Kemény, *La Società d'esplorazione Commerciale*, esp. pp. 14–54; J. Wright, 'The Cyrenaican expedition of Giuseppe Haimann', *Libyan Studies*, 14 (1983), pp. 1–3.
15 J. Wright, *The Cyrenaican expedition*; Milanini Kemény, *La Società d'esplorazione Commerciale*, esp. p. 102–12.
16 Gambi, *Una Storia*; Gambi, 'Geography and imperialism', esp. pp. 80–1. For Battisti and Biasutti's initial statement in *La Cultura Geografica*, see: La redazione, 'Ai lettori', *La Cultura Geografica*, 1, 1, 15 Gennaio (1899), pp. 1–2. Biasutti, as I note later, was more ready to praise the relationship between the state and geography in the 1940s.
17 F. Surdich, 'La spedizione Stefanini-Puccioni in Somalia (1924)', *Bollettino della Società Geografica Italiana*, Serie XI, vol. IX (1992), pp. 125–40.
18 Gambi, 'Geography and imperialism', p. 82; on the wider Italian intervention in Libya, see A. Del Boca, *Gli italiani in Libia. Dal Fascismo a Gheddafi*, Bari, 1988; and on Italian imperialism more generally, J. L. Miège, *L'imperialismo coloniale italiano, dal 1870 ai giorni nostri*, Milan, 1976.
19 Gambi, 'Geography and imperialism', p. 82.
20 C. G. Segré, *'Fourth Shore', the Italian colonization of Libya*, Chicago, 1974, esp. pp. 46–81.
21 On Mussolini's visit to the SGI in 1924, at which he promised the society increased government funding, see: 'Atti della Società, seduta straordinaria del 27 Maggio, Ricevimento di S. E. il Presidente del Consiglio dei Ministri, On. Benito Mussolini', *Bollettino Della Reale Società Geografica Italiana*, Serie VI, vol. I, 5–6 (1924), pp. 221–4.
22 M. Fuller, 'Building power: Italian architecture and urbanism in Libya and Ethiopia', in N. AlSayyad (ed.), *Forms of Dominance, on the Architecture and Urbanism of the Colonial Enterprise*, Aldershot, 1992, pp. 211–39.
23 L. Rusca, 'Problemi Coloniali al X Congresso Geografico Italiano', *Rivista delle Colonie Italiana*, 1, 1 (1928), pp. 177–84.
24 N. Vachelli, 'Coscienza Geografica', in *L'Oltremare*, 2, 4 (1928), p. 159, cited in Fuller, *Building Power*, p. 216.
25 C. Zoli, 'Presentazione dell'opera', in Reale Società Geografica Italiana, *Il Sahara Italiano, Parte Prima. Frezzan e Oasi di Gat*, Rome, 1937, p. 9.
26 E. Scarin, 'I confini della Libia', *Rivista Geografica Italiana*, 42 (1935), pp. 77–102; Scarin, 'Le condizione altrimetriche della Libia', *Rivista Geografica Italiana*, 43 (1936), pp. 29–35; and Scarin, 'La zona della Gara Mullata', *Rivista Geografica Italiana*, 44 (1938), pp. 122–34.
27 See his report: E. Scarin, 'Insediamenti e tipi di dimore', in Reale Società Geografica Italiana, *Il Sahara Italiano*, pp. 603–44.
28 F. Lando, 'Geografie di casa altrui: l'Africa negli studi geografici italiani durante il ventennio fascista', *Terra d'Africa*, 1933, pp. 73–124, the quotation is from p. 84.
29 Ministere des Colonies, *Le Sahara Italien, Guide officiel de la section Italienne*, Rome, 1934.
30 R. Bagnold, *Libyan Sands*, London, 1926. Whilst planning an expedition at the London

RGS in the 1920s, Bagnold recalls how 'There were three areas of seducing blankness, omitting Italian Libya (which it would now be only a matter of time until it was explored and mapped).' pp. 208–9.

31 For an example of the swift mapping of new Italian territories see A. Dardano, 'Areometria dell'Oltre Giuba (Giubaland Italiano)', *Bollettino della Reale Società Geografica Italiana*, Serie VI, vol. I, 5–6 (1924), pp. 268–70.

32 On the history of Italian cartography in Libya, see E. Casti Moreschi, 'Nomi e segni per l'Africa italiana: la carta geografica nel progetto coloniale', *Terra d'Africa*, 1992, pp. 13–60; C. Taversi, *L'Italia in Africa, Storia della Cartografia coloniale Italiana*, Rome, 1964; and, from the period, E. De Agostini, 'La cartografia', Reale Società Geografica Italiana, *Il Sahara Italiano*, pp. 645–58; on the range of maps available, even during the *reconquista*, see, 'Ufficio Studie propaganda della Cirenaica', in *Rivista delle Colonie Italiane*, 1, 3 (1928), pp. 367–8; also see Sindicato Nazionale Fascista dei Geometri, *Corso per Geometri Coloniali*, Roma, vol I., 1937, on the institutionalization of 'colonial-surveying' in this period.

33 I. Balbo, 'La colonizzazione in Libia', *L'Agricoltura Coloniale*, August, 1939; and P. D'Agostini-Orsini di Camerota, *La colonizzazione Africana nel Sistema Fascista*, Milan, 1941, pp. 101–7.

34 Istituto Geografico Militare, *L'Istituto Geografico Militare in Africa Orientale, 1885–1937*, Florence, 1939; for a lengthy review of this work, see; G. Peisino, 'Rassegna Cartografica', *Geopolitica*, 2, 2 (1940), pp. 133–5. Other such public and unabashed statements of cartography's role in the conquest of the colonies include: A. Dardano, 'Cartografia Coloniale', *Rivista delle Colonie Italiano*, 1, 2 (1928), p. 265–72.

35 The *Istituto Coloniale Fascista* was created from the *Istituto Coloniale Italiano* which was founded in 1906 by eminent geographers of the day amongst others. With the foundation of Italy's East African empire in the late 1930s, it was renamed the *Istituto dell'Africa Italiana*. On something of the *Istituto Coloniale Italiano*, and its origins in the pro-colonial lobby of late-Liberal Italy, see: R. J. B. Bosworth, 'The opening of the Victor Emmanuel Monument', *Italian Quarterly*, 16 (1975), pp. 78–87.

36 See G. Dainelli, 'Scuola Nazionale di Geografia', *Bollettino della Reale Società Geografica Italiana*, Serie VII, vol. IX, 2 (1941), pp. 81–9; Consociazione Touristica Italiana, *Atlante fisico ed economico d'Italia*, Florence, 1940, and on some reaction to the Atlas, 'L'Atlante fisico-economico d'Italia della CTI', in *Rivista Geografica Italiana*, 46, 1–3 (1939), pp. 206–7.

37 On the regime's use of the TCI and the 1937 occasion when Mussolini personally ordered that the TCI's name be Italicised into the 'Consociazione Touristica Italiana', see L. Di Mauro, 'L'Italia e le guide turistiche dall'Unità ad oggi', in C. De Seta (ed.) *Storia d'Italia, Annals V, Il Paesaggio*, Turin, 1980, pp. 367–428, esp. pp. 406–12.

38 Consociazione Touristica Italiana, *Guida dell'Africa Orientale Italiana*, Milan, 1938, (*Guide to Italian East Africa*); which received a positive review from London's *Geographical Journal*, 93, 5 (1939), pp. 448–9.

39 L. F. De Magistris and G. C. Pico, *L'impero d'Italia*, Verona, 1938. The co-authorship attributed to Pico, as national director of education, may have been honorific. De Magistris, though, was involved with some of the more Fascistised institutions of the period, including the ICF.

40 The *Paesi d'Attualità* series was edited by Elio Migliorini, the editor of the *Bollettino* of the SGI. Migliorini penned the first study, *La Siria*, Rome, 1941; and the fourth, *La Tunisia*, Rome, 1941. Other contributions from this period included works upon Egypt, Savoy, Ljubljana, Dalmatia, Montenegro and Corsica. All began with a regional geography of the area in question, then outlined its contemporary importance and ended with an account of Italian interests in the region. The *Bollettino della Reale Società Geografica Italiana*, Serie VII, vol. VII, 1 (1942), pp. 36–7, announced that the series was to be continued and re-titled *Paesi dell'ordine nuovo*, with the assistance of additional government funding.

41 Contemporary comments from R. Biasutti, 'Della nuova 'Geopolitica', del rinnovate 'Bollettino della R. Società Geografica Italiana, e di altre cose', *Rivista Geografica Italiana*, 46, 1–3 (1939), pp. 64–5; and again from The *Geographical Journal*, 93, 4

(1939), p. 372.

42 *Coscienza geografica* translates as 'geographical consciousness', a phrase that enjoyed an increasing currency during the 'imperial phase' of Fascist Italy. On Bottai see S. Cassese, 'Giuseppe Bottai', in *Dizionario Biografico degli Italiani*, Rome, 1971, pp. 389–404; and on his cultural project see, A. J. De Grand, *Bottai e la cultura Fascista*, Bari, 1978; Italian geographers were pleased by the attention of this influential figure, see R. Biasutti, 'Della nuova "Geopolitica", del rinnovate Bollettino della R. Società Geografica Italiana, e di altre cose', *Rivista Geografica Italiana*, 46, 1–3 (1939), pp. 64–9, in which he makes mention of Geographers referring to Bottai as 'their' minister.

43 G. Bottai, Mete ai Geografi, *Bollettino della Reale Società Geografica Italiana*, Serie VII, vol. IV, 1 (1939), pp. 1–3. Bottai also appeared to acknowledge the value of geography as a broad-ranging and synthesizing science, and as one that could be used to widen the horizons of the Italian people.

44 R. Biasutti, 'I geografi Italiani convocati dal Ministro dell'Educazione Nazionale', *Rivista Geografica Italiana*, 48, 1–2 (1941), pp. 75–9, quotation taken from p. 79.

45 E. Massi, 'Römische und italienische Mittelmeergeopolitik', *Zeitschrift für Geopolitik*, 16, 8–9 (1939), pp. 551–66; for a later, post-war account, in which he traces Italian geopolitics as emerging from a historical dialogue between geography and politics, see E. Massi, 'Geopolitica, Della teoria originale ai nuovi orientamenti', *Bollettino della Società Geografica Italiana*, Serie XI, vol. III, 1–6 (1986), pp. 3–45. On the appointment of Roletto to Trieste, and on the specific culture of that border city, see A. Vinci, ' "Geopolitica" e Balcani: l'esperienze di un gruppo di intellettuali in un ateneo di confine', *Società e storia*, 47 (1990), pp. 87–127, esp. pp. 89–103.

46 On the commitment of Massi and Roletto to geography, see Vinci, *'Geopolitica' e Balcani*; also Roletto's inaugural lecture, G. Roletto, *La Geografia come scienza utilitaria*, Trieste, 1929; G. Valussi, 'L'Opera scientifica di Giorgio Roletto', *Bollettino della Società Geografica Italiana*, Serie IX, vol. VI., 7–8 (1965) pp. 313–26; M. Lo Monaco, 'Ernesto Massi: Mezzo secolo di analisi geografiche per la sintesi economica', in Dipartimento di studi Geoeconomici Statistici e storici per l'analisi regionale, Università di Roma, *Scritti in onore di Ernesto Massi*, Bologna, 1987, pp. 7–11; on Massi as particularly interested in the growth and development of a distinctive 'geopolitical' discourse in inter-war Europe, see: E. Massi, 'Geografia politica e geopolitica', *La cultura Geografica*, 6 (1931), pp. 1–11; E. Massi, 'Lo Stato quale oggetto Geografico', *Rivista di geografia*, 7, 5 (1932) pp. 169–76; and E. Massi, 'Nuovi indirizzi della geografia politica in Francia', *Rivista internazionale di Scienze Sociale*, Serie III, vol. IX, 2 (1938), pp. 194–208.

47 See Massi's contributions: E. Massi, 'Il Mediterraneo e l'Italia', and 'Il problema economico della colonizzazione', in Gruppo Rionale Fascista "E. Tonioli", Ufficio coloniale dell'Istituto Coloniale Fascista, Milano, *Corso di Cultura Coloniale*, Milan, 1937, pp. 7–12, and 63–70. The series was reported in *Impero Italiano*, 2, 2 (1937), p. 7.

48 'Un ufficio istituto in sede', *Impero Italiano*, 2, 8 (1937), p. 6; and on some of the courses, see note 44.

49 E. Massi, *La participazione delle colonie alla produzione delle Materie prime*, Milano, 1937, (a second edition was published in 1939); the maps were also reproduced upon a full-page in *Impero Italiano*, 2, 8 (1937), p. 3, in the launch of the subsection 'Collana di studi Coloniali'. This work is typical of Massi's interests in the second half of the 1930s on African resources, their distribution, transportation and political control, for instance, E. Massi, 'L'Africa nell'economia mondiale', *Conferenza di Alta Cultura Coloniale*, Fasc. 3, 1937 and E. Massi, 'Il problema coloniale in Germania', Estratto dalla Rivista *Vita e Pensiero*, February 1937, pp. 3–12.

50 M. Antonsich, 'La coscienza geografico imperiale del regime Fascista "Geopolitica" (1939–1942)', unpublished Tesi di Laurea, Catholic University of the Sacred Heart, Milan, 1991, pp. 78–80.

51 A. J. De Grand, *Bottai*, p. 275n.

52 G. Bottai, 'S. E. Bottai alla 'Geopolitica', *Geopolitica*, 1, 1 (1939), pp. 3–4, the quotation is from page 4.

53 G. Roletto and E. Massi, 'Per una Geopolitica Italiana', *Geopolitica*, 1, 1 (1939), pp.

5–11.

54 The other 'official' cartographer was Dante Lunder, a Triestene journalist, Fascist organiser, and sometime lecturer on Fascist colonialism. Although his maps were also loaded with 'persuasive' graphics and shot through with sentiments similar to Morandi's, they did not appear as frequently and were not as sophisticated as his colleague's. See *Geopolitica*, 3, 12 (1941), p. 610, for Morandi's acknowledgements at the end of his year as Milanese co-editor.

55 *Sintesi Geopolitiche* means 'geopolitical synthesis'; this form of mapping is considered in greater depth in my forthcoming thesis, 'Geography and geopolitics in Fascist Italy', University of Loughborough, in preparation.

56 C. G. Segré, 'Colonization in the French Maghreb: model for Italy's Fourth Shore?', *The Maghreb Review*, 12, 1–2 (1987), pp. 34–7; see also. G. L. Fowler, 'Italian colonization of Tripolitania', *Annals of the Association of American Geographers*, 62, 4 (1972), pp. 627–40.

57 E. Massi, *L'Africa economica*, vol. 1, Milan, 1941; P. D'Agostino-Orsini di Camerota, *Che cosa é l'Africa*, 5 vols, Rome, 1937.

58 See, on Libya for example, E. Migliorini, 'Risorse economiche', Reale Società Geografica Italiana, *Il Sahara Italiano*, pp. 561–90.

59 Libya's function as a bridge into Africa, and on the SGI's exploratory activity that was helping to realise this, see Corrado Zoli in his report to the 12th Italian Geographical Congress (Sardinia), 1935, C. Zoli, 'L'attività delle Reale Società Geografica Italiana dal 1930 al 1934', in *Atti del XII Congresso Geografico Italiano*, 1935, pp. 9–14.

60 On the importance of a colonial profile to Italy in this period, see D. Mack Smith, *Mussolini's Roman Empire*, London, 1976; I. Balbo, 'Coloni in Libia', *Nuova Antologia*, 1 November 1938; J. Wright, 'Mussolini and the Sword of Islam', *The Maghreb Review*, 12, 1–2 (1987), pp. 29–33; on some of the historical background to Italian ambitions in Africa, R. L. Hess, 'Italy and Africa: colonial ambitions in the First World War', *Journal of African History*, 4, 1 (1963), pp. 105–26.

61 M. Monmonier, *How to Lie with maps*, Chicago, 1991; S. Toniolo, 'Cartografia come strumento d'informazione e disinformazione geografica', *Bollettino della Società Geografica Italiana*, Serie XI, vol I. (1984), pp. 519–24.

62 Richard von Schumacher had created a series of geopolitical mapping symbols in 'Zur Theorie der Geopolitischen Signatur', *Zeitschrift für Geopolitik*, 12 (1935), pp. 247–65. Massi read *Zeitschrift für Geopolitik*, and may have passed this information to Morandi; regardless, this system was reproduced in Italy in U. Toschi, *Appunti di Geografia Politica*, Bari, 1940, second edition, p. 66, and Morandi obviously knew of this schema.

63 G. O'Tuathail and J. A. Agnew, 'Geopolitics and discourse: practical geopolitical reasoning in United States Foreign Policy', *Political Geography*, 11, 2 (1992), pp. 190–204. Also, G. O'Tuathail, 'The effacement of place: US Foreign Policy and the spatiality of the Gulf Crisis', *Antipode*, 25, 1 (1993), pp. 4–11.

64 See J. B. Harley, 'Silences and secrecy: the hidden agenda of cartography in early-modern Europe', *Imago Mundi*, 40 (1988), pp. 57–76; and J. B. Harley, *Maps, Knowledge and Power*.

65 *Geopolitica*, 2, 10, (1940), pp. 420–421; the first 'Sintesi Geopolitiche' (1, 1 (1939), pp. 58–9) and the third (Figure 10.4) were the only two of the series to boast a reader's guide. These guides introduced geopolitical mapping, pointed out the main themes of each geopolitical synthesis, and were intended to be tutors in the reading of geopolitical maps.

66 Richard von Schumacher was responding to this perceived requirement with his: 'Zur Theorie der Geopolitischen Signatur'. For an overview of the maps of German *Geopolitik*, see G. Herb, 'Persuasive cartography in Geopolitik and national socialism', *Political Geography Quarterly*, 8, 3 (1989), pp. 289–303. However, such techniques were applied far more frequently in Italy's geopolitical journal.

67 See E. Massi, *Il valore economico dei 'Mandati' Africani*, Estratto dal periodico 'Africa', 56 (1938), pp. 1–62; E. Massi, *L'Africa economica*, esp. pp. 115–24.

68 *Geopolitica*, 2, 6–7 (1940), pp. 300–301.

69 Within academic geopolitics 'Eurafrica' was first proposed by Haushofer in: *Geopolitik der Panideen*, Berlin, 1931, and, 'Die weltpolitische Machtverlagerung seit 1914 und die internationalen Fronten der Panideen', in K. Haushofer and K. Trampler (eds) *Deutschlands Weg an der Zeitenwende*, 1931, pp. 208–23. See also J. D. Trunzo, *Eurafrica*, Ann Arbor, 1973; and C. R. Ageron, 'L'idée d'eurafrique et le débat colonial franco-allemand de l'entre-deux-guerres', *Revue d'Histoire Moderne et contemporaine*, 22 (1975), pp. 447–75. In Italy the concept was adopted widely, both within and without *Geopolitica*: P. De Agostino-Orsini di Camerota, 'Note geoeconomiche sull'Eurafrica', *Geopolitica*, 2, 2 (1941), pp. 90–6; B. Francolini, 'Il Mediterraneo nella guerra e nella geografia eurafricana', in *Geopolitica*, 4, 6 (1942), pp. 288–90; P. D'Agostino-Orsini di Camerota, *Eurafrica: L'Africa per L'Europe, L'Europe per L'Africa*, Rome 1934; A. Sestini, 'Per la cartografia geopolitica dell'Eurafrica', *Rivista Geografica Italiana*, 50, 1–3 (1943), pp. 8–11.

70 Reflected by projects such as the British ambition for a Cairo-Cape railway, and the earlier French interest in a Trans-Saharan railway. The 'penetration' of Africa by western colonial powers, in concert with the conquest of the Sahara and the 'oriental' Maghreb via modern science, are all implicated in the gendered representation of Africa as female and submissive to the west's paternalistic masculinity: a western-imperial script that was reinforced by these theories and maps. On these wider issues see E. W. Said, *Orientalism*, London, 1978.

71 G. Roletto, *Le tendenze geopolitiche continentali e l'asse Eurafrica*, Milan, 1937.

72 J. O'Loughlin and H. van der Wusten, 'Political geography of pan-regions', *The Geographical Review*, 80, 1 (1990), pp. 1–20.

73 R. King, *The Industrial geography of Italy*, London, 1985, esp. pp. 1–21.

74 The 'Battle' for autarky was a response to the 1935 League of Nations sanctions against Italy, it was not insignificant in furthering popular consent for Fascist imperialism.

75 Note that the French 'Axis' does not originate from Paris, as the arrow would obscure the scale of the circle that represents the Italian community in Tunisia. Also note, after C. G. Segré, *Colonization in the French Maghreb*, p. 34, the reality that Italians totalled only a small proportion of the Europeans in North Africa.

76 A. Biondo, 'La Transafricana Italiano', *Geopolitica*, 3, 12 (1941), pp. 569–75. The map is on page 571.

77 A. Biondo, 'Studi e progetti, La Transafircana Italiana', *Impero Italiano*, 5, 10 (1940), p. 1. There was apparently no equivalent editorial control over the standard of maps in *Zeitschrift für Geopolitik*, Haushofer's journal sometimes published very unclear maps, some by Haushofer himself. During the early 1940s, the German journal sometimes reproduced the maps of *Geopolitica*, including this one by Biondo, *Zeitschrift für Geopolitik*, 19, 5 (1942), pp. 252–3.

78 King, *The Industrial Geography of Italy*, pp. 123–8; R. Mazzuca, 'L'energia idroelettrica: Vecchi e nuovi modi di uso e di produzione in Valle di Susa', in S. Conti and G. Lusso (eds) *Aree e problemi di un regione in transizione, Escurzione geografiche in Piemonte*, Bologna, 1986, pp. 69–77, esp. pp. 69–70.

79 Although organic metaphors such as this were rare in the *Sintesi Geopolitiche*.

80 *Geopolitica*, 1, 3 (1939), pp. 160–1. The 'reader's guide' is on page 159.

81 This ties into a further point. The arrows that indicate migratory-shifts leave Italy to colonise other territory, whereas France apparently has no emigration, but is subject to immigration from several nations. This ties into the Fascist refrain that France, with a declining birth rate, was a 'decadent' and diminishing power in contrast to Italy.

82 This curving arrow implies dynamism and movement just as straight arrows suggest strength.

83 In Italy, the Mediterranean was indeed considered to be Italy's natural 'spazio vitale' (living-space), and was described in these exact terms by Hitler upon his May 1938 visit to Rome, see E. Wiskemann, *The Rome-Berlin Axis*, London, 1949, p. 137.

84 *Geopolitica*, 2, 1 (1940), p. 24–5.

85 Whilst this idea of 'zones of influence' was common in *Geopolitica* (for example, the Mediterranean or 'Eurafrica'), the actual term 'spazio-vitale' was not used uncritically in the journal and it was the focus of a definitional debate. See, for example, A. Fossatti,

'Sulla formazione degli spazi vitali', *Geopolitica*, 4 (1942), pp. 211–12. Beyond *Geopolitica*, the term was also discussed and employed, at times without any reference to academic geopolitics. For example: B. Nice, 'Sul concetto geografico di spazio vitale e di grande spazio', *Rivista di studi politici internazionale*, (1943), pp. 359–75; or contrast A. Messineo, *Spazio vitale e grande spazio*, Rome, 1942, and, D. Soprano, *Spazio vitale*, Milan, 1942. The term 'Geopolitics' was also contested at this time, and 'officially' defined in: E. Migliorini, 'Geopolitica', in Partito Nazionale Fascista (ed), *Dizionario di Politica*, Rome, 1940, pp. 250–1.

86 Antonsich, *La coscienza geografico imperiale*, pp. 78–82; A. Vinci, *'Geopolitica' e Balcani*, pp. 112–13.

87 For example, Morandi contributed twenty-one maps to Egidio Moleti di Sant'Andrea, *Mare Nostrum*, Rome, 1939.

88 'The School of Mystical Fascism', 'The National Institute of Fascist Culture', and 'The Institute for the Study of International Politics'; on some of these bodies, and their interests in geography, see, A. Montenegro, 'Popoli'; un'esperienza di divulgazione storico-geografica negli anni della guerra fascista', *Italia contemporanea*, 145 (1981), p. 3–37.

Acknowledgements

This paper developed from my doctoral research upon geography and geopolitics in Fascist Italy. For research periods in Italy I am grateful to the British School at Rome for a grant in aid of research, 1991–2, and to the Sir Dudley Stamp Memorial Fund. I am especially indebted to the Italian Government Foreign Office for a long-term scholarship, 1992–3. I thank Mike Heffernan, Denis Cosgrove, and Morag Bell for invaluable guidance.

CHAPTER ELEVEN

The cartography of colonialism and decolonisation: the case of Swaziland

Jeffrey C. Stone

Introduction

The fourteen-power Berlin conference on Africa in 1884–85 failed to avert the threat to 'the old system of free trade imperialism' in Africa.[1] The collaborative arrangements between European powers which characterised the previous four centuries of European imperialism in Africa gave way to the separate assertion of the authority of individual European nations in Africa. Colonial rule briefly replaced imperial infleunce for the first five or six decades of this century, before colonial authority was removed almost as rapidly as it had been imposed. The post-colonial relationship between Europe and Africa bears comparison with that of the pre-colonial period, so that twentieth-century European colonialism in Africa may be seen as a brief aberration in the long-standing and evolving imperial connection between a continent which is relatively rich and powerful and one which is weak and disorganised. The cartography of Africa from the fifteenth century to the present day has been interpreted within that framework of events.[2]

If this is a valid interpretation of both the diplomatic history and the evolving cartography of Africa, then the cartographic evidence should lie principally in the phases of transition from imperialism to colonialism and back to imperialism, particularly the first phase from imperial influence to colonial rule which occurred with relatively little preparation, so that functional change was particularly abrupt and the need for maps appropriate to the new function was immediate. The change should be starkly apparent by comparison with the very long preceding period of continuity of imperial purpose among European nations. By comparison, the transition from colonial control to a more imperialistic relationship between unequal partners in the post-independence period was not achieved without some foresight and even preparation, although decolonisation was not planned to happen in the way that it did. Within the colonial period, the origins of self-government in former British Africa can be traced back at least as far as 1923, when the Secretary of State for the Colonies 'enunicated the principle of paramountcy of African interests'.[3] Historians who have analysed the subsequent process of decolonisation in Africa offer an explanatory context for the evolution of the cartography of the extended phase of decolonisation.

The decolonisation model

In what Clapham has described as 'a masterly survey of African decolon-ization',[4] Professor Hargreaves explicitly declines to match his readers' expectations of a closing chapter which briefly epitomises his analysis of events. However, his interpretation of the changing nature of colonial rule provides a potential framework for understanding the artefacts of colonial government. In basing my interpretation of the cartography of a part of African decolonisation on Professor Hargreaves's analysis, I must therefore try to summarise the analysis even though such attemnpts are 'rarely good history'.[5] I must at least seek to identify those phases or events in the decolonisation process which seem particularly relevant to an understanding of the cartography of decolonisation.

The first threats to the colonial tranquility of sub-Saharan Africa in the 1920s arose from the great depression of 1929, followed by the first hesitant steps towards a policy of active development. Development and mapping became synonymous from the first. Ground survey and mapping were forms of development in their own right and means to facilitate economic development. They were part of the 'experiment with new techniques for state intervention and planning of economic development',[6] in order to foster an imperial economy as a bulwark against depression in the international economy. In British Africa, the process was set in motion by the Colonial Development Act of 1929. Within the Colonial Office, specialist advisory committees were increasingly employed,[7] witness the Colonial Survey Committee which first met in 1905 but whose work was much reduced in World War One, before becoming increasingly active in the inter-war years, particularly in co-ordination between territories.[8] The feeling of a need to survey and appraise the resources of the imperial domain extended to academic bodies such as the British Association for the Advancement of Science. The work of Section E (Geography) of the BA on the 'furtherance and development of geographical knowledge of tropical Africa' dates to 1926,[9] although not yet with an adequate cartographic base for plotting its findings.

Imperial policy remained a subject for debate in the 1930s, of course, as is evident from the publication in 1938 of Lord Hailey's monumental *African Survey*. Hailey had become identified with 'the theme that the object and justification of colonialism was its contribution to the material betterment of backward peoples'.[10] He recognised that massive aid would be required to improve health, nutritional and living standards. The Colonial Office in the 1930s was not unreceptive to such suggestions although there were political dimensions to the material betterment of colonial peoples, as can be exemplified in the case of Swaziland. By 1935, the Colonial Secretary had recognised that Britain would have to adopt a welfare-development approach with more vigour, if it were to justify its refusal to transfer the three High Commission territories to South Africa.[11] There was evolution in policy formulation for colonial territories in the 1930s, even though implementation may have been muted in British colonial Africa.

Cartography was to some extent a barometer of the development process on the ground and evidence of the changing nature of the colonial relationship in the aftermath of the depression of 1929. It was indicative of the first tentative steps towards development, described in the context of the end of the 1930s as no more than the 'various impulses towards reform'[12] which made themselves felt unevenly across British Africa. Thus, despite what in British colonial territories was described in 1956 as a 'preoccupation with mapping for development projects',[13] Swaziland was one of four territories that up to about 1944 had no official topographic maps. Origins of the sort of developments which were likely to place map making in high profile in British colonial Africa can indeed be traced back to 1929, but progress was partly determined by the economic and internationalist orientation of development as a buffer against depression in the world economy. Consequently, it was both modest and geographically uneven. The latter characteristic arose from the different constitutional relationships between Whitehall and individual British African territories, and because of the high order of territorial autonomy.

With the ultimate goal of self-government re-affirmed by the Colonial Secretary in 1938, the Colonial Development and Welfare Bill was enacted in 1940, to promote development planning more vigorously in the colonies. The Act marked the beginning of more active involvement by the colonial state in both economic and social well being. War lessened the immediate consequence of the Act, but in 1945 it was amended by making provision for the funding of central services. In 1945 it funded the establishment of a central Directorate of Colonial Surveys with major remits in the fields of geodetic and topographic survey.[14] In the broader context, the economic aftermath in Britain of World War II and the new imperative of the cold war gave rise to the concept of a pool of Comomonwealth resources harnessed to the cause of European revival and enhanced international influence. This further twist in the nature of the imperial relationship was in turn to be reappraised following the events in Suez in 1956. The consequent contraction in Britain's commitments overseas meant a reduction in total Treasury expenditure. Development assistance in colonial territories increasingly took the form of expertise, a shift in the emphasis of policy implementation which was potentially to the advantage of the increaingly high technology fields of photogrammetry and tellurometry as the means for providing rapid and accurate topographic map cover.

Swaziland: colonialism in practice

If the evolution of British colonial policy during decolonisation can thus be simplified for purposes of providing a framework for understanding the evolution of topographic map cover in British colonial Africa, the different circumstances of individual British colonial territories unfortunately add complexity. British colonial decolonisation policy may be simplified ultimately to the pre-war and post-war phases (with the sub-division of the latter), but the response of particular territories depended to some extent on their abilities to match the objectives of Whitehall policy in the construction of a larger Commonwealth economic and political entity. Furthermore, the allocation of Treasury resources and British expertise within the Commonwealth included political considerations, particularly in the case of a territory bordering South Africa. Hence, the substantial differences in progress in topographic mapping in the later decades in British colonial Africa, despite the existence of an overall policy framework during the decolonisation phase.

The political and administrative history of colonial Swaziland is significantly different to other British colonial territories in southern Africa, let alone in Africa as a whole, (although British African colonial territories were typically varied in this respect). Britain was not the first external administration to assume at least limited authority over Swaziland. Despite conventions in 1881 and 1884 between Great Britain and the South African Republic (SAR), guaranteeing the independence of Swaziland, the great number of land concessions granted in Swaziland to Europeans gave rise to disputes into which neighbouring governments were drawn as arbitrators without any authority. Consequently, further conventions were drafted in 1890, 1893 and 1894, initially establishing a triumvirate government representative of Britain, SAR and the Swazis, but latterly conceding 'rights of protection, legislation, jurisdiction and administration' but not incorporation to

[301]

the SAR government,[15] and without control of the internal affairs of the Swazis. In 1895, a Resident Special Commissioner was appointed from the SAR to administer Swaziland, whilst Britain was represented by a Consul. In 1899, The Resident Special Commissioner withdrew from Swaziland and in 1902, after the South African War, Britain assumed the authority of the SAR in Swaziland.[16] A British Special Commissioner was sent to administer Swaziland, with full powers of legislation and jurisdiction defined by Order-in-Council in 1903. The Special Commissioner was answerable to the Governor of Transvaal, now a British colony, but with the granting of responsible government to the Transvaal in 1906, the Governor's authority over Swaziland was transferred to the High Commissioner for South Africa. This was an important change in the chain of command, because in consequence, when a separate Dominions Office was established in 1925 to take over from the Colonial Office the business relating to the self-governing Dominions, the new office also took over responsibility for the three territories administered by the South Africa High Commission. Hence Swaziland affairs were dealt with in Whitehall by the Dominions Office (renamed the Commonswealth Relations Office in 1947) throughout almost the entire period of colonial development and decolonisation, a phase which was masterminded by the Colonial Office. Only in 1961, when South Africa left the Commonwealth, was responsibility for the High Commission Territories transferred to the Colonial Office.[17] However, in 1966, the Colonial Office was absorbed into the Commonwealth Office, which reassumed responsibility for Swaziland until internal self-government in 1967 and continued briefly to take residual responsibility for the territory until independence in September 1968.

Swaziland may therefore be seen as a particularly severe test of the utility of any broad framework of British colonial policy and practice for understanding the evolution of British colonial topographic mapping in Africa. On the other hand, explanation is not necessarily being sought for an identical cartographic evolution between territories. The delegation of authority and the financial autonomy of British colonial territories resulted in different cartographic histories and it is the differences as much as the similarities which have to be understood in the context of evolving British colonial policy in Africa. Nevertheless, if the decolonisation model of colonial policy evolution proves useful as an explanatory device in the case of Swaziland, then it deserves to be put to the test elsewhere in former British Africa.

Four phases are envisaged within the total colonial period. Characteristically, there is an initial short phase of imposition of colonial rule, urgently necessitating very different cartography by comparison with the prolonged pre-colonial phase, by virtue of the change in the nature of the functional relationship from imperialism to colonialism. There-after, there follows a phase of relative inactivity, if not tranquility. The two remaining phases fall within the decades of decolonisation. They led to the restoration of a more imperialistic relationship after independence. They are epitomised firstly by the phrase 'impulses towards reform' which Professor Hargreaves[18] applies to the 1930s and secondly by the word 'development', as it was variously interpreted and implemented in the post-war period.

The cartography of colonial Swaziland: colonialism imposed

The major change in cartographic requirements as a result of the imposition of colonial rule, which has been exemplified in the case of northern Rhodesia,[19] seems improbable in the case of the imposition of British rule in Swaziland. That first phase of change should presumably have occurred with the imposition of the authority of the Transvaal government in the 1890s. In fact, it did not happen then. From 1895 to 1899, the SAR exercised powers of protectorate over Swaziland, but those powers did not include the management of the internal affairs of the Swazis. Moreover, the ruling authorities were very much more familiar with the small neighbouring country than were most representatives of European powers who were establishing the rule of colonial law on the ground in larger territories eleswhere in Africa. From 1866 to the South African War and beyond, the boundaries of Swaziland were the subjects of numerous commissions involving the Swazi nation and the governments of Portugal, Britain and the South African Republic, with subsequent demarcation on the ground recorded cartographically.[20] More important, concessions of land in the heart of Swaziland had been granted to Europeans from as early as 1860. The number of concessions increased greatly in the 1880s, whilst the SAR government which had ambitions of an outlet to the sea through Swaziland, indirectly secured some infrastructural rights through the country. The government of the SAR was not administering unfamiliar territory in the 1890s. However, the imposition of the British administration did result in an immediate outburst of significant cartographic activity which is in contrast to the form of earlier map making. To demonstrate the contrast between pre-colonial and post-colonial cartography, the earlier work must be touched upon.

Despite the fact that a late nineteenth-century external authority preceded British administration in Swaziland, and despite the earlier decades when boundaries were beaconed and the beneficiaries of concessions were present on the ground within loosely defined areas, there

[303]

is nevertheless some evidence of what is the characteristic pre-colonial cartographic phase. The evidence is perhaps less substantial than in some other countries of southern Africa such as Zambia or Botswana, but it exists. It takes the form of maps in the mould of the nineteenth-century European explorers, whose journeys had little direct connection with the initiation of colonialism, but was a manifestation of what Professor Bridges[21] has termed the unofficial mind of imperialism. Travellers' sketches were compiled[22] including the map provided by W. C. Harris in 1837 to accompany his account of a hunting expedition, which speaks of little else but the thrills of the chase. However, he does describe his method of locating himself on the map which he was carrying and hence his method of recording features observed on the ground.[23] His map is of interest, as the first to locate both the Swazis[24] and also a few of the physical features of the region. Much more important in this category of pre-colonial map, is the outstanding sketch map complied by Allister Mitchell Miller in 1896,[25] at a time when Britain was primarily concerned to sustain the integrity of Swaziland against encroachment from the Transvaal. Miller had substantial commercial interests in land concessions in Swaziland including direct involvement in the drawing up of the controversial Unallotted Lands Concession in 1889. He sought to promote the economic potential of Swaziland particularly through European agricultural settlement, as is shown by his paper to the Royal Colonial Institute, extolling the virtues of the agricultural potential of Swaziland.[26] Miller's essentially commerical motivation typifies the pre-colonial unofficial mind of imperialism and his map has been the subject of a separate study by John Masson.[27] Another well known depiction of Swaziland, on the 1899 map of the Transvaal (Sheet 6) at 1:476,000 by Jeppe (1899), falls into the same category since it is derived from Miller.[28]

The usual tendency for the characteristic cartography of imperialism to continue briefly into the colonial period[29] seemed not to have happened in the case of Swaziland. Indeed, the reverse was true in that cartographic content which was appropriate to the establishment of an effective colonial administration actually preceded both the Swaziland Order in Council of 25 June 1903 and the arrival in Swaziland in 1902 of the Special Commissioner sent by the British Governor of the Transvaal to take administrative charge of the country. Major Jackson's Transvaal and Natal Series, 1901-04 at 1:148,752 (1,000 Cape roods to 1 inch) had been compiled by the Survey and Mapping Section, Field Intelligence Department, Army Headquarters of British forces in Pretoria, for military purposes. The neutrality of Swaziland during the South African War meant that it would not have been a priority area for military purposes, but it was included in six sheets of the series (15, 17B, 33, 34,

35 and 41). The sheets were revised several times between 1900 and 1902, and revision of the depiction of Swaziland did occur between editions. The sources are described as Transvaal farm surveys, concession and boundary surveys and Miller's sketch map, so that as long as it may be thought of as a topographic map, it seems to conform to the characteristic early type of colonial topographic map, the compilation map.[30] If a pre-colonial compilation map seems to be something of a paradox, (making compilations before the sketches and traverses of the initial phase of administrative imposition have become available) then the explanation lies not solely in the peculiar circumstances of war, but also in the fact that this was not the first external administration of Swaziland. Boundary surveys, travellers sketches and concessions diagrams existed from the period of SAR rule and earlier, as a result of the numerous land concessions of the late 1880s[31]. The result was a compilation and depiction of Swaziland which was less detailed than neighbouring Transvaal, but included drainage, form lines, routes, concessions and some named settlements.

Although the sort of rudimentary cover which may result from the initial phase of imposition of colonial rule on the ground, was already available in Swaziland, the first few years of British administration nevertheless did see a remarkable exercise in mensuration which in 1908 lead to the expectation that 'Swaziland will be one of the best – if not the best – and most completely surveyed territories in south Africa.'[32] The origins of this extraordinary cartographic episode are contained in Swaziland Administration Proclamation No. 3 of 1904 which indicated the policy which the High Commissioner proposed to adopt with regard to the land concessions granted mainly during the reign of Dlamini IV [Mbandzeni] almost two decades previously. The number and the complexity of the concessions was 'a serious embarrassment for the new Administration',[33] arising from a land scramble beyond all previous proportions in southern Africa.[34] The circumstances in which British colonial administration was initially established in Swaziland were clearly quite atypical, something to be born in mind when comparing the cartography of the colonial administration with that of other British territories, or when matching it against any broad framework of the evolution of British colonial policy.

The Proclamation not only indicated the principles which the Governor would adopt in unscrambling the problem. It also established the mechanisms for implementing his resolution of the problem. A Commission was to be appointed to effect the policy decisions and a survey was 'to be made of the boundaries of all lands affected by any concessions' of which notice had been given. The Report of the Swaziland Concessions Commission dated 1906, contains a plan of

Bremersdorp township plus five documents entitled 'Sketch map of Swaziland' at a scale of 4 miles to 1 inch, showing the mineral, land and grazing concessions plotted in a generalised manner, thus identifying the concessions to be surveyed. It is, perhaps, a reflection of the inadequacy of the then existing geodetic and cadastral data base that the Swaziland National Archives library copy of the report contains a copy of sheet 6 of 'Jeppe's Map of the Transvaal' (1899), with manuscript additions in colour. The concessions are located on this single small-scale map, presumably because this was the best map available for the purpose. However, such compilation mapping, and in particular, the sketch map of Swaziland by A. M. Miller, was about to become out-moded in spectacular fashion.

The Swaziland Concessions Commission was appointed on 1 October 1904. On 11 February 1905, it appointed G. C. Murray as Surveyor to the Commission and he began to plan the division of the country into sections for survey. The country was divided into five sections or strips extending from west to east across the country including both high and low veld, bounded largely by features visible on the ground. In June 1905 instructions were issued to five surveyors: R. Pizzighelli, A. H. F. Duncan, J. F. I. Curlewis, W. R. Lanham and F. S. Watermeyer. They were provided with copies of all land and mineral concessions and instructed to survey them without reference to any sub-division, so that on the face of it, they were to implement a solely cadastral exercise. They began work in the following month and by June 1907 the last general plan was received, except that survey of boundaries of concessions within the very extensive Vermaak concession in the southeast of the country was incomplete. The work was hindered by the climate in the low veld during summer and by rain and mists in the high veld during the same season. The President of the Commission reported in August 1907 that 'the surveys have been very carefully and efficiently carried out'.[35] By October 1906, preliminary general plans of the concession boundaries had been opened for inspection and protests could be lodged in the three months following. There was also a question of provision for what were described as 'native interests', so that further survey would still be required before the plans could be advanced beyond their preliminary state.

The evidence of what happened subsequently, derives in part from the few maps which have so far been located, as well as from archival sources. In February 1908, the then High Commissioner, Lord Selbourne, asked that the native area which had been provisionally marked out in the Pizzighelli survey be referred by the Special Commissioner to the Assistant Commissioners of the districts concerned, for their scrutiny.[36] Selbourne had appointed George Grey as Special

Commissioner in 1907, to demarcate the Swazi area by cutting one-third from all except mineral concessions. Grey indicated that he was arranging to make maps of each Assistant Commissioners' District from the five latitudinal survey strips as he thought that the administration would be better able to consider the native area with maps in this form. This arrangement may explain the provenance of four manuscript maps (lodged in the Office of the Surveyor-General, Mbabane), each in two sections on blue tracing film, of Mbabane District North, Mbabane District South, Ubombo District and Hlatikulu District, although in addition to the five surveyors initially appointed by the Commission, the maps are also credited to A. Joubert, who held the post of Government Land Surveyor and was the Special Commissioner's Assistant. The provenance of the four maps may also be related to the fact that Grey himself used the district maps in his examination of the country.[37] The extant maps do not comprise complete coverage of the country and are in excellent condition, so that these particular manuscript copies were probably not sent out to the Districts concerned. They are not dated but they all appear to derive from the same pen. A great deal of work went into their original compilation. For example, the Hlatikulu map includes the following topographical features: form lines, named rivers, water furrow, store, numbered roads, outspans, settlements, Nzama royal graves, farm houses, missions, and paths. They are drawn to the same scale as the plans presented by the Commission's surveyors.

The next phase of the Commission's work seems to have been the insertion of proposed native areas on maps of the five surveyed sections of the country. Between February and September 1908, Grey dispatched to the High Commissioner sections I, II, IIIA, IIIB, IV and V, with the proposed areas marked off. The task gave rise to 'a kraal map of most of the high country of the Ingwempisi River' by Joubert,[38] which would be an interesting historical record, if it survives. Although 'detailed population distribution maps were drawn',[39] they have proved difficult to locate. It is interesting, nevertheless, that this seemingly cadastral exercise incorporated careful location of the population, the prime requirement of the administration in the first phase of colonial topographic map making elsewhere in former British Africa.[40] The field work on what was described as 'the preliminary stage of the Swaziland partition',[41] was completed in August 1908.

One further period of work on the ground by the Commission's surveyors is apparent from cartographic and documentary evidence. Grey's measurements were only approximate. His report refers to the survey of the native areas which was to be made, so that the next step was a detailed survey of the native areas. The Boundary Commission, which had reported in 1908 was reconstituted in 1909, probably to reconcile the results of the survey of native areas with the observations of the earlier concessions boundary survey.[42] This last phase of work by the Commission resulted in further map compilation. The Surveyor-General, Mbabane, holds what are probably unique documents in the form of four finely executed manuscript maps. They appear to be four of the five final signed products of the Commission's surveyors and are uniformly entitled 'GENERAL PLAN OF SECTION – SWAZILAND', each with the relevant section number, 1, 2, 3 or 4. The plan of Section 5 was not seen by this author in 1985, although it was seen a few years previously.[43] Three of the plans indicate when the country depicted was surveyed. Section 1 reads 'Surveyed in June to December 1905, July to November 1910'. Section 2 was 'surveyed in August–December 1905, April–September 1906, September 1908, August–October 1909'. Section 3 is not so informative but Section 4 was 'surveyed in 1905 to 1910'. The further work was therefore carried out on the ground in the period 1909–10.

The four manuscript maps are extremely large, each comprising several sheets mounted on linen and measuring respectively 120 × 183 cm, 99 × 202 cm, 133 × 283 cm and 115 × 200 cm. They bear the signatures of Pizzighelli (1), Duncan (3) and Lanham (4), with Section 2 bearing Joubert's signature, not Curlewis who did the early work on that Section. Their content, as indicated by the 'References' (or key) on each, is of particular interest, showing them to be much more than cadastral plans. They include district boundaries, roads, footpaths, rights of way, police or government stations, stores, homesteads, mines, missions, kraals of important chiefs, as well as native areas, compensation areas and the various categories of concessions: mineral, perpetual and long period land, short period land, grazing, timber, unallotted land unencumbered, unallotted land encumbered, unallotted mineral and lapsed. They include significant amounts of topographical detail and as such, they are valuable historical, as well as legal records. Incidentally, the Surveyor-General also holds nine large manuscript traces of portions or extracts of the Sections. The contents of these traces are not identical to the General Plans of Sections and their relationship to the Plans has not been established. They may have been compiled in the course of preparation for the published compilation of 1932.

The large dimensions of the General Plans of Sections are products of their scale, '400 Cape roods to 1 English inch' (1:59,500). The geodetic framework for each Section was constructed independent one of another, each surveyor starting from his origin which was not at the same time related to those of other surveyors. A DCS Surveyor commented to local staff in 1949 on longitudinal and latitudinal discrepancies noticeable over considerable distances, although not apparent in localised areas.[44] The work of the Commission did not necessitate putting the Sections together, although when this was attempted (probably for the 1932 compilation), it was done by matching up the five Sections along their latitudinal topographical boundaries rather than by the more accurate use of co-ordinates.

Significantly for topographic map provision in Swaziland, the sectional maps of the Commission were quickly put to use in compiling a smaller scale map of the territory, replacing the single sheet 'Sketch map of Swaziland' at 1:633,600 (TSGS 2035 c. 1905), which had been compiled from Jeppe's map. In other words, the cartographic heritage of Miller finally came to an end in 1914, with the publication by the Surveyor-General, Pretoria, of a 'Map of Swaziland' in two parts (North and South) at 1,000 Cape roods to 1 inch. Comparison of a sample area around Mbabane, as shown on the published map, with the corresponding cover as shown on Section 3 of the Concessions Commission manuscript map, shows that the published map was derived very largely from the larger-scale manuscript. Of the twenty-four names within the sample area on the manuscript, twenty-one appear on the printed map, mainly rivers and streams but including a footpath, a waterfall, a fort and the Residency. There are no names on the corresponding part of the printed which may not derive from the manuscript Section. Cadastral information is also largely carried forward, although beacons tend not to be numbered on the printed map. However, comparison of the keys (or 'References') on the two maps at once suggests that there are additional items of topographic information on the printed map, located by means of symbols or letter codes, if not by local place names. These include Police Posts, Post Offices, Post & Telegraph Offices and Telegraph Lines. Game Reserve is also shown. The printed map contains an amalgam of cadastral and topographic information, and although the origins of the map are most unusual, the contents of the map, and indeed to some extent, the contents of its large-scale forerunner, are therefore comparable with the sort of compilation cover which characterises the imposition phase of colonial rule. Add this to the earlier military maps, and Swaziland can be seen as having the customary phase of compilation mapping, at the outset of British colonial rule.

The cartography of colonial Swaziland: from tranquillity to decolonisation

The publication in 1932 of a series of sixteen sheets at 1:59,000, dated 1932 and providing complete cover of Swaziland may seem to be something of a landmark in the cartographic evolution of the territory.

However, its cartographic importance perhaps belies its appearance. It incorporated relatively little new information and corresponds only to a modest development in governmental function. Rather, it facilitated access in Swaziland and elsewhere to the work of the Concessions Commissions' surveyors from 1905 to 1911, and to new diagrams surveyed in conjunction with Portuguese surveyors by H. K. Matthews in 1926, in the border area in the northeast of the country.[45] The dispute was resolved in 1920 by an Anglo-Portuguese border commission, which found that Pizzighelli's work in the area in 1906, shown on Section 1 of the Swaziland Concessions Commission's General Plan was topographically inaccurate.[46] Otherwise, the 1932 compilation is based on the five Sections of the General Plan of 1905–11, including a sketch of the location of the five sections in relation to the sixteeen sheets of the compilation and crediting the five sections to Pizzighelli, Joubert, Duncan, Lanham and Watermeyer with the dates of their surveys. Matthew's survey is added in manuscript to the printed 'Authority' on the compilation set presently held by the Surveyor-General, Mbabane. Otherwise, the Authority makes only the following additions to the five Sections of the General Plan of 1905–11: 'and subsequent diagrams registered and filed in the Surveyor General's Office, Pretoria, Union of South Africa, prior to the first of November 1932. Road classification and native spelling, Swaziland Administration'. The maps suggest continuity and consolidation in the development of colonial administration, rather than a watershed in any respect.

The series seems to have emanated almost by accident, from the Public Works Department in Mbabane. In January 1932, the Government Engineer wrote to the Assistant Government Secretary,[47] enclosing a sketch map of Swaziland with the country divided into eight sections and asking him to obtain these separate sections drawn to a scale of 1:59500. The purpose was to plot the roads of Swaziland, paying particular attention to the relationship of farm and concession boundaries. The need was for plans at a sufficiently large scale to allow the Government Engineer to mark sections of roads according to the nature of their surfaces and the requirement seems to have been triggered off by the intention to reclassify the roads of Swaziland, adopting 'new designations of main, trunk and branch roads and bridle paths',[48] although the finished map only differentiated between main roads, branch roads and bridle paths. The Government Engineer presumably would have been aware of the Commissions' sectional plans. Indeed the Government Secretary, in passing on the request to the Surveyor-General, Pretoria, specifically referred to similarity with the original surveyors' plans. Hence the choice of scale. The novelty in the request was that the original five sections should be recompiled into eight sections, with only two sets of eight sections required.

Thereafter, the task seemed to expand and to assume new dimensions which had initially not been intended. Almost at once, the Surveyor-General was discussing the means of printing the finished compilation. The prior compilation of the new plans was a major task of draughtsmanship, but it provided an opportunity to catch up on the Surveyor-General's compilation work which had been put off previously. The information which the new compilation might contain was dependent to some extent on which reproduction process was selected, and the Surveyor-General sought the advice of the Government Secretary about any additional information which would make the new copilation more useful. In this way, the preparation of the compilation gathered its own momentum rather than being an exercise which was planned in detail at the outset in response to recognisable stimuli. In part, it also seems to have been the product of the chance factor of a particularly competent draughtsman by the name of Collander, who was probably influential in the decision to increase the number of sheets to cover the same area from eight to sixteen. Topographic details apart from roads were not given high priority in the continuing discussion of content. The potential value of the new compilation was expressly seen 'in connection with all land work' and in 'locating correctly the public roads throughout the territory.'[49] Apart from the increased attention to roads, the other function of the series was very much a continuing one, as is emphasised by the arrangements which were made subsequently with the Surveyor-General in Pretoria to upgrade the legal information on his master copies on a regular basis. The printed series was delivered to the Government Secretary in Spetember 1933.

A minor cartographic consequence of the 1932 Compilation Series, was the use of the index map at 1:750,000 in the preparation of a revised small-scale single-sheet map of Swaziland. After protracted correspondence between the Government Secretary, Mbabane, and the Surveyor-General, Pretoria, about content and cost, 500 copies and this small map were received in Mbabane in May 1935.[50] It can be identified by the imprint: 'Mapping Section Surveyor-General's Office: Pretoria, 1935.' It was out of stock by 1938 and further correspondence ensued about a replacement. In this instance, the Director of Public Works, Mbabane, was anxious to take account of concern recently expressed in Downing Street about lack of topographical information when schemes were under consideration by 'the C.D.F. Committee'[51] and consequently he suggested that the recently completed cover at 1:500,000 by the Irrigation Department in South Africa might be utilised as the basis of an improved small-scale map. The developmental attitudes of

Whitchall were having their first effects on the cartography of Swaziland. At the suggestion of the Surveyor-General, Pretoria, it was eventually decided to compile a single sheet map at 1:500,000, utilising not only the Irrigation Department's topographical map but also the cadastral survey of the Union currently under preparation at 1:250,000. The result was a single sheet map of Swaziland in four colours at 1:500,000 with a part of the imprint reading 'Drawn in the Trig Survey Office, Pretoria, for the Swaziland Administration'. The map is not dated but 500 copies were printed in March 1939. In the course of compilation, further small failings were found with the work of the 1905–11 survey, this time in connection with heights of peaks on the northwest boundary. The map remained in use for some time, however. For example, it was appended to the Liversage Report published in 1948. In content, it was a modest recompilation and it was largely the fortuitous product of cartographic work in the Union of South Africa and the special relationship between Swaziland and the Surveyor-General, Pretoria, who continued to record all deeds in Swaziland until independence. Land registration was effected in the Swaziland Deeds Office which was under the charge of the Registrar of Deeds for the Transvaal, whilst surveys were controlled by the Surveyor-General for the Transvaal, also in Pretoria.[52] As part of the arrangement, compilations were constructed by the Surveyor-General's draughtsmen at no cost, although any subsequent printing costs had to be defrayed by the Swaziland administration.

Cartographic developments in South Africa in the 1930s and in the war years affected map provision in Swaziland. In 1935 the High Commissioner was informed of the intention of the Union Department of Irrigation to prepare a contour map of the Union including Swaziland, in 10 sheets. They sought the co-operation of the authorities in Swaziland[53] and the field survey parties extended their work into Swaziland in 1935. Contoured sheets nos 2 and 3 covering Swaziland at 1:500,000 were received in Swaziland in July 1936. In 1942, sheet 13 (Swaziland) of the 1:500,000 South Africa (Air) series was published by the Survey Depot, SAEC, Johannesburg. Unpublished cover at 1:250,000 was also prepared for military or aeronautical use including a sheet dated 1942, entitled 'Mbabane'. Then after the war, the work which had begun in 1936 on the SA 1:250,000 Topocadastral Series[54] came to fruition, with the publication in 1947 of sheets 14 and 21 covering Swaziland and including inset plans of Swaziland towns. Swaziland had acquired further map cover with little effort on the part of the Swaziland administration. However, it remained compilation mapping, much of the recent data being cadastral rather than topographic. The 1:250,000 map was based on the cadastral map of the

Territory, supplemented by a rapid reconnaissance survey carried out in 1942 by military engineers. It was acknowledged as 'not suitable for engineering, geological, irrigation or other purposes for which a reliable topographical map is required'.[55] As is emphasised to the point of irony by the extremely rudimentary map endorsed 'P.W.D. 1929' in the published Colonial Office Report on Swaziland for 1948, the territory still awaited a uniform topographic survey which would finally supplant the persistent cadastral heritage dating back to the Concessions Commission.

The cartography of colonial Swaziland: from decolonisation to independence

The Directorate of Colonial Surveys became involved in Swaziland soon after its establishment in 1946. An application was submitted for a topographical survey of Swaziland in 1948 and an initial grant from Colonial Development and Welfare funds was awarded. Thereafter, Swaziland was a regular beneficiary from that part of C D & W funds allocated for schemes directly administered by the UK, in this case, issued in accordance with the overall programme of the DCS. Swaziland also received C D & W funds issued in a different way, as a territorial allocation for specific schemes proposed by Swaziland and approved by the UK, such as education, medical services and public works. However, topographic survey was funded in the first of these two ways, so that it was not in competition with other requirements identified within the administration.

An aerial survey was flown by the RAF in June 1947, as part of a larger programme covering parts of east, central and southern Africa. Provision of ground control resulted in re-examination of the pre-existing system of beacons including those of the Concessions Commission, which was not seen as adequate.[56] These beacons had been supplemented by geodetic work in the 1930s and 1940s,[57] but further trigonometrical and height control was necessitated by the photography. This was provided by a surveyor sent to Swaziland by DCS in 1948, who apparently confirmed the lack of overall sympathy of the first survey co-ordinates,[58] although the quality of his own work was to be seriously criticised by his superiors in London.[59] It was not until 1955, that a composite triangulation was constructed, again by a DCS survey party, incorporating 28 primary stations (9 new ones and 19 established South African stations), plus 36 secondary points.[60]

Work commenced on mapping at DCS Headquarters, Teddington, in 1949, following discussions with the Trigonometrical Survey Office in Cape Town, who strongly recommended the inclusion of cadastral information to increase the value of the maps.[61] However, the initial concern of DCS was the production of preliminary plots, without any information of that nature, so that the cadastral continuity was at last broken.[62] Work was suspended for more urgent tasks before much was achieved. It proceeded from 1950 erratically, with other work sometimes given priority, but by 1951, ten preliminary plots covering a priority area in Swaziland had been issued. In 1953, the High Commissioner proposed that progress be speeded up by putting the work out to a private company, but DCS acknowledged the over-riding urgency of Swaziland and detached surveyors from Tanganyika to start work in Swaziland on further ground control so that the contoured edition could be prepared.[63] By 1954, twenty-two preliminary plots had been issued, the remaining eleven were nearing completion and contouring was receiving high priority. The heighting of Swaziland, together with a trigonometrical survey to cover the whole territory with a triangulation network tied to the Union system, took longer than estimated because of adverse observational conditions, but it was completed by December 1956.[64] By 1957, the uncontoured cover was complete and publication of the contoured edition had commenced. By 1958, only nine contoured sheets were outstanding and by 1962, coverage at 1:50,000 with a 50-foot contour interval was complete.

If there were any initial uncertainties about the ability of the High Commission territories to benefit from a central organisation for geodetic and topographical surveys funded from Colonial Development and Welfare, because of the chain of responsibility through the Dominions Office rather than the Colonial Office, then these were quickly dispelled. The Dominions Office proposed to the High Commissioner as early as 1944 that they be included and the offer was accepted. The only question was the problem of maintaining adequate liaison with the survey authority in South Africa.[65] Once work commenced, progress was erratic, because DCS was under conflicting pressures from several territorial administrations who sought topographic cover adequate for proposed development projects. Amongst the High Commission territories, the Chief Geologist, Mbabane, was pressing in 1946 for the inclusion of the whole of Swaziland in the scheme, on the grounds of the relative smallness of the territory, the total lack of accurate topographic maps and of the belief that 'reliable maps would be invaluable in furthering the work, not only of the geological survey, but of other departments (e.g. livestock and agriculture, native settlement etc) and of district officers as well'.[66] Here was an enthusiast for maps! Perhaps in consequence, Swaziland was accorded priority among the three High Commission territories, particularly with the possibility of mineral developments in mind. Nevertheless, when work commenced in 1950,

Swaziland 'had a lowish priority, and not first in the area in which your geologists are interested in'.[67] By 1952, the sheets covering the Swaziland Ranches area had been given high priority at the urgent request of the administration, but far from speeding up efforts on the remaining sheets, there was a suggestion of diverting existing effort in Swaziland to get some work going in Bechuanaland;[68] indeed work had stopped in Swaziland by 1953, in favour of more urgent mapping elsewhere.[69] Then came the High Commissioner's suggestion to have the work done by a private company, as a result of the views of consulting civil engineers dealing with a hydro-electric proposal. This led to DCS giving greater priority to contouring the catchment of the Great Usutu River and its tributaries and an area of low-lying country to the east, and while the surveyors were in Swaziland, to the completion of height control over the territory.[70] There was also a question of the adequacy of maps in preparation for the preliminary railway survey which would provide access to the Usutu forests for the Commonwealth Development Corporation. This led to acceleration of mapping in the Usutu-Umpilusi area. Five sheets of the 1:50,000 contoured series were printed in 1955, followed by 15 in 1956. Having met the urgent need, progress slowed down, with cover complete by 1962. The substantial impact of the development ethic of the period upon a hard pressed DCS and on the production and form of topographic maps in Swaziland is evident.

Post-independence cartography

Continuity in the production of topographic maps into the post-colonial period is remarkable. Independence had no obvious effect on map production, which continued under the auspices of DOS with only minor administrative changes in Mbabane. The revision of the 1:50,000 topographic series began prior to independence and continued steadily thereafter, at a rate of three or four sheets per year, so that by 1976, the final revised sheet had been published and a new revised edition was already in preparation.

If there was any change in the rate of map production after independence, it was to enhance output, though whether the change in the status of the country was the direct cause is even less clear. There may well be an indirect connection in the sense that the increase in developmental activity which characteristically follows the achievement of independence by a former colony, generates increased demand for maps. There is some evidence to suggest that this happened in Swaziland. The evidence is in the form of mapping for particular needs, some of it in the form of specialist mapping. For example, new air photography was obtained by 1973, to facilitate a new cycle of revision of the 1:50,000 series which was required 'because of recent development in many parts of the country.[71] At the same time, large-scale photography was taken of five Rural Development Areas, of Mbabane and Manzini, and of the national Sugar Farm. The consequent large-scale mapping of the Mbabane and Manzini areas was contracted out, but it was monitored by DOS and financed from British aid funds.[72] However, other work was undertaken by DOS both on the ground and in Tolworth, for example, 'precise levelling work which with the rapidly increasing development in agriculture, mining and irrigation, was badly needed'.[73] Partially contracted work was also undertaken on 'an urgent new project to map the Swaziland Coalfield',[74] which was to result in eighty-eight contoured sheets at 1:10,000 by 1976. Within a decade of independence, not only was a further revision of the basic 1:50,000 topographic cover being published and much larger-scale contoured mapping made available, but a new layered 1:250,000 single sheet of the entire country had been produced and revised editions of the geological series were under way.[75] The evolution of map production seemed little affected by the change from colonial status to independence, except perhaps for an increase in the amount of cover published, commensurate with the rising pace of development. If maps are a barometer, the nature of the post-colonial relationship with Britain, as expressed through the DOS arm of ODA had been securely rooted in the later phase of decolonisation. This is perhaps not surprising in a territory which was relatively late in achieving independence.

The specialist map requirements of colonialism in Swaziland

Before attempting to match the above sequence of cartographic events with the generalised model of the functional phases of British colonialism in Africa, in the hope of explaining the cartographic evolution of colonial Swaziland, it would be helpful to note any further requirements or uses of maps specific to Swaziland, in addition to those already noted in the 1940s and 1950s. As has been shown, the 1906–11 Section Plans of the Concessions Commission were redrawn into district maps during the first phase of the work, but this was to aid the process of identifying native areas rather than to meet subsequent needs of the administrative staff at district level. They were an integral part of the first phase of imposition of colonial rule on the ground in Swaziland, but there is no evidence that they were used in the same way that district maps were used elsewhere in British colonial Africa, that is, in asserting and maintaining administrative control in the early decades of colonial rule.[76] Indeed, there is some evidence to the contrary. By 1939, the term *tax camp* had become firmly established in Swaziland but the Government Secretary indicated to district staff that he wished to see it replaced by administrative tours.[77] He was concerned to increase the emphasis on agricultural, veterinary, educational, medical, public works and forestry activities, as well as carrying out what was seen as the essential duty of tax collection, plus legal work. The responses show that the standard procedure was a winter tour in which subjects other than tax collection were dealt with. A report of an Assistant District Commissioner who carried out a winter administrative tour in southern Swaziland in 1939 is divided into the following heads: new policy as regards tax collection; improved methods of agriculture with special reference to: soil erosion; dairies; improvement of stock, sale of bulls by government; improvement in sheep breeding; cattle sales erection of an auction camp; noxious weeds; cruelty to animals; outbreak of war. However, it also emerged that tours were constrained by the necessity to base tours around regularly utilised camp sites, which were presumably few in number, because of the difficulty in transporting 'a large tax tent'. The District Commissioner, Central District, reported his intention (in 1939) 'to visit as far as possible native areas not usually visited and a lighter tent will be necessary'. Horses were considered as a means of enabling District Officers to tour more widely and freely. It seems that touring until 1939 involved temporary residence at a few fixed sites, quite unlike the much more mobile tradition of touring by bicycle or on foot with frequent changes in camp sites by use of temporarily employed carriers, visiting every village in an area without exception, which was the revered administrative tradition in some other British African colonial territories. In the Swaziland circumstance, there would have been much less of a need for detailed maps of African settlements and paths or touring routes, as was the case for example in Northern Rhodesia where tour reports were regularly accompanied by a freshly sketched tour report map, or the Sudan where the district map was a standard item on the DC's office wall.[78]

The absence of a demand for maps in the normal course of administration is also apparent in the 1930s, from correspondence between the High Commissioner's Office and the Under-Secretary for South African External Affairs in Pretoria,[79] on the subject of the contour map covering Swaziland which was being prepared by the Union Department of Irrigation. Swaziland fell between two sheets and so the Under-Secretary suggested in 1935 that the Swaziland administration might like to have a separate edition compiled, showing the territory on a single sheet, the additional cost to be met by Swaziland. The Resident Commissioner accepted the proposal, agreeing that it would be an advantage to have the territory on a single sheet. However, he sought to place an order of only twenty copies, which was insufficient to justify the cost of a special printing block, but was presumably indicative of the real demand for maps, despite the Resident Commissioner's acknowledgment of the advantages of the proposed cover over the 1932 compilation which 'does not show mountain topography or altitudes.'

The more dispersed nature of Swazi rural settlement by comparison with more nucleated forms in some other parts of Africa, may have mitigated against the comprehensive village-to-village form of touring, which led elsewhere to the use of maps as normal practice. Probably more significant was the lack of foci for touring. The chiefs' homesteads might have provided them, but the administration did not have authority over chiefs.[80] Also, war intervened to prevent the implementation of the reforms proposed in 1939. Under-staffing made it difficult to conduct even the existing form of touring. Nevertheless, the change in attitude towards the nature of rural administration persisted. Enquiries initiated in connection with the Socio-Economic Survey of Swaziland, conducted in 1947 by V. Liversage (an agricultural economist with experience in Kenya) provides evidence of agreement amongst administrative staff on the need for more time to be spent on administrative tours, separated from tax tours which then took up two months of the year or twice as long as administrative tours.[81] However, there is no suggestion that any form of touring might have been facilitated by better maps of the territory, although this may be too fine a point of detail to be anticipated in this particular context.

To quote an officer who served in the post-war period: 'there was little demand for maps in day-to-day district administration since the districts were small and we were very familiar with the terrain and distribution of population'.[82] In the 1950s, single days were spent away from headquarters at police posts where cases were tried, taxes taken and a variety of problems brought by individuals were dealt with. Developmental questions could be attended to by specific visits, returning home at night because of the short distances involved. Other factors which had a bearing on the conduct of touring was the surviving influence of the older generation in the administration who saw maintenance of law and order as their prime function, and also the very indirect nature of colonial rule in Swaziland in relation to any matters outside the District Officer's judicial powers. Despite movements towards change in the nature and frequency of touring, there was in fact little change at any time, and hence little need for maps to facilitate touring.

There is one piece of evidence of cartographic usage in rural affairs in the form of an undated map (without reference) in the National Archives of Swaziland. The map is a dye-line reproduction (96 × 52 cm) of eight sections of a manuscript tracing, which have been taped together. It shows large parts of Mankaiana and Hlatikulu Districts, drafted by D. H. Harvey, Assistant Commissioner, probably in the late 1920s or early 1930s.[83] There is no scale, but the key lists rivers, bridle paths and old roads, wagon roads, stores, posts, missions, kraals and chiefs' boundaries. It seems to be a thematic sketch map designed to locate overlaps of the areas of jurisdiction of chiefs. It is both a valuable historical record and evidence of a rural administrative problem which lent itself to cartographic portrayal. It is an unusual example of the use of maps in rural administration in Swaziland, although as early as 1915, the Acting Resident Commissioner had asked for information on the carrying capacity of native Areas. He asked for the information to be mapped and was partially successful in obtaining the information in cartographic form, although not until 1921.[84]

There is also some evidence of maps being constructed, supplemented, or put to use in tasks of government other than rural administration. The consulting engineers (Hawkins, Jeffares & Green) to the C D & W funded Hydrographic Survey constructed a map at 1:250,000 in 1949 ('Drawing No. S/H/136/1') concerned primarily with drainage, and reproduced in dye-line. By 1962, the Department of Land Utilisation was plotting its own data on the 1:50,000 series, including dip tanks, cream separating depots and field offices. The maps were said to have proved invaluable in planning.[85] Economic and demographic mapping based on the 1960 random sample survey and the 1962 agricultural census was in preparation. An economic atlas of Swaziland was planned and was extended in the planning stage to a National Atlas in three parts (physical, agricultural and minerals, and demography), to be funded as part of the C D & W programme for 1965–68, but never completed.[86] The same office reported that maps and diagrams were prepared for the Livestock Commission, FAO, UNICEF, District staff and others. Cartography was seemingly a measure of the pace of developmental activity, albeit still without a Surveyor-General or any equivalent professional office within the Territory. In the last six years prior to independence, DOS also published soil, geological and minerals maps of Swaziland, as well as the first seven sheets of a revision of the 1:50,000 series, incorporating information supplied by the Department of Land Utilisation.

The cartographic evolution of Swaziland reviewed

Swaziland's cartography evolution within and around the period of British rule from 1902 to 1968, particularly the decolonisation phase, can now be set against the changing nature of British colonial policy in

Africa, as a possible explanatory framework for the evolution of topographic mapping. The policy phases which historians identify, begin with the prolonged period of free trade imperialism, characterised by mutuality of European free-trading interests and lack of direct control over the internal affairs of Africa. Colonialism provides the great functional contrast with imperialism. The initial phase of imposition or establishment of an effective administration, was followed by a period of relative tranquility verging on *laissez-faire*, before the decade of the 1930s which saw the first impulses towards reform, following the great depression. The post-war phase of accelerated development began with direct investment in colonial resources by Whitehall, but after Suez in 1956, experts were increasingly substituted for cash. The post-war developmental phase of activity paved the way for the reformulated relationship between Britain and Africa of the post-independence decades, and a return to something more akin to the long-standing imperial relationship between Africa and Britain as a part of Europe.

It is beyond the scope of this article to examine the four centuries of pre-colonial European cartography of southern Africa, to search for detailed evidence of continuity. The evidence of this phase has been tentatively offered elsewhere for the continent as a whole.[87] Suffice it to say that the particular case of Swaziland is complicated by the earlier period of colonial rule by a non-European power in the 1890s, although one influential map from the pre-colonial period in Swaziland (by A. M. Miller) has been identified, and its characteristic influence into the early colonial period has been recognised.

The imposition of British colonial rule did result in a remarkable outburst in cartographic activity which was a direct consequence of the change in the nature of the relationship between Britain as the now ruling authority and the pre-colonial authority within the territorial area which had come to be defined as Swaziland. However, it was a rather different need to that experienced elsewhere in British Africa, where the imperative was to locate the people to be ruled. In Swaziland, the imperative was to resolve the problem of the excessive number of pre-colonial land concessions although the location of the Swazi inhabitants gave rise to a related problem. Much prior contact between European and African in Swaziland meant that the territory was nowhere *terra incognita* and there is little evidence during the early decades of colonial rule of the sort of cartographic requirement for rural administration which has been identified elsewhere in British Africa. The immediate need to resolve the administrative problems of the concessions resulted in cartography of such high order that its data persisted into the post-war period. In consequence of their purpose, the Sectional Plans were not primarily topographic, although topographic information was inserted within the cadastral survey to the extent that it could act as a surrogate for a topographic map. With this particular phase of map making complete by 1914, tranquility did indeed descend, as far as further cartography was concerned.

The 1930s period of impulses towards reform coincided in Swaziland with the origins of the 1932 compilation (strictly speaking, not the first example of compilation mapping in Swaziland), arising from modest engineering requirements, but allowing improved control of day-to-day land registration functions, as well as facilitating incidental cartographic needs such as the improved small-scale territorial map. Impulses continued to be felt in the field of cartography, however, through the work of the Colonial Development Fund committee in London and the cartographic work of the South African authorities, which provided modest cartographic benefit to Swaziland.

Swaziland was a significant post-war beneficiary of the Colonial Development and Welfare Act, with direct grants not only for topographical survey. Indeed, the topographical survey work was in consequence of developments in the field of mineral and hydrological exploration, livestock husbandry, afforestation, and communications. By comparison with most larger British colonial territories in Africa, Swaziland was not only well provided with complete high-quality topographic cover by a relatively early date; it was also provided with some additional land resource mapping. There is clear cartographic evidence in Swaziland, of the initial post-war phase of funded development by Britain. Although cover was not complete by 1958, the momentum was great enough to ensure completion in the post-Suez decade of the expert in Africa, when grant aid for development projects was moderated. Topographic cover was complete by 1962. Nevertheless, a modest degree of momentum was maintained in the last few years of the colonial period with the completion of some specialist mapping and the commencement of revision of the basic topographic series.

Moving into the period of independence, there is characteristically much greater continuity in the cartographic transition, than at the outset of colonial rule. Swaziland's transition to independence was more protracted than most former British African countries and the high degree of preparation, at least in terms of cartographic provision, was as unusual as were the consequences of colonial imposition in 1902. However, the later phase of transition conforms to the general pattern, as decolonisation slowly led to the restoration of a more imperialistic relationship, with the modest momentum in cartographic production maintained or even enhanced after independence.

Conclusion

Despite the atypical origins and constitutional form of colonial rule in Swaziland, and the very unusual form which colonial cartography initially assumed, the subsequent evolution of its cartography accords well with the most recent thinking of historians about the changing nature of British colonial rule in Africa. The unusual form and content of some of its colonial cartography might tend to suggest that Swaziland was something of a unique case in the history of British colonial mapping in Africa. In fact, recent historiography helps us to identify common denominators with the colonial cartography of other British territories in Africa, and to understand the cartographic history of Swaziland.

Notes

1 J. D. Hargreaves, 'The making of the boundaries focus on West Africa', in A. I. Asiwaju (ed.), *Partitioned Africans*, London, 1985, p. 21.
2 J. C. Stone, 'Imperialism, colonialism and cartography,' *Transactions Institute British Geographers*, N.S. 13 (1988), pp. 57–64.
3 Sir A. Cohen, *British Policy in Changing Africa*, London, 1959, p. 21.
4 C. Clapham, 'The end of the affair', *Times Higher Education Supplement*, 829, 23 September 1988, p. 21.
5 J. D. Hargreaves, *Decolonization in Africa*, London, 1988, p. 229.
6 Hargreaves, *Decolonization*, p. 42.
7 C. Jeffries, *The Colonial Office*, London, 1956.
8 G. McGrath, 'The surveying and mapping of British East Africa 1890–1946', *Cartographica Monograph*, 18, (1976).
9 J. C. Stone, 'The British Association essays on the human geography of Northern Rhodesia 1931–35', *Zambian Geographical Journal*, 33/34 (1979), pp. 31–48.
10 J. W. Cell, 'Lord Hailey and the making of the African Survey', *African Affairs*, 88 (1989), pp. 481–505.
11 Cell, 'Lord Hailey', p. 505.
12 Hargreaves, *Decolonization*, p. 45.
13 Lord Hailey, *An African Survey*, London, 1957.
14 G. McGrath, 'Mapping for development. The contributions of the Directorate of Overseas Surveys', *Cartographica Monograph*, 29–30 (1983).
15 Lord Hailey, *The Republic of South Africa and the High Commission Territories*, London, 1963. A. J. Van Wyk, 'Swaziland. A political study', *Communications Africa Institute of South Africa, 9 (1969)*.
16 A. R. Booth, *Swaziland. Tradition and Change in a Southern African Kingdom*, Boulder, Colorado, 1983.
17 J. Garner, *The Commonwealth Office 1925–1968*, London, 1978.
18 Hargreaves, *Decolonization*.
19 J. C. Stone, 'The district map: an episode in British colonial cartography in Africa, with particular reference to Northern Rhodesia', *Cartographic Journal*, 19, 2 (1982), pp. 104–14. J. C. Stone, 'The compilation map: a technique for topographic mapping by British colonial surveys', *Cartographic Journal*, 22 (1984), pp. 121–8.
20 J. S. M. Matsebula, *A History of Swaziland*, Cape Town, 1976.
21 R. C. Bridges, 'The historical role of British explorers in East Africa', *Terrae Incognitae*, 14 (1982), pp. 1–21.
22 Public Record Office CO885/5 Item 58. List of maps etc. inserted in Colonial Office Papers 1875 to 1886.

23 W. C. Harris, *Narrative of an Expedition into Southern Africa during the years 1836, and 1837*, Bombay, 1938, p. 240

24 J. R. Masson, 'The first map of Swaziland, and matters incidental thereto', *Geographical Journal*, 155, 3, (1989), p. 337.

25 A. M. Miller, 'Sketch map of Swaziland made for the Umbandine Swaziland Concessions Syndicate, Limited by Allister M. Miller', 1:190,000, 94 × 122 cm, Edward Standford, London, 1896.

26 A. M. Miller, 'Swaziland', *Journal Royal Colonial Institute*, 31, 7 (1900), pp. 468–99.

27 Masson, 'The first map of Swaziland', pp. 335–41.

28 Mr J. R. Masson, personal communication.

29 Stone, 'Imperiaism'.

30 Stone, 'The compilation map', p. 121.

31 For example, see item 3318 in P. A. Penfold (ed.), *Maps and Plans in the Public Record Office 3 Africa*, London, 1982, p. 357, entitled 'Sketch map of Swazie-Land Concessions 1889 . . .'.

32 Public Record Office DO 119/818. President of Swaziland Concessions to Imperial Secretary, Johannesburg, 14.1.08.

33 Hailey, *African Survey*, p. 372.

34 A. J. Christopher, *The Crown Lands of British South Africa 1853–1914*, Kingston, Ontario, 1984.

35 Public Record Office DO119/800. Summary of the work of the Swaziland Concessions Commission.

36 Public Record Office DO119/608. Correspondence between the Special Commissioner for Swaziland and the High Commissioner, on Swaziland partition.

37 J. Crush, 'The colonial division of space: the significance of the Swaziland land partition', *International Journal of African Historical Studies*, 13, 1 (1980), pp. 71–86.

38 Public Record Office DO119/608.

39 Crush, 'The colonial division of space', p. 81.

40 Stone, 'The district map'.

41 Public Record Office DO119/608.

42 Mr H. M. Jones, personal communication.

43 Dr A. J. Christopher, personal communication.

44 Swaziland National Archives 767. Survey Regulations: Government Engineer to Government Secretary, 20 September 1949.

45 Swaziland National Archives RCS 138/20. Surveyor-General, Pretoria, to Government Secretary.

46 Public Record Office DO 119/944. Anglo-Portuguese commission for definition of Swaziland–Portuguese boundary.

47 Swaziland National Archives RCS 161/22. Government Engineer to Government Secretary, 27 January 1932.

48 Swaziland National Archives RCS 161/22. Surveyor-General, Pretoria, to Government Secretary, 10 November 1932.

49 Swaziland National Archives RCS 161/22. Acting Government Secretary to Surveyor-General, Pretoria, 19 October 1932.

50 Swaziland National Archives RCS 161/22. Government Secretary to Surveyor-General, Pretoria, 2 May 1935.

51 Swaziland National Archives RCS 266/38. Director of Public Works to Government Secretary, 29 April 1938.

52 *Colonial Reports Annual. No. 1404 Swaziland Report for 1927*, London.

53 Swaziland National Archives RCS 219/35. Under Secretary for South African External Affairs to Administrative Secretary, High Commissioner's Office, 15 February 1935.

54 E. Liebenberg, 'Topographical maps of South Africa, 1879–1979', paper presented to the International Map Seminar, CSIR Conference Centre, Pretoria, 1979.

55 Public Record Office OD6/437. High Commissioner, Pretoria to Dominions Office, London, 27 June 1945.

56 Public Record Office OD6/437. DCS to Trigonometrical Survey, Cape Town, 12 May 1947.

57 Public Record Office OD6/559. DCS to Sutton, 1 February 1955.
58 Swaziland National Archives 767. Government Engineer to Government Secretary, 20 September 1949.
59 Public Record Office OD6/445. DCS to H. Cotton, Swaziland Survey Party, 3 January 1950.
60 Public Record Office OD6/559. Surveyor Furmiston, Swaziland, to DCS, 30 June 1945.
61 *Directorate of Colonial (Geodetic and Topographical) Surveys. Annual Report for Year Ending 31 March 1949*, London, 1949.
62 Public Record Office OD6/285. DCS to Trigonometrical Survey, Cape Town, 8 August 1951.
63 Public Record Office OD6/286. Swaziland Topographical Survey. Note of a meeting at the Commonwealth Relations Office, 16 April 1953.
64 Public Record Office OD6/560. Letter of 13 December 1956.
65 Public Record Office OD6/437.
66 Public Record Office OD6/437. Chief Geologist, Mbabane, to Government Secretary, 29 June 1946.
67 Public Record Office OD6/445. DCS Memo, 23 September 1950.
68 Public Record Office OD6/285. CDC to Commonwealth Relations Office, 14 July 1952.
69 Public Record Office OD6/286. DCS to CDC, 3 February 1953.
70 Public Record Office OD6/446. DCS to Director of Public Works, Mbabane, 27 April 1953.
71 *Directorate of Overseas Surveys. Annual Report for the year ended 31 March 1973*, London, 1973.
72 *Directorate of Overseas Surveys. Annual Report for the year ended 31 March 1974*, London, 1974.
73 *Directorate of Overseas Surveys. Annual Report for the year ended 31 March 1975*, London, 1975.
74 *Ibid.*
75 *Directorate of Overseas Surveys. Annual Report for the year ended 31 March 1978*, London, 1978.
76 Stone, 'The district map'.
77 Swaziland National Archives 393/39. Government Secretary to all Heads of Departments, District Commissioners and Assistant District Commissioners, 20 April 1939.
78 K. D. D. Henderson, *Set Under Authority*, London, 1987, p. 33.
79 Swaziland National Archives RCS 219/35. Acting Resident Commissioner to Under Secretary for South African External Affairs, 27 February 1935.
80 Jones, personal communication.
81 Swaziland National Archives 1216. Socio-Economic Survey. Collection of Data.
82 Jones, personal communication.
83 Jones, personal communication.
84 Jones, personal communication.
85 Annual Report Department of Land Utilization for the year ended December 1962, Mbabane, 1962.
86 Masson, personal communication.
87 Stone, 'Imperialism'.

Acknowledgments

The assistance of the Director, Swaziland National Archives and of the Surveyor General, Mbabane, is gratefully acknowledge. I am greatly indebted to Mr H. M. Jones and Mr J. R. Masson OBE for advice, guidance and hospitality. I am also indebted to the Carnegie Trust for the Universities of Scotland and to Aberdeen University for financial support.

INDEX

Aberdare, Lord 99
Aberdeen 101, 102, 109, 126
Aberdeen, Earl of 101
Aborigines Protection Society 177
Abyssinia 54, 63–6, 68, 246, 269, 272, 273, 277
Académie des Sciences 38, 39
Accra 166
Aceh 81, 82, 84, 85, 89
Adana 242
Addis Ababa 289
Aden, Gulf of 289
Admiralty 15–7, 24–7, 237, 241
Adowa 269
Adriatic 246
Aegean 269
Africa 4, 43–5, 57–61, 62–3, 82, 97–101, 122, 126–9, 132, 137–8, 146, 160–1, 172, 175–9, 189, 192–3, 197–9, 221–6, 231–2, 240, 249, 269–73, 275–8, 280, 287, 290, 298–9, 301, 313–15, 320
  Central 58, 61, 98, 161, 249, 281–4, 313
  East 58, 61, 97, 127, 159, 161, 176, 179, 237, 249, 268, 277, 282, 287–9, 313
  North 5, 93, 173, 222, 271, 280–2
  Southern 43, 85, 137, 146, 159–61, 176, 304–5, 313
  West 60, 62, 99, 126–30, 132–5, 137–9, 140–7, 159, 165, 222, 239
African Association 57
African Trade Society (Afrikaansche Handelsvereeniging) 85
Africans 60, 142, 192, 204
Afrikaners 203
Akpap 133
Albany 20
Albany Settlement 193
Albert Hall 212
alcoholism 126–7
Aldershot 170
Alexandretta 247
Alexandria 3
Algeria 7, 226–7, 247

Almeida, Pierre Camena d' 234
Alpine Club 169
Alpine Journal 169
Alps 74, 169, 277
Alsace-Lorraine 222, 228, 235, 245–6
American Geographical Society 251
Americas 104
Amsterdam 83
Anatolia 246–7
Anderson, Alexander 39, 40
  Delugia 40
  Geography and History of St Vincent 40
Andes 74
Anglocentrism 152
Annales de Géographie 228
Annesley Bay 64
Anson 14
Antarctic 102, 107, 120, 122
Anthropological Institute 112, 161, 177, 179, 183
Anthropological Society of London 177
anthropology 2, 83, 177–8
Arab nationalism 244
Arafura Sea 25
archaeology 2–3, 194
Archer, Frederick Scott 56
Arctic 102, 107, 122
Argentina 93
Armenia 229
Arnold, Thomas 154
Aro country 130, 137
Arochuka 130, 133
Art Union Journal 56
Ashanti War 82
Asia 4, 116, 196, 221, 224, 231, 240–1
Asquith, H.H. 245
astronomy 2
Atlantic 13
Australasia 160–3, 249
Australia 13, 18–9, 28, 43, 102, 157–9, 164–6, 172, 174–5, 180–2, 193, 229, 241, 249
Austro-Hungarian empire 229
Axis 281, 287, 291

[325]

cultural imperialism 182
Cunningham, Allan 16, 18, 26
Curlewis, J.F.I. 306, 308
Currie, Sir Donald 101, 111
Cyprus 172
Cyrenaica 269, 272

Daguerre, Louis 54, 56
daguerreotype 57
Dahomey 230, 239
Dakar 227
Darby, Henry Clifford 180
Dardanelles 241
Darwin, Charles 42, 45, 47–8, 177
  Origin of Species 42, 48
Darwinism 84
Daughters of the Empire Guild 210
Dawkins, Boyd 99
De Beers Company 204–5
deforestation 36–9, 41, 45–8
Dehra Dun 42, 47
Delagoa Bay 17
Demangeon, Albert 234, 251
Denison, Sir William 45
Derby, Earl of 106
Derbyshire 195, 203, 214
Description d'Egypte 3
desiccation 37–2, 47
diamonds 203–6
Dictionary of National Biography
  169
Diligentia Club 83
Disraeli, Benjamin 64
Dixie, Florence 195, 204
Djibouti 227, 250, 287
Dominions Office 302, 314
Donaldson, James 102
Doumergue, Gaston 235–7, 248
Downing Street 311
Dresch, Jean 6
Duala 232, 234, 237
Dubois, Marcel 223, 228
Duncan, A.H.F. 306, 308, 310
Dundas, Henry 20
Dundee 101, 109, 126–9, 137, 146
Durham 114
Dutch East Indian Company 193
Dutch East Indies 20, 81–90, 93

Earl, George Windsor 25
East Anglia 180

East India Company 15, 17, 20, 24, 37,
  40, 46
East Midlands 190
Eastern Province 206
Edem, King of Ekenge 144
Edinburgh 69, 100–3, 109, 115
Efik 127, 141, 145, 147
Egerton, H.E. 159, 165, 168
Egypt 3–4, 166, 196, 249, 287, 289
Ekpe 127, 128
El Niño 40
Eliot, George 130
Elmina, Ghana 82
Emigrants Information Office 156,
  162
emigration 84, 164, 190
Enciclopedia Italiana 266
Encyclopaedia Britannica 16
endemism 37
engineering 64
English 158, 224
English Channel 39
English Ethnological Society 177
Enlightenment 3
environmental concern 37
environmental degradation 49
environmentalism 37, 42
Enyong 132, 138, 141
Eritrea 268, 272, 289
Erysipelas 138
Esposition du Sahara 271
Ethiopia 272, 289
ethnography 3, 83, 194
Ethnological Society 69, 179
ethnology 59, 96
Etienne, Eugène 224, 247
eucalypti 45
'Eurafrica' 280–2, 287–9
Eurocentrism 7, 152, 168, 198
evolution 48
exotic scenes 56
expedition artists 58
extinctions 37, 48

Falkland Islands 172
Far East 162
'fattening houses' 140
female domesticity 189
Ferry, Jules 224
Fezzan 271, 272
Fidel, Camille 246–8, 251
Fisher, A. Hugh 182

gold 203, 205–6
Gold Coast 182, 230, 239
Goldburn, Henry 16
Goldie, Sir George 102, 106
Goltstein, W. van 87
Grant, William 19
Great Barrier Reef 22
Great Usutu River 315
Greek colonies 158
Green, J.R. 154
greenhouse effect 43
Greenland 197
Gregory, Augustus 32
Gregory, J. 182
    *The Geography of Victoria,
    Historical, Physical and Political*
    182
    *The Imperial Geography for New
    Zealand Schools* 182
Grenada 39–40
Gresham College, London 38
Grey, Earl 61, 62
Grey, Sir Edward 237, 241
Grey, Sir George 13–15, 17–19, 22,
    24–32, 63
Grove, George 157
Guadeloupe 227
Guild of Loyal Women 210
Guinea 227
Gulf War 2
Guyana 17

Hailey, Lord 300
Haimann, Angela 269
Haimann, Giuseppe 269
Hales, Stephen 38
    Vegetable Staticks 38
Halley, Edmund 37
Hanbury-Williams, Colonel John 201
Hanoi 226–7
Hanover Bay 13
Harcourt, Freda 64
Harcourt, Viscount 237
Harmand, Jules 240
Harmattan 138
Harold, Sergeant John 65
Harris, W.C. 304
Harvey, D.H. 319
Haushofer, Karl 253, 274
Hay, R. 16, 17, 22
Hedin, Sven 106, 116, 120, 228
Hegel, Georg Wilhelm Friedrich 154

Heidelburg 231
Helm, Elijah 98
Herodotus 30
Herschel, Sir John 56, 57
Hettner, Alfred 231
Hex River Mountains 200
Hexham 113, 114
hill stations 179
Himalayas 74
Historical Association 184
Hobson, J.A. 1
Hogarth, David George 210
Holy Crusade 64, 243
Holy Land 243
Holy Roman Empire 173
Hong Kong 69, 70, 166
Hooker, Sir William 40
Horton, W. 16
Hovell 15
Hudson Bay 162
Hull 93, 242
Hulot, Baron Étienne 227, 229–32,
    234–6, 244–5, 248
human sacrifice 128, 141, 146
Humbert, Charles 240
Hutton, J.F. 98–9
Huxley, Thomas Henry 179, 207
Hydrographic Office 25
hydrology 47

Ikpe 138
Illustrated London News 182
Imperial Colonist 194
Imperial Education Conference (1923)
    167
Imperial Forestry Institute 42, 47
Imperial Institute 96, 112
Imperial School of Forestry 47
Imperial Studies Movement 183
imperialism
    and culture 54
    economic 1
    universal 3
India 31, 40–2, 45–7, 64–5, 69, 74, 127,
    159, 162, 170, 172, 179, 193, 195,
    212, 237, 249
Indian Army 43
Indian Forest Service 42
Indian Ocean 38, 227, 237, 277, 289
Indian Trigonometric Survey 68
indigenous culture 133
Indonesia 25